1^{00}

LINEAR
MATHEMATICS

UNIVERSITY MATHEMATICS SERIES

Fred Brauer and John A. Nohel, Editors
University of Wisconsin

ELEMENTARY DIFFERENTIAL EQUATIONS: PRINCIPLES, PROBLEMS, AND SOLUTIONS
Fred Brauer and John A. Nohel, University of Wisconsin

ORDINARY DIFFERENTIAL EQUATIONS: A FIRST COURSE
Fred Brauer and John A. Nohel, University of Wisconsin

THE QUALITATIVE THEORY OF ORDINARY DIFFERENTIAL EQUATIONS
Fred Brauer and John A. Nohel, University of Wisconsin

MATHEMATICAL LOGIC: A FIRST COURSE
Joel W. Robbin, University of Wisconsin

LINEAR MATHEMATICS: AN INTRODUCTION TO LINEAR ALGEBRA AND LINEAR DIFFERENTIAL EQUATIONS
Fred Brauer, John A. Nohel, and Hans Schneider, University of Wisconsin

LINEAR MATHEMATICS

AN INTRODUCTION TO LINEAR ALGEBRA AND LINEAR DIFFERENTIAL EQUATIONS

Fred Brauer
John A. Nohel
Hans Schneider

University of Wisconsin

W. A. BENJAMIN, INC.
Menlo Park, California · Reading, Massachusetts
London · Amsterdam · Don Mills, Ontario · Sydney

Second printing, August 1974

Copyright © 1970 by W. A. Benjamin, Inc. Philippines copyright 1970 by W. A. Benjamin, Inc.

All rights reserved. No part of this publication may be reproduced, stored in a retrieval system, or transmitted, in any form or by any means, electronic, mechanical, photocopying, recording, or otherwise, without the prior written permission of the publisher. Printed in the United States of America. Published simultaneously in Canada. Library of Congress Catalog Card No. 79-183667.

ISBN 0-8053-1206-4
ABCDEFGHIJ-CO-7987654

*To Audrey
 Barbara
 David
 Deborah
 Dick
 Michael
 Mike
 Peter
 Tom*

C.2

PROPERTY OF UNITED STATES

CSEC
QA251
.B73

PREFACE

Linear algebra and linear differential equations are interesting mathematical topics fully deserving study on their merit. Both have a long history and ample literature that clearly establish deep connections between them, and both have been central in the development of modern mathematics. For this reason they have been studied for some time by students of mathematics at the undergraduate and graduate levels.

Engineers and physical scientists have long recognized the need for studying linear differential equations. However, in recent years it has become apparent that an understanding of some of the fundamental ideas of linear algebra is also of great value to them. There are at least two reasons why this is true. The first is that many physical problems lead to mathematical models whose solution involves linear algebra and (or) linear differential equations directly. The second is that many physical problems lead to mathematical models, the solution of which requires an understanding of a mathematical theory formulated in terms of linear algebra. This is illustrated vividly by the study of linear systems of differential equations. The purpose of this book is to develop some ideas of linear algebra and linear differential equations, and to exploit the interplay between them. Through examples we also attempt to show to the interested student why these topics are of interest to engineers as well as to pure and applied scientists.

The authors believe that students should be introduced to linear algebra early in their careers but not so early that they are unprepared for applications to other branches of mathematics. Therefore, *we suggest that the material in this book be taught to students who have had a standard course in calculus.* Over the past three years, we have taught the material of the first seven chapters with reasonable success to students at the University of Wisconsin in their fourth semester, following a three-semester sequence of courses in calculus with a brief introduction to elementary differential equations (seven to eight weeks). We have found that this material gives students adequate background in linear algebra and linear differential equations to enable them to proceed to courses in abstract algebra, advanced calculus (specifically those topics in the study of functions of several variables for which some knowledge of linear transformations is essential), applied matrix analysis, or nonlinear ordinary differential equations. At the same time, this course provides students not desiring further mathematical training with an interesting and useful terminal course.

In Chapter 1, we present a variety of physical problems, the solutions of which involve concepts of linear algebra and differential equations. These problems are presented to provide motivation for those students who need to be convinced that the material in the book is useful even for nonmathematicians. They are suggested as light reading, rather than as objects of concentrated study. Students who are not particulary interested in applications can omit Chapter 1 without loss of continuity.

The main part of the book begins in Chapter 2. We believe that it is wise to introduce linear algebra on an elementary level by beginning with fairly easy concrete ideas before proceeding to more abstract concepts. In accordance with this belief, we begin with matrix operations, linear systems of algebraic equations, and determinants in Chapters 2, 3, and 4. However, we also believe that this rather easy material should be done quickly. In Chapter 5, we develop the more abstract theory of vector spaces that is needed in Chapter 6; the development builds on examples already studied in Chapter 3.

Chapters 6 and 7 contain the theory and methods of solutions of linear systems of differential equations; linear scalar differential equations become a simple special case. The general theory developed in Chapter 6 is a nontrivial application of linear algebra that should solidify the abstract concepts introduced in Chapter 5. The study of linear systems of differential equations with constant coefficients in Chapter 7 motivates the introduction of the concept of eigenvalues of a matrix—an important part of linear algebra. In this chapter we again study an algebraic theory and then solidify our understanding by using it to develop a complete theory of systems of linear differential equations with constant coefficients.

For instructors who wish to give a little more emphasis to differential equations and a little less to linear algebra we have provided, in Chapter 8, an alternate way of studying linear differential equations with constant coefficients by means of the Laplace transform. In this connection, we mention that instructors should, of course, be guided in selection of material by the previous training of their students. Many calculus courses include a study of determinants, and there is no need to give a thorough treatment of Chapter 4 to students who are already familiar with determinants. We stress again that Chapters 2, 3, and 4 can be covered quickly, since most students have some acquaintance with linear algebraic equations. On the other hand, Chapters 5, 6, and 7 represent a different level of abstraction, and should not be rushed.

Three appendixes are included to make the book essentially self-contained. They contain proofs of results stated and used without proof in the text, and may be of interest to the ambitious student. We do not, however, consider them essential for the development of the course.

We point out to the reader that this book is only an introduction to linear algebra and differential equations. Many important areas are omitted or mentioned only briefly in this book. The interested reader is referred to advanced courses and books for such topics.

We would like to thank several persons for their invaluable assistance in the preparation of this book. Drs. George Barker, David Ferguson, William Hintzman, and Jack Williamson prepared notes based on lectures; Dr. Robert L. Wilson prepared supplementary problems as well as answers to selected exercises. We acknowledge with pleasure discussions with our colleague Professor Ben Noble, concerning applications of linear mathematics to various physical problems. A valuable source for such problems is his book, *Applications of Undergraduate Mathematics in Engineering* (Mathematics Association of America and Macmillan, New York, 1967). We have made frequent use of this material, particularly in Chapter 1. The reviews by Professors E. Barston, M. Hausner, and D. Sherbert proved helpful. We are particularly grateful to Professor Marvin Marcus for his painstaking reading of the entire manuscript and for his critical remarks which helped to improve it. As usual, Mrs. Phyllis J. Rickli deserves our thanks for deciphering our handwriting and typing the manuscript.

Fred Brauer
John A. Nohel
Hans Schneider

Madison, Wisconsin

NOTE TO THE STUDENT

We hope that we have provided you with a readable introduction to linear algebra and linear differential equations. We suggest that any mathematics book is to be read with paper and pencil at hand. There will inevitably be places where you must fill in some details. Scattered throughout the text you will find numerous exercises which are designed to help you to follow the argument and to reach a better understanding of the subject. If you cannot carry out an exercise, assume its validity and continue reading, but remember to fill the gap at the earliest opportunity. The exercises at the ends of sections are intended to give practice in applying the material in the text. The miscellaneous exercises at the end of each chapter are intended to help you review the chapter as a whole, as well as to supplement some of the material in the chapter.

Answers to some exercises may be found at the back of the book. If your answers do not agree with ours, recheck your work. If you still do not agree, your answer may be equivalent to ours but written in a different form. We will appreciate hearing comments from our readers.

F.B., J.N., H.S.

CONTENTS

Chapter 1	**Some Examples**		1
Chapter 2	**Matrix Operations**		17
	2.1	Basic Definitions	17
	2.2	Addition and Multiplication of Matrices	19
	2.3	Matrix Multiplication	21
	2.4	Nonsingular Matrices and Inverses	28
Chapter 3	**Linear Systems of Algebraic Equations**		35
	3.1	Three Examples	35
	3.2	Solutions and Equivalent Systems	38
	3.3	Elementary Row Operations and Elementary Matrices	41
	3.4	Row Equivalence and Equivalent Systems of Equations	46
	3.5	Row Echelon Form	49
	3.6	Linear Homogeneous Systems of Algebraic Equations	55
	3.7	General Linear Systems of Algebraic Equations	60
	3.8	The Inverse of a Nonsingular Matrix	65

Contents

Chapter 4 Determinants — **71**

- 4.1 Definition and Notation — 71
- 4.2 Some Properties of Determinants — 75
- 4.3 The Determinant of the Transposed Matrix — 82
- 4.4 Expansion by Cofactors of Any Row or Column — 87
- 4.5 Inverses and Cramer's Rule — 89
- 4.6 Solution of Example 2, Chapter 1—A Plane Pin-Jointed Framework — 94
- 4.7 Another Way of Evaluating Determinants — 97

Chapter 5 Vector Spaces — **103**

- 5.1 Real Euclidean n-Dimensional Space — 103
- 5.2 Abstract Vector Spaces and Subspaces — 107
- 5.3 Subspaces — 112
- 5.4 Span, Linear Dependence, and Linear Independence — 115
- 5.5 Basis and Dimension of a Vector Space — 126
- 5.6 Linear Systems of Algebraic Equations Revisited — 136
- 5.7 Linear Transformations — 146

Chapter 6 Linear Systems of Differential Equations — **151**

- 6.1 Introduction — 151
- 6.2 The Existence and Uniqueness Theorem — 165
- 6.3 Linear Homogeneous Systems — 170
- 6.4 Solutions of Scalar Linear Differential Equations with Constant Coefficients — 183
- 6.5 Linear Nonhomogeneous Systems — 195

Chapter 7 Eigenvalues, Eigenvectors, and Linear Systems of Differential Equations with Constant Coefficient — **208**

- 7.1 The Exponential of a Matrix — 208
- 7.2 Eigenvalues and Eigenvectors of Matrices — 213

		Contents	*xiii*

	7.3	Calculation of a Fundamental Matrix	224
	7.4	Two-Dimensional Linear Systems	230
	7.5	The General Case	245
	7.6	Solution of Example 8, Chapter 1. An Electric Circuit	254
Chapter 8	**The Laplace Transform**		**265**
	8.1	Introduction	265
	8.2	Basic Properties of the Laplace Transform	266
	8.3	The Inverse Transform	275
	8.4	Applications to Linear Equations with Constant Coefficients	281
	8.5	Applications to Linear Systems	282
	8.6	A Table of Laplace Transforms	301
APPENDIX	*1*	*The Exponential Matrix*	305
	2	*The Existence and Uniqueness Theorem for Linear Systems of Differential Equations*	308
	3	*Generalized Eigenvectors, Invariant Subspaces, and Canonical Forms of Matrices*	324
		SOLUTIONS TO SELECTED EXERCISES	335
		INDEX	343

LINEAR
MATHEMATICS

Chapter *1* ‖ SOME EXAMPLES

A number of physical problems that lead to mathematical questions concerning linear algebra or linear differential equations are formulated in this chapter. These problems can be solved by techniques that we explore throughout the book; we return to the solutions of the problems posed here in appropriate places in the text. However, the reader who is not interested in applications can proceed immediately to Chapter 2 without any loss of mathematical continuity and omit all subsequent references to these problems.

The reader will have already encountered a number of problems that lead to the solution of a linear system of algebraic equations (one of the central problems of linear algebra). We mention just one of these.

Example 1. Consider the problem of finding an equation of the plane passing through three given points P_1, P_2, P_3 in three-dimensional Euclidean space that do not lie on a straight line (see Figure 1.1, where we have represented the points P_1, P_2, P_3 as lying in the first octant). A plane in three-dimensional space can be represented by an equation of the form

$$Ax + By + Cz = D,$$

where A, B, C, D are constants, and x, y, z are the coordinate axes. If the points P_1, P_2, P_3 have coordinates $(x_1, y_1, z_1), (x_2, y_2, z_2), (x_3, y_3, z_3)$, respectively, this can be formulated as follows: Find numbers A, B, C, D, not

2   *Some Examples*

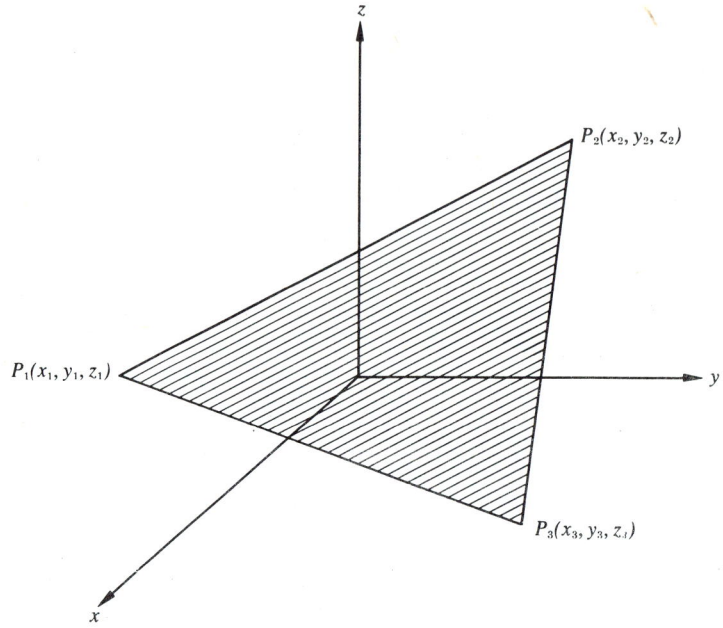

Figure 1.1

all zero, such that

$$Ax_1 + By_1 + Cz_1 = D$$
$$Ax_2 + By_2 + Cz_2 = D \qquad (1.1)$$
$$Ax_3 + By_3 + Cz_3 = D.$$

This is a system of three linear equations in the four unknowns A, B, C, D. Since not all four of the quantities A, B, C, D can be zero, we may divide by one of them to obtain a system of three equations in three unknowns. For example, if the plane does not pass through the origin, then D is different from zero, and we may divide by D. We then determine the three quantities $A' = A/D$, $B' = B/D$, $C' = C/D$ such that

$$A'x_1 + B'y_1 + C'z_1 = 1$$
$$A'x_2 + B'y_2 + C'z_2 = 1 \qquad (1.2)$$
$$A'x_3 + B'y_3 + C'z_3 = 1.$$

Once this has been done (if it is possible), the equation of the desired plane can be written in the form

$$A'x + B'y + C'z = 1$$

If the plane does pass through the origin, then D is zero and the system (1.1) becomes

$$\begin{aligned} Ax_1 + By_1 + Cz_1 &= 0 \\ Ax_2 + By_2 + Cz_2 &= 0 \\ Ax_3 + By_3 + Cz_3 &= 0. \end{aligned} \qquad (1.3)$$

However, we cannot tell in advance whether, given the points $P_1(x_1, y_1, z_1)$, $P_2(x_2, y_2, z_2)$, $P_3(x_3, y_3, z_3)$, the plane through these points passes through the origin. If it does, we solve the system (1.3); if the plane does not pass through the origin, we solve the system (1.2).

The remaining examples in this chapter will come from various physical situations. The mathematical formulation of physical problems depends on a set of basic physical laws and physical assumptions. We stress the fact that a different set of physical assumptions in a particular problem may lead to a different mathematical model. Further, the accuracy of a particular model depends primarily on how reasonable the physical assumptions are. Every mathematical model represents an idealization; physical laws are only approximations to reality.

We now consider a very simple problem in statics.

Example 2. Calculate the stresses and strains in the plane mechanical framework shown in Figure 1.2, subject to given external forces F_1 and F_2 acting at the point A as shown. Each of the members of the framework, labeled (1) to (5), makes a fixed angle θ_1 to θ_5, respectively, with the horizontal. (Note that in Figure 1.2, the angles θ_3, θ_4, θ_5 are negative.) We make the following assumptions:

   (i) The frame is pin-jointed; that is, its members are joined together at A and joined to the wall by pins, in such a way that the ends are free to rotate.
  (ii) Forces at the joints are applied along the five members, and no bending moment is transmitted.
 (iii) The weights of all members are negligible.
 (iv) If the forces F_1 and F_2 are both zero, then there are no stresses in the members.
  (v) Each member obeys **Hooke's law**: When a force is applied along its length, the extension e of the member is directly proportional to the

4 *Some Examples*

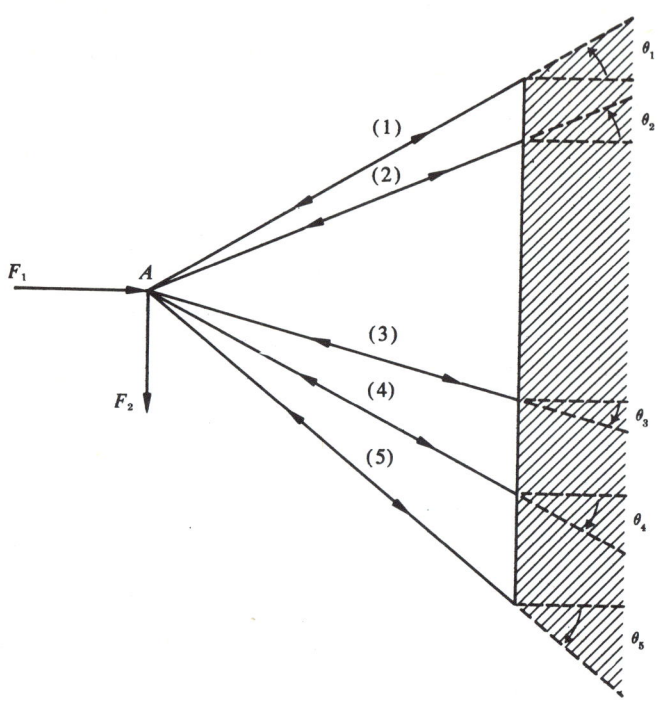

Figure 1.2

force; that is, $e = kT$, where T is the force (tension) and k is a constant of proportionality called the **flexibility** (extension per unit force).

(vi) The changes in the angles θ_i ($i = 1, \ldots, 5$) produced by application of the external forces are negligible.

To solve this problem, let T_i be the tension, e_i the extension, and k_i the flexibility of the ith member ($i = 1, 2, \ldots, 5$) when the forces F_1 and F_2 are applied at A. (The double arrows in Figure 1.2 indicate the two possibilities that the ith rod can be extended ($T_i > 0$) or compressed ($T_i < 0$).)

To formulate the problem mathematically, we first write down the equilibrium equations by resolving all forces horizontally and vertically:

$$-T_1 \cos \theta_1 - T_2 \cos \theta_2 - \cdots - T_5 \cos \theta_5 = F_1$$
$$T_1 \sin \theta_1 + T_2 \sin \theta_2 + \cdots + T_5 \sin \theta_5 = F_2. \quad (1.4)$$

These equations relate the internal and external forces in the framework. Here, we are considering horizontal forces to the left as positive. In Figure

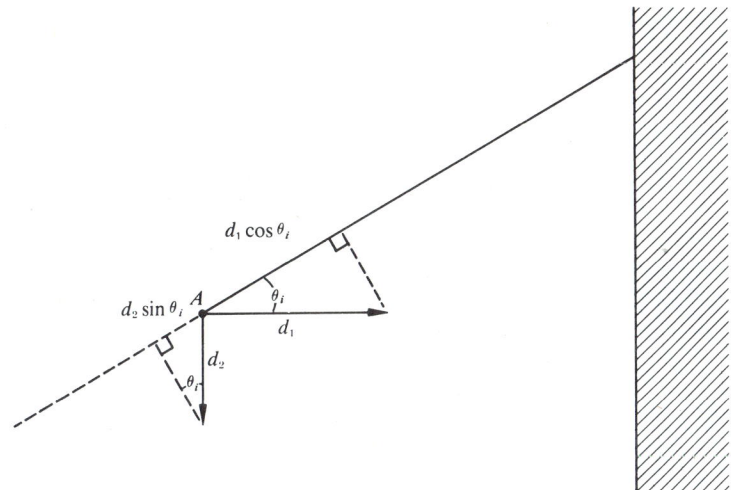

Figure 1.3

1.2, the first two vertical components are positive since $\theta_1 > 0$, $\theta_2 > 0$, and the last three vertical components are negative since $\theta_3 < 0$, $\theta_4 < 0$, $\theta_5 < 0$.

We now wish to relate internal and external displacements in the framework. Suppose that when the forces F_1 and F_2 are applied, the point A moves d_1 units horizontally and d_2 units vertically, measured **in the same direction** as the respective forces F_1 and F_2. Since e_i is the extension of the ith member, when we resolve d_1 and d_2 along the rods, we obtain

$$e_i = -d_1 \cos \theta_i + d_2 \sin \theta_i \qquad (i = 1, 2, \ldots, 5), \tag{1.5}$$

as can be seen from Figure 1.3. Notice that the component $d_1 \cos \theta_i$ decreases the extension e_i, whereas the component $d_2 \sin \theta_i$ increases e_i if $\theta_i > 0$ and decreases e_i if $\theta_i < 0$. From Hooke's law (v), we also have that for each member of the framework

$$e_i = k_i T_i \qquad (i = 1, 2, \ldots, 5), \tag{1.6}$$

where the k_i are the known flexibilities. The mathematical problem in this case is to solve simultaneously the linear system of algebraic equations consisting of (1.4), (1.5), and (1.6). This is a system of 12 equations in the 12 unknown quantities $T_1, T_2, \ldots, T_5, e_1, e_2, \ldots, e_5, d_1, d_2$.

We now consider a problem in dimensional analysis; such problems arise frequently in science and engineering.

Some Examples

Example 3. A fluid flow situation depends on the following physical quantities: the velocity V, the density ρ, the diameter D, the gravity g, and the viscosity μ. In terms of the fundamental physical quantities, mass M, length L, and time T, the dimensions of these quantities, regardless of which consistent units are employed, are given in the following table.

Quantity	V	ρ	D	g	μ
Dimension	LT^{-1}	ML^{-3}	L	LT^{-2}	$ML^{-1}T^{-1}$

For example, V is measured in length per unit time; hence, the dimension LT^{-1}. We are interested in two problems.

(i) Is it possible to formulate **dimensionless products** of the form

$$V^a \rho^b D^c g^d \mu^e \ ? \tag{1.7}$$

(ii) If so, how many such distinct (independent) products are possible?

To formulate these as mathematical problems, we must define what we mean by the statement that the product (1.7) is dimensionless. The product (1.7) is said to be dimensionless if and only if there are real numbers a, b, c, d, e such that

$$(LT^{-1})^a (ML^{-3})^b L^c (LT^{-2})^d (ML^{-1}T^{-1})^e = M^0 L^0 T^0 \ ;$$

that is,

$$M^{b+e} L^{a-3b+c+d-e} T^{-a-2d-e} = M^0 L^0 T^0. \tag{1.8}$$

From (1.8), we see immediately that the product (1.7) is dimensionless if and only if the numbers, a, b, c, d, e are chosen in such a way that

$$\begin{aligned} b \qquad\qquad\ + e &= 0 \\ a - 3b + c + \ d - e &= 0 \\ -a \qquad\quad - 2d - e &= 0. \end{aligned} \tag{1.9}$$

The mathematical problems are now the following:

(i) Does the system (1.9) of three linear algebraic equations in five unknowns a, b, c, d, e have a solution other than $a = 0, b = 0, c = 0, d = 0, e = 0$? (We will see in Section 3.6 that the answer to this equation is yes.)

(ii) How many independent solutions does system (1.9) possess? (Independence will be defined precisely in Section 5.4, and this problem can be solved after Section 5.6.)

An important problem in chemistry leads to questions that can easily be answered by methods of linear algebra. A single molecule of a substance has a definite mass called the **molecular weight**. The number of grams of a substance equal to its molecular weight is called 1 **gram-mole**. (Thus, 1 mole of any substance has the same number of molecules.) A chemical reaction between substances A, B, C, D might be

$$A + 2B \Rightarrow C + D,$$

and this may be read as: 1 mole of substance A reacting with 2 moles of substance B produces 1 mole each of substance C and D. It is convenient to write the above reaction in the symbolic form

$$A + 2B - C - D = 0.$$

Example 4. Consider a mixture of CO, H_2, and CH_4 that is fed into a furnace and burned with oxygen to produce CO, CO_2, and H_2O. From chemistry we know that the possible reactions taking place are

$$CO + \tfrac{1}{2}O_2 \Rightarrow CO_2$$
$$H_2 + \tfrac{1}{2}O_2 \Rightarrow H_2O$$
$$CH_4 + \tfrac{1}{2}O_2 \Rightarrow CO_2 + 2H_2O$$
$$CH_4 + \tfrac{3}{2}O_2 \Rightarrow CO + 2H_2O,$$

and there are no others. The problem is to determine whether these reactions are independent of one another or whether any of these are combinations of some of the others.

To formulate this mathematically let $A_1 = CO$, $A_2 = O_2$, $A_3 = CO_2$, $A_4 = H_2$, $A_5 = H_2O$, $A_6 = CH_4$. Then we may write the above reactions as the linear equations

$$\begin{aligned} A_1 + \tfrac{1}{2}A_2 - A_3 &= 0 \\ A_4 + \tfrac{1}{2}A_2 - A_5 &= 0 \\ A_6 + \tfrac{1}{2}A_2 - A_3 - 2A_5 &= 0 \\ A_6 + \tfrac{3}{2}A_2 - A_1 - 2A_5 &= 0. \end{aligned} \qquad (1.10)$$

This is a linear system of four equations and six unknowns, and the mathematical problems are: (i) How many independent solutions does the system (1.10) possess (see the explanation at the end of Example 3)? (ii) If the equations in system (1.10) are not independent, which ones can be omitted?

8 Some Examples

Many problems in the applications are time dependent. (The problems treated so far have been assumed to be time independent.) We will now examine several time-dependent problems whose solution will lead us naturally to the study of linear ordinary differential equations. Although the examples that we will study are very simple, an understanding of them is essential for the analysis of complex physical systems.

Example 5. A weight of mass m is connected to a rigid wall by means of a spring with spring constant $k > 0$. An external force $F(t) = A \cos \omega t$ acts on the weight, and the system slides in a straight line on a frictionless table as shown in Figure 1.4. Let $y(t)$ denote the displacement of the weight from equilibrium of the system, with $y > 0$, whenever the displacement of the weight stretches the spring and $y < 0$ whenever the displacement t of the weight compresses the spring. At equilibrium, $y = 0$ and the spring is unstretched. Then we may pose the following problems:

(i) If the system starts from rest at time $t = 0$ with an initial displacement y_0, determine the motion of the system for $t > 0$.

(ii) If $F(t) \equiv 0$ and if the system starts from some given initial state $y(0) = a$, $y'(0) = b$, where a, b are not both zero, show that the system executes a simple harmonic motion (here $'$ denotes the derivative with respect to t). Also determine the **natural frequency** of the system (that is, the frequency when $F(t) \equiv 0$).

(iii) Show that if $\omega = (k/m)^{1/2}$ (that is, if the natural frequency of the system is equal to the **applied frequency**), then the solutions of the system become unbounded at $t \to \infty$. (This is the so-called case of **resonance**.)

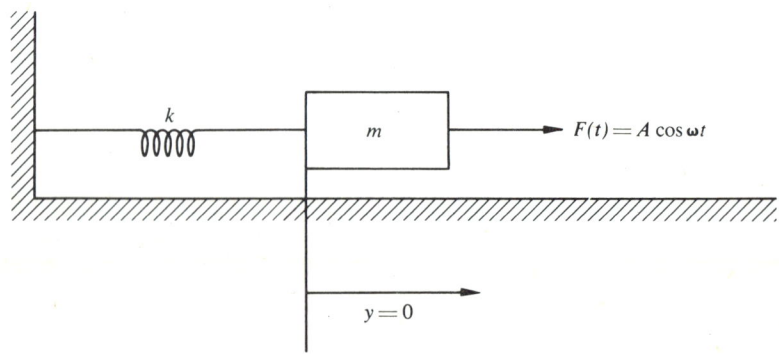

Figure 1.4

In order to formulate this problem mathematically, we make the following assumptions:

(i) The spring has zero mass.
(ii) The weight can be treated as though it were a point mass of mass m.
(iii) The spring satisfies **Hooke's law**, which in the present context states that the spring exerts a restoring force toward the (unstretched) equilibrium position; the magnitude of this force is proportional to the displacement of the spring from equilibrium, and the constant of proportionality $k > 0$ is called the **spring constant**.
(iv) Neither the table nor the surrounding medium offer any resistance to the motion of the system.
(v) The system obeys Newton's second law of motion.

Suppose that at any arbitrary time t the system is in the position shown in Figure 1.4 $y(t)$ units from equilibrium. We list **all the forces** acting on the particle of mass m at time t:

(i) The restoring force of the spring is $-ky(t)$ by Hooke's law. (Note that if $y(t) > 0$ this restoring force acts to decrease the displacement, whereas if $y(t) < 0$ the restoring force acts to increase the displacement $y(t)$ of the particle of mass m.)
(ii) The external force $F(t) = A \cos \omega t$.

Thus, the sum of all forces acting on the particle at time t is $-ky(t) + A \cos \omega t$. Now, **Newton's second law of motion** applied to the moving particle of mass m states that

$$\frac{d}{dt}(m\mathbf{v}) = \text{sum of the forces acting on the particle.} \qquad (1.11)$$

Here, $\mathbf{v}$ is the **velocity vector** and $m\mathbf{v}$ is the momentum. (Newton's law is stated in terms of vectors. Since the motion of the present system is restricted to a line, the vectors involved are one dimensional and vector notation can be eliminated.) Because the mass m is constant and because in this case $\mathbf{v} = y'(t)$, we obtain, using (1.11),

$$my''(t) = -ky(t) + A \cos \omega t,$$

or the displacement $y(t)$ satisfies the equation

$$y''(t) + \frac{k}{m} y(t) = \frac{A}{m} \cos \omega t,$$

10 Some Examples

which we write symbolically as

$$y'' + \frac{k}{m} y = \frac{A}{m} \cos \omega t \tag{1.12}$$

Equation 1.12 is a **differential equation**. It is **linear** (of first algebraic degree in y and its derivatives) and of **second order** (since y'' is the highest order derivative present). The mathematical problem is to find a function ϕ, called a **solution** of (1.12) defined in the interval $0 \leq t < \infty$ such that ϕ' and ϕ'' exist and such that for $0 < t < \infty$,

$$\phi''(t) + \frac{k}{m} \phi(t) = \frac{A}{m} \cos \omega t.$$

Since the system starts from rest at $t = 0$ with an initial displacement y_0, we shall also require that the function ϕ satisfy

$$\phi(0) = y_0, \qquad \phi'(0) = 0; \tag{1.13}$$

these are the **initial conditions**. The problem consisting of the differential equation (1.12) and the initial conditions (1.13) is called an **initial value problem**. Once such a function has been found, the remaining questions can be answered. This function ϕ (it can be shown that there is only one such function) completely determines the motion of the system under assumptions (i)–(v). We stress the fact that a different set of assumptions will lead to a mathematical model different from the one described by Eq. 1.12 and the initial conditions (1.13).

Example 6. Consider the system described in Example 5 and in Figure 1.4, but drop the assumption that the system slides on a frictionless table. Instead, assume that the table offers a force of friction proportional to the velocity with constant of proportionality $b > 0$.

  (i) Assuming that assumptions (i)–(v) hold, determine the motion of the system with the same initial state as that of the system described in Example 5.
  (ii) If the external force $F(t) \equiv 0$, discuss the behavior of the system in each of the cases $b^2 < 4km$, $b^2 > 4km$, and $b^2 = 4km$.

Here the formulation of the mathematical model is similar to the one in Example 5. We merely add $-by'(t)$ to the forces considered previously; this is the force caused by friction. Thus, Newton's second law says in this case that

$$my''(t) = -ky(t) - by'(t) + A\cos\omega t,$$

or (1.14)

$$y'' + \frac{b}{m}y' + \frac{k}{m}y = \frac{A}{m}\cos\omega t.$$

Equation 1.14 is another differential equation; like (1.12) it is linear and of the second order. The mathematical problem is now one of determining a function ψ, defined for $0 \leq t < \infty$, such that $\psi'(t)$ and $\psi''(t)$ exist for $0 < t < \infty$ and such that

$$\psi''(t) + \frac{b}{m}\psi'(t) + \frac{k}{m}\psi(t) = \frac{A}{m}\cos\omega t,$$

for $0 < t < \infty$. Such a function ψ is called a **solution** of Equation 1.14 on $0 < t < \infty$. We also require that ψ should satisfy initial conditions

$$\psi(0) = y_0, \qquad \psi'(0) = 0. \tag{1.15}$$

Equations 1.14 and 1.15 constitute another **initial value problem**.

We now consider a mechanical system that is made up of two or more "mass springs."

Example 7. A weight of mass m_1 is connected to a rigid wall by a spring having spring constant $k_1 > 0$. A second weight of mass m_2 is connected to the weight of mass m_1 by means of a spring having spring constant $k_2 > 0$. An external face $F(t)$ is applied to the second weight. The whole system slides in a straight line on a frictionless table, as shown in Figure 1.5. Let $y_1(t)$ denote the displacement of the first weight from its rest position (equilibrium)

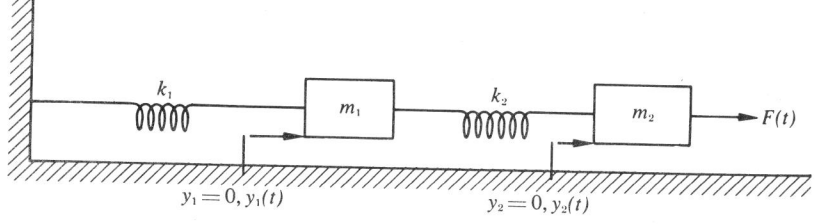

Figure 1.5

and $y_2(t)$ the displacement of the second weight from equilibrium. At equilibrium $y_1 = y_2 = 0$ and both springs are unstretched.

(a) If at time $t = 0$ the system starts from rest with initial displacements $y_1(0) = y_{10}$, $y_2(0) = y_{20}$, determine the motion of the system.

(b) If $m_1 = m_2 = m$, $k_1 = k_2 = k$, and $F(t) \equiv 0$, show that the motion of the system is a superposition of two simple harmonic motions with natural frequencies

$$\frac{1}{2\pi}\left(\frac{3+\sqrt{5}}{2}\right)^{1/2}\left(\frac{k}{m}\right)^{1/2} \quad \text{and} \quad \frac{1}{2\pi}\left(\frac{3-\sqrt{5}}{2}\right)^{1/2}\left(\frac{k}{m}\right)^{1/2}.$$

In order to formulate this problem mathematically, we assume (i)–(v) of Example 5 and apply Newton's second law of motion to each moving weight which we regard as a point mass. Suppose that at time t the system if in the position shown in Figure 1.5; then the only forces acting on the particle of mass m_1 are the restoring force of the first spring $-k_1 y_1(t)$ (by Hooke's law); and the restoring force of the second spring whose net extension is $y_2 - y_1$ (because it is stretched y_2 units by the second weight and compressed y_1 units by the first weight. Thus, the restoring force of the second spring is $k_2[y_2(t) - y_1(t)]$. Applying Newton's second law to the particle of mass m_1, we therefore obtain

$$m_1 y_1''(t) = -k_1 y_1(t) + k_2(y_2(t) - y_1(t)). \tag{1.16}$$

The only forces acting on the particle of mass m_2 are the restoring force $-k_2(y_2(t) - y_1(t))$ of the second spring whose net extension is $y_2 - y_1$ units, and the external force $F(t)$. Thus, Newton's second law applied to the second particle yields

$$m_2 y_2''(t) = -k_2[y_2(t) - y_1(t)] + F(t). \tag{1.17}$$

From the statement of the problem we also have the initial conditions $y_1(0) = y_{10}$, $y_1'(0) = 0$, $y_2(0) = y_{20}$, $y_2'(0) = 0$. Equations 1.16 and 1.17 govern the motion and the mathematical problem is to find a pair of functions ϕ_1, ϕ_2 defined for $t \geq 0$, such that $\phi_1', \phi_1'', \phi_2', \phi_2''$ exist for $t > 0$, such that

$$\begin{cases} m_1 \phi_1''(t) = -k_1 \phi_1(t) + k_2(\phi_2(t) - \phi_1(t)) \\ m_2 \phi_2''(t) = -k_2(\phi_2(t) - \phi_1(t)) + F(t), \end{cases}$$

for every $t > 0$ and such that

$$\phi_1(0) = y_{10}, \quad \phi_1'(0) = 0, \quad \phi_2(0) = y_{20}, \quad \phi_2'(0) = 0.$$

If such functions have been found (it will be seen later (Section 6.2) that there is only one such pair of functions), we say that we have found a solution of the **initial value problem** consisting of the **system of linear differential equations** of second order

$$\begin{cases} m_1 y_1'' = -k_1 y_1 + k_2(y_2 - y_1) \\ m_2 y_2'' = -k_2(y_2 - y_1) + F(t) \end{cases} \qquad (1.18)$$

and the **initial conditions**

$$y_1(0) = y_{10}, \qquad y_2(0) = y_{20}, \qquad y_1'(0) = 0, \qquad y_2'(0). \qquad (1.19)$$

In order to determine the motion and answer question (b), we will have to learn how to solve the initial value problems (1.18) and (1.19).

Closely related to mechanical systems are electrical systems that also lead to linear systems of ordinary differential equations.

Example 8. Consider the electrical circuit shown in Figure 1.6 in which the known time-varying source current $i_s(t)$ is connected to nodes A and C; in the circuit $v_1(t)$ is the variable voltage (unknown) across the 5/3-farad condenser, $i_1(t)$ is the variable current through the 3/5-henry inductor. The polarities are indicated in Figure 1.6. When $v_1(t) > 0$, the potential of node A is larger than that of node C (measured with respect to a common reference). When $i_1(t) > 0$, the current flows from node A to node B. Suppose that at $t = 0$ we are given

$$v_1(0) = 0.6 \text{ volts}, \qquad i_1(0) = 1.0 \text{ ampere}, \qquad v_2(0) = 1.2 \text{ volts}.$$

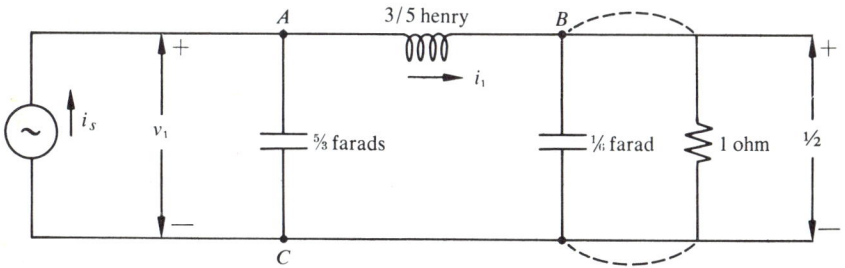

Figure 1.6

14   Some Examples

It is desired to determine the voltages $v_1(t)$, $v_2(t)$, and the current $i_1(t)$ as functions of time (in terms of the given source current $i_s(t)$).

To set this problem up mathematically, we shall use **Ohm's law** ($v = iR$, relating voltage, current, and resistance) and the formulas $i = Cv'(t)$ and $Li'(t)$ (relating the "current through" to the "voltage across" condensers and inductors; here C is the capacitance, L is the inductance, and $' = d/dt$). We shall also need **Kirchhoff's law of currents**: The sum of the currents entering and leaving a given node is zero.

Suppose that at a time t the source current $i_s(t)$ leaves node C and enters node A as shown in Figure 1.6. At the same time the current i_1 leaves node A (through the inductance of 3/5 henry) and a current $(5/3)v_1'(t)$ leaves node A through the capacitance of 5/3 farads. Thus, by Kirchhoff's law applied to node A we have

$$i_s(t) - i_1(t) - \tfrac{5}{3}v_1'(t) = 0. \tag{1.20}$$

Similarly, a current i_1 enters node B, while currents $\tfrac{1}{6}v_2'(t)$ and $v_2(t)/1$ leave node B through the $\tfrac{1}{6}$-farad condenser and 1-ohm resistor, respectively. Thus, Kirchhoff's law applied to node B gives

$$i_1(t) - \tfrac{1}{6}v_2'(t) - v_1(t) = 0. \tag{1.21}$$

Then in the middle loop of the circuit shown in Figure 1.6, since the sum of the voltage drops must also be zero (another one of Kirchhoff's laws), we have

$$v_1(t) - \tfrac{3}{5}i_1'(t) - v_2(t) = 0. \tag{1.22}$$

Solving each Eq. 1.20, 1.22, and 1.21 for the quantities v_1', i_1', v_2', respectively, we see that the mathematical problem is to solve the initial value problem

$$\begin{cases} v_1' = -\tfrac{3}{5}i_1 + \tfrac{3}{5}i_s(t), & v_1(0) = 0.6 \text{ volt} \\ i_1' = \tfrac{5}{3}v_1 - \tfrac{5}{3}v_2, & i_1(0) = 1 \text{ ampere} \\ v_2' = 6i_1 - 6v_2, & v_2(0) = 1.2 \text{ volts} \end{cases} \tag{1.23}$$

for the unknown functions v_1, i_1, v_2, where $i_s(t)$ is a given source current, that is, a given function defined for $0 \leq t < \infty$. This initial value problem consists of a system of three linear first-order differential equations. By its **solution,** we shall mean a set of three functions ϕ_1, ϕ_2, ϕ_3 defined for $t \geq 0$, such that ϕ_1', ϕ_2', ϕ_3' and ϕ_1, ϕ_2, ϕ_3 are continuous for $t > 0$ and such that

$$\phi_1'(t) = -\tfrac{3}{5}\phi_2(t) + \tfrac{3}{5}i_s(t), \qquad \phi_1(0) = 0.6$$
$$\phi_2'(t) = \tfrac{5}{3}\phi_1(t) - \tfrac{5}{3}\phi_2(t), \qquad \phi_2(0) = 1$$
$$\phi_3'(t) = 6\phi_2(t) - 6\phi_3(t), \qquad \phi_3(0) = 1.2$$

We shall be able to solve this problem completely in Section 7.6.

We have now formulated eight applied problems in mathematical terms. The first four problems lead to linear systems of algebraic equations, whereas the last four lead to linear systems of differential equations. We point out that in Examples 2, 5, 6, 7, and 8 we are attempting to predict the behavior of a physical system. If our mathematical model is to have any validity, we must know that the mathematical problem has a solution (because the physical system does something), and that the mathematical problem has only one solution (because it if had more than one, we could not tell which one would describe the physical system). These two properties are called **existence of solutions** and **uniqueness of solutions**, respectively. We shall see that the mathematical problems in Examples 2, 5, 6, 7, and 8 do indeed have these properties.

Also, the physical laws used in setting up the mathematical problems are only approximations and not absolute truths. A more accurate approximation to physical reality would lead to a more complicated, nonlinear, mathematical problem that is usually impossible to solve explicitly. One of the central problems of theoretical physics (or applied mathematics) is to attempt to describe the behavior of solutions of such problems. Any discussion of this problem is beyond the scope of this book, but we refer the interested reader to *Qualitative Theory of Ordinary Differential Equations*, by F. Brauer and J. A. Nohel (W. A. Benjamin, Inc., New York, 1969), after he has finished this book. It turns out that a thorough understanding of the linear approximation is essential for any study of nonlinear systems. The purpose of this book is to study two important classes of linear problems: linear systems of algebraic equations and linear systems of differential equations.

Examples 1–4, 7, and 8 in this chapter are most conveniently analyzed in the setting of linear algebra. For this purpose, we introduce vector and matrix notation, which enables us to rewrite these problems in a more convenient form, in Chapter 2. In Chapter 3, linear systems of algebraic equations are studied in this setting. In Chapter 4 we give a brief study of determinants; these are sometimes a convenient tool in linear algebra and linear differential equations. Example 1 can be conveniently solved by Cramer's rule (see Exercise 6, Section 4.5); Example 2 is solved completely in Section 4.6. In Chapter 5, we abstract the ideas of Chapter 3 to the theory of vector spaces, and we give a large variety of examples that are essential in the systematic study of linear systems of differential equations. At the same time, using the results of Chapter 5 we can solve Examples 3 and 4 completely

16   *Some Examples*

(see Exercises 4 and 6, Section 5.6). In Chapter 6 we develop the general theory of linear systems of differential equations and make extensive use of the results of Chapter 5. We obtain the standard results on scalar linear differential equations as special cases. In order to provide the reader with a supply of examples, we solve the scalar linear homogeneous differential equation with constant coefficients. Using these results, the reader will be able to solve Examples 5 and 6 (see Exercises 11 and 12, Section 6.5). Chapter 7 contains the solution of general linear systems of first-order ordinary differential equations with constant coefficients. This requires a development of the **theory of eigenvalues and eigenvectors of matrices** (which is of considerable interest in itself). Examples 7 and 8 may be solved using these results. (See Section 7.6, where Example 8 is solved completely and Example 7 is given as an Exercise 7.) Finally, Chapter 8 discusses the Laplace transformation and how it may be used to solve scalar linear differential equations and systems of linear differential equations with constant coefficients. This chapter is largely independent of the other chapters, and provides an alternate means for solution of Examples 5, 6, 7, and 8.

Chapter 2 | MATRIX OPERATIONS

In Chapter 1 we formulated a variety of problems. On order to deal with these problems we require a more compact notation. For this reason we introduce the notion of a **matrix**, and discuss the elementary properties of matrices. This will make it possible for us to study the solution of linear systems of algebraic equations in Chapter 3. It is at the same time very convenient for the systematic study of systems of differential equations in Chapters 6 and 7.

2.1 Basic Definitions

We begin with a formal definition.

Definition 1. *A matrix is a rectangular array of complex numbers. These numbers are called entries of the matrix.*

If all the entries of a matrix are real numbers, we say that the matrix is real.

Example 1. The following are examples of matrices:

$$A = [1 + i \quad 2 - 3i \quad 7], \qquad B = \begin{bmatrix} 1 + i \\ 2 - 3i \\ 7 \end{bmatrix},$$

$$C = \begin{bmatrix} 1 & 0 & -1 \\ 2 & 7 & 3 \\ \sqrt{2} & \pi & \ln 2 \end{bmatrix}, \qquad D = \begin{bmatrix} 1 & -1 \\ -3 & 2 \\ \sqrt{3} & 7+i \end{bmatrix}.$$

Of these, only C is real. A horizontal array such as [2 7 3] in C is called a **row** of C and a vertical array such as B itself is called a **column** of B. Thus, the matrix A has 1 row and 3 columns and we will say that A is a 1×3 matrix; the matrix B is a 3×1 matrix with 3 rows and one column; similarly, C is a 3×3 matrix and D is a 3×2 matrix.

In our discussion of matrices we shall employ the following notation:

$\mathscr{R}$ will denote the real numbers;

$\mathscr{C}$ will denote the complex numbers.

In many situations we will not need to specify whether we are dealing with $\mathscr{R}$ or $\mathscr{C}$; in such cases we will use $\mathscr{F}$ to denote either $\mathscr{R}$ or $\mathscr{C}$ with the understanding that $\mathscr{F}$ is to be replaced by one of these consistently in any particular argument.

In general we can write a matrix with entries a_{ij} in the form

$$A = \begin{bmatrix} a_{11} & a_{12} & a_{13} & \cdots & a_{1n} \\ a_{21} & a_{22} & a_{23} & \cdots & a_{2n} \\ \vdots & \vdots & \vdots & & \vdots \\ a_{m1} & a_{m2} & a_{m3} & \cdots & a_{mn} \end{bmatrix}$$

with m rows and n columns, and we say that A is an $m \times n$ matrix over $\mathscr{F}$. The array $[a_{i1} \quad a_{i2} \quad \cdots \quad a_{in}]$ is called the ith row of A and similarly the array

$$\begin{bmatrix} a_{1j} \\ a_{2j} \\ \vdots \\ a_{nj} \end{bmatrix}$$

is called the jth column of A. Thus, a_{ij} is the entry in the ith row and jth column of A. If $m = n$ we say that A is a square matrix. It is customary to call a matrix with only one column a column vector (usually denoted by boldface lower-case letters, (for example, **a**). Similarly, a matrix with only one row is called a row vector.

Normally, the entries of a matrix A are denoted by a_{ij}, those of B by b_{ij}, and so on. The entries of a column or row vector **a** are normally denoted by a_i.

We shall denote by $\mathscr{F}_{mn}$ the set of all $m \times n$ matrices with entries in $\mathscr{F}$. Thus, $\mathscr{R}_{mn}$ is the set of all real $m \times n$ matrices and $\mathscr{C}_{mn}$ is the set of all complex $m \times n$ matrices. In Example 1, $A \in \mathscr{C}_{13}$ ($\in$ is a symbol for "is a member of")

$B \in \mathscr{C}_{31}$, $C \in \mathscr{R}_{33}$, $D \in \mathscr{C}_{32}$. (Of course, C could also be regarded as a member of $\mathscr{C}_{33}$.)

Definition 2. *Let $A \in \mathscr{F}_{mn}$, $B \in \mathscr{F}_{pq}$. Then $A = B$ if and only if $m = p$, $n = q$ and $a_{ij} = b_{ij}$ for $i = 1, \ldots, m$, $j = 1, \ldots, n$.*

The reader should note that we are only comparing matrices with the same number of rows and the same number of columns. Thus in Example 1, $A \neq B$.

● **EXERCISES**

   1. Is the 2×3 matrix each of whose entries is zero equal to the 3×3 matrix, each of whose entries is zero?
   2. Is the 2×3 matrix each of whose entries is zero equal to the 2×3 matrix, each of whose entries is one?

2.2 Addition and Multiplication of Matrices

We shall now define the basic operations on matrices. The first of these is addition.

Definition 1. *Let $A \in \mathscr{F}_{mn}$ and $B \in \mathscr{F}_{pq}$. Then $A + B$ is defined if and only if $m = p$ and $n = q$. In this case, the matrix $C \in \mathscr{F}_{mn}$ with entries $c_{ij} = a_{ij} + b_{ij}$ for $i = 1, \ldots, m$, $j = 1, \ldots, n$, is called the sum of A and B, and is written $C = A + B$.*

For example, if

$$A = \begin{bmatrix} 2 & 3 & 1 \\ 3 & 1 & 2 \end{bmatrix}, \quad B = \begin{bmatrix} 1 & 1 & i \\ 1 & 0 & -2 \end{bmatrix}, \quad C = \begin{bmatrix} 1 & 3 \\ 3 & 1 \end{bmatrix},$$

then

$$A + B = \begin{bmatrix} 3 & 4 & 1+i \\ 4 & 1 & 0 \end{bmatrix},$$

while the sums of A and C and of B and C are not defined.

The essential properties of matrix addition are given in the following theorem.

Matrix Operations

Theorem 1.

(i) $\mathscr{F}_{mn}$ *is closed under addition; that is, the sum of two $m \times n$ matrices is an $m \times n$ matrix.*

(ii) *Addition is commutative; that is $A + B = B + A$ for all $A, B \in \mathscr{F}_{mn}$.*

(iii) *Addition is associative; that is $(A + B) + C = A + (B + C)$ for all $A, B, C \in \mathscr{F}_{mn}$.*

(iv) *There exists an $m \times n$ zero matrix, denoted by 0 (all of whose entries are zero) such that $A + 0 = A$ for every $A \in \mathscr{F}_{mn}$.*

(v) *For every matrix $A \in \mathscr{F}_{mn}$, there exists a unique matrix, denoted by $-A$, such that $A + (-A) = 0$. If a_{ij} are the entries of A, then $-a_{ij}$ are the entries of $-A$ for $i = 1, \ldots, m$, and $j = 1, \ldots, n$.*

Proof. The statement (i) is merely a reformulation of part of the definition of addition. To prove (ii), let $C = A + B$ and $D = B + A$. Since addition in $\mathscr{F}$ is commutative, we have $c_{ij} = a_{ij} + b_{ij} = b_{ij} + a_{ij} = d_{ij}$, where $i = 1, \ldots, m$, and $j = 1, \ldots, n$, and thus, $C = D$. The proof of (iii) is similar to that of (ii). The proofs of (iv) and (v) follow immediately from inspection of the definitions of 0 and $-A$, respectively. To prove uniqueness in (v), let $A + X = 0$. Then $a_{ij} + x_{ij} = 0$, whence $x_{ij} = -a_{ij}$ for $i = 1, \ldots, m$ and $j = 1, \ldots, n$, and thus $X = -A$. ∎

The second basic operation on matrices is multiplication by scalars. *A scalar is simply an element of $\mathscr{F}$.*

Definition 2. *If $A \in \mathscr{F}_{mn}$ and $\alpha \in \mathscr{F}$, then the matrix B with entries $b_{ij} = \alpha a_{ij}$ ($i = 1, \ldots, m_j$; $j = 1, \ldots, n$) is called the product of the scalar α and the matrix A, and is denoted by*

$$B = \alpha A = A\alpha.$$

Multiplication by scalars has the following basic properties.

Theorem 2. *For all $\alpha, \beta \in \mathscr{F}$ and all $A, B \in \mathscr{F}_{mn}$ we have the following properties:*

(i) $\alpha(A + B) = \alpha A + \alpha B$.

(ii) $(\alpha + \beta)A = \alpha A + \beta A$.

(iii) $(\alpha\beta)A = \alpha(\beta A)$.

(iv) $1 \cdot A = A$.

The proof is left as an easy exercise for the reader.

2.3 Matrix Multiplication

● **EXERCISES**

1. Form all possible sums of pairs of the following matrices:

$$A = \begin{bmatrix} 2 & 1 & 3 \\ 9 & 4 & 6 \\ -1 & 2 & 0 \end{bmatrix} \qquad B = \begin{bmatrix} 1 & 1 & 1 \\ 2 & 2 & 2 \\ -1 & -1 & -1 \end{bmatrix}$$

$$C = \begin{bmatrix} 1 & 0 & 0 & 1 \\ 0 & 1 & 0 & 1 \\ 0 & 0 & 1 & 1 \end{bmatrix} \qquad D = \begin{bmatrix} 2 & 9 & 0 & 0 \\ 1 & 6 & 1 & 1 \\ 6 & 3 & 2 & -7 \end{bmatrix}$$

$$F = \begin{bmatrix} 0 & 0 & 0 & 0 \\ 1 & 0 & 0 & 0 \\ 0 & -1 & 0 & 0 \end{bmatrix} \qquad G = \begin{bmatrix} 2 & 0 & 4 \\ 9 & 1 & 5 \end{bmatrix}$$

2. Using the matrices of Exercise 1, evaluate each of the expressions $3A$, $A - 2B$, $2C + 5D$, $-G$, $C + D - F$.

3. Theorem 1 says that an operation (called addition) is defined on a certain collection of objects (called matrices) in such a way that the operation satisfies properties (i) through (v). The following is a list of different collections of objects. Is it possible to define in some way an addition operation for each collection so that the operation satisfies properties (i) through (v)?

(a) Real numbers.
(b) Complex numbers.
(c) Integers.
(d) Continuous functions.
(e) Infinite sequences.
(f) Convergent sequences.
(g) Differentiable functions.
(h) Vectors (i.e., the vectors of physics, force vectors, etc.).
(i) Polynomials.

[*Remark:* Any collection of objects with an addition defined satisfying (i) through (v) is called a commutative group. In other words, Theorem 1 simply says that $\mathscr{F}_{mn}$ is a group with respect to addition.]

4. Consider the collections in Exercise 3. Is it possible to define multiplication by scalars from $\mathscr{R}$ or from $\mathscr{C}$ in such a way that, with the addition defined in Exercise 3, Theorem 2 is satisfied.

2.3 Matrix Multiplication

Matrices arise in a natural way from a consideration of systems of linear (algebraic) equations. To introduce matrix multiplication we look at such a system.

22 Matrix Operations

Consider the linear system of equations,

$$z_1 = b_{11}y_1 + b_{12}y_2$$
$$z_2 = b_{21}y_1 + b_{22}y_2$$
$$z_3 = b_{31}y_1 + b_{32}y_2,$$

where the b_{ij}'s denote given constants. Symbolically, we can represent this system by the equation $\mathbf{z} = B\mathbf{y}$, where

$$\mathbf{z} = \begin{bmatrix} z_1 \\ z_2 \\ z_3 \end{bmatrix}; \quad B = \begin{bmatrix} b_{11} & b_{12} \\ b_{21} & b_{22} \\ b_{31} & b_{32} \end{bmatrix}; \quad \mathbf{y} = \begin{bmatrix} y_1 \\ y_2 \end{bmatrix}.$$

Suppose we also have the linear system

$$y_1 = a_{11}x_1 + a_{12}x_2$$
$$y_2 = a_{21}x_1 + a_{22}x_2,$$

where the a_{ij}'s are given constants. Again, symbolically, $\mathbf{y} = A\mathbf{x}$, where

$$\mathbf{y} = \begin{bmatrix} y_1 \\ y_2 \end{bmatrix}; \quad A = \begin{bmatrix} a_{11} & a_{12} \\ a_{21} & a_{22} \end{bmatrix}; \quad \mathbf{x} = \begin{bmatrix} x_1 \\ x_2 \end{bmatrix}.$$

For these two sets of systems we can express $\mathbf{z}$ in terms of $\mathbf{x}$ by the equations

$$z_1 = b_{11}(a_{11}x_1 + a_{12}x_2) + b_{12}(a_{21}x_1 + a_{22}x_2)$$
$$z_2 = b_{21}(a_{11}x_1 + a_{12}x_2) + b_{22}(a_{21}x_1 + a_{22}x_2)$$
$$z_3 = b_{31}(a_{11}x_1 + a_{12}x_2) + b_{32}(a_{21}x_1 + a_{22}x_2)$$

or

$$z_1 = (b_{11}a_{11} + b_{12}a_{21})x_1 + (b_{11}a_{12} + b_{12}a_{22})x_2$$
$$z_2 = (b_{21}a_{11} + b_{22}a_{21})x_1 + (b_{21}a_{12} + b_{22}a_{22})x_2$$
$$z_3 = (b_{31}a_{11} + b_{32}a_{21})x_1 + (b_{31}a_{12} + b_{32}a_{22})x_2.$$

Thus, symbolically, $\mathbf{z} = C\mathbf{x}$, where

$$C = \begin{bmatrix} b_{11}a_{11} + b_{12}a_{21} & b_{11}a_{12} + b_{12}a_{22} \\ b_{21}a_{11} + b_{22}a_{21} & b_{21}a_{12} + b_{22}a_{22} \\ b_{31}a_{11} + b_{32}a_{21} & b_{31}a_{12} + b_{32}a_{22} \end{bmatrix}. \tag{2.1}$$

2.3 Matrix Multiplication

But we also have symbolically,

$$\mathbf{z} = B\mathbf{y} = B(A\mathbf{x}).$$

Now, obviously we would wish to write

$$\mathbf{z} = (BA)\mathbf{x}.$$

In order to do this we must define $BA = C$, where C is given by (2.1). This definition leads to the following evaluation of the matrix product:

$$BA = \begin{bmatrix} b_{11} & b_{12} \\ b_{21} & b_{22} \\ b_{31} & b_{32} \end{bmatrix} \begin{bmatrix} a_{11} & a_{12} \\ a_{21} & a_{22} \end{bmatrix} = \begin{bmatrix} b_{11}a_{11} + b_{12}a_{21} & b_{11}a_{12} + b_{12}a_{22} \\ b_{21}a_{11} + b_{22}a_{21} & b_{21}a_{12} + b_{22}a_{22} \\ b_{31}a_{11} + b_{32}a_{21} & b_{31}a_{12} + b_{32}a_{22} \end{bmatrix}.$$

Note that B and A are not of the same size; however, the number of columns of B is equal to the number of rows of A.

With this special case as motivation, we make the following definition: *If B is an $m \times n$ matrix and A is an $n \times p$ matrix, then the matrix product $C = BA$ is an $m \times p$ matrix with entries c_{ij}, where*

$$c_{ij} = \sum_{k=1}^{n} b_{ik} a_{kj} \qquad (i = 1, \ldots, m; j = 1, \ldots, p).$$

Note that we do not define the product BA unless the number of columns of B is equal to the number of rows of A.

Example 1. If

$$B = \begin{bmatrix} 1 & 3 \\ 0 & -1 \\ 5 & 1 \end{bmatrix}, \quad A = \begin{bmatrix} 3 & -5 \\ 5 & 3 \end{bmatrix} \quad \text{then } BA = \begin{bmatrix} 18 & 4 \\ -5 & -3 \\ 20 & -22 \end{bmatrix}.$$

The symbolic equation $\mathbf{z} = B\mathbf{y}$ introduced above now gains mathematical meaning when we interpret the right-hand side as the matrix product $B\mathbf{y}$, where

$$B = \begin{bmatrix} b_{11} & b_{12} \\ b_{21} & b_{22} \\ b_{31} & b_{32} \end{bmatrix}, \quad \mathbf{y} = \begin{bmatrix} y_1 \\ y_2 \end{bmatrix}.$$

It is useful to observe that if B is in $\mathscr{F}_{mn}$ and A is in $\mathscr{F}_{np}$, then $BA = C$

has

$$c_{ij} = [b_{i1}, \ldots, b_{in}] \begin{bmatrix} a_{1j} \\ \vdots \\ a_{nj} \end{bmatrix} \quad (i = 1, \ldots, m;\ j = 1, \ldots, p).$$

In other words, the entry c_{ij} of the matrix product BA is the matrix product of the ith row vector of B by the jth column vector of A; we shall abbreviate this by the equation

$$c_{ij} = b_{i*} a_{*j},$$

where b_{i*} represents the ith row of B and a_{*j} represents the jth column of A.

Let the matrices B, A, and $C = BA$ be as in Example 1. Observe that one has $[18 \quad 4] = 1[3 \quad -5] + 3[5 \quad 3]$. Thus,

$$c_{1*} = b_{11} a_{1*} + b_{12} a_{2*}.$$

Similarly, we obtain

$$c_{2*} = [-5 \quad -3] = 0[3 \quad -5] + -1[5 \quad 3] = b_{21} a_{1*} + b_{22} a_{2*}$$
$$c_{3*} = [20 \quad -22] = 5[3 \quad -5] + 1[5 \quad 3] = b_{31} a_{1*} + b_{32} a_{2*}.$$

The reader can also check that

$$c_{*1} = b_{*1} a_{11} + b_{*2} a_{21}$$
$$c_{*2} = b_{*1} a_{12} + b_{*2} a_{22}$$

In general, if $B \in \mathscr{F}_{mn}$ and $A \in \mathscr{F}_{np}$ the ith row and jth column of the product matrix $C = BA$ are given by

$$\begin{aligned} c_{i*} &= b_{i1} a_{1*} + b_{i2} a_{2*} + \cdots + b_{in} a_{n*} = b_{i*} A \\ c_{*j} &= b_{*1} a_{1j} + b_{*2} a_{2j} + \cdots + b_{*n} a_{nj} = B a_{*j}, \end{aligned} \quad (2.2)$$

respectively. Moreover, we also have

$$C = b_{*1} a_{1*} + b_{*2} a_{2*} + \cdots + b_{*n} a_{n*} = \sum_{k=1}^{n} b_{*k} a_{k*}. \quad (2.3)$$

2.3 Matrix Multiplication

We check this formula for Example 1 as follows:

$$b_{*1}a_{1*} + b_{*2}a_{2*} = \begin{bmatrix} 1 \\ 0 \\ 5 \end{bmatrix}[3 \ -5] + \begin{bmatrix} 3 \\ -1 \\ 1 \end{bmatrix}[5 \ 3]$$

$$= \begin{bmatrix} 3 & -5 \\ 0 & 0 \\ 15 & -25 \end{bmatrix} + \begin{bmatrix} 15 & 9 \\ -5 & -3 \\ 5 & 3 \end{bmatrix}$$

$$= \begin{bmatrix} 18 & 4 \\ -5 & -3 \\ 20 & -22 \end{bmatrix} = BA.$$

• EXERCISES

1. Calculate the following products. Also, verify the validity of (2.2) and (2.3) in each case.

(a) $5\begin{bmatrix} 1 & 2 \\ 3 & 4 \end{bmatrix}$. $\begin{bmatrix} 5 & 10 \\ 15 & 20 \end{bmatrix}$

(b) $\begin{bmatrix} 5 & 0 \\ 0 & 5 \end{bmatrix}\begin{bmatrix} 1 & 2 \\ 3 & 4 \end{bmatrix} = \begin{bmatrix} 5 & 10 \\ 15 & 20 \end{bmatrix}$

(c) $[1 \ 2 \ 3]\begin{bmatrix} 4 \\ 5 \\ 6 \end{bmatrix}$. $= [32]$

(d) $\begin{bmatrix} 4 \\ 5 \\ 6 \end{bmatrix}[1 \ 2 \ 3]$. undefined

(e) $\begin{bmatrix} 1 & 0 & 3 \\ 2 & -1 & 0 \end{bmatrix}\begin{bmatrix} 7 \\ -3 \\ 2 \end{bmatrix} = \begin{bmatrix} 13 \\ 17 \end{bmatrix}$

(f) $\begin{bmatrix} 1 & 0 \\ 0 & 1 \end{bmatrix}\begin{bmatrix} 18 & 72 \\ -3 & 5 \end{bmatrix} = \begin{bmatrix} 18 & 72 \\ -3 & 5 \end{bmatrix}$

(g) $\begin{bmatrix} 0 & 1 \\ 1 & 0 \end{bmatrix}\begin{bmatrix} 0 & 1 \\ 1 & 0 \end{bmatrix} = \begin{bmatrix} 1 & 0 \\ 0 & 1 \end{bmatrix} = I_{22}$

(h) $\begin{bmatrix} 2 & 1 \\ 1 & 1 \end{bmatrix}\begin{bmatrix} 1 & -1 \\ -1 & 2 \end{bmatrix} = \begin{bmatrix} 1 & 0 \\ 0 & 1 \end{bmatrix} = I_{22}$

(i) $\begin{bmatrix} 1 & 1 \\ -1 & 1 \end{bmatrix}\begin{bmatrix} 2 & 1 \\ -1 & 2 \end{bmatrix} = \begin{bmatrix} 1 & 3 \\ -1 & 1 \end{bmatrix}$

(j) $\begin{bmatrix} 2 & 3 \\ -3 & 2 \end{bmatrix}\begin{bmatrix} 8 & 2 \\ -2 & 8 \end{bmatrix} = \begin{bmatrix} 10 & 28 \\ -28 & 10 \end{bmatrix}$

(k) $\begin{bmatrix} \cos\theta & \sin\theta \\ -\sin\theta & \cos\theta \end{bmatrix}\begin{bmatrix} \cos\theta & -\sin\theta \\ \sin\theta & \cos\theta \end{bmatrix}$, $= \begin{bmatrix} 1 & 0 \\ 0 & 1 \end{bmatrix} = I_2$

(l) $\begin{bmatrix} 0 & 1 \\ 0 & 3 \end{bmatrix}\begin{bmatrix} 4 & 7 \\ 0 & 0 \end{bmatrix} = \begin{bmatrix} 0 & 0 \\ 0 & 0 \end{bmatrix} = 0_{22}$

where θ is a given constant.

2. Let $A \in \mathscr{F}_{mn}$ and let 0 be the zero matrix in $\mathscr{F}_{np}$. Show that $A0$ is the zero matrix in $\mathscr{F}_{mp}$. State and prove an analogous result for the product $0A$.

→**3.** If A, B are in $\mathscr{F}_{nn}$, is $AB = BA$? Explain why.

4. (a) Let $\mathcal{M}$ be the collection of all matrices of the form

$$\begin{bmatrix} a & b \\ -b & a \end{bmatrix}$$

where a, b are real numbers. Show that $\mathcal{M}$ is closed with respect to

  (i) matrix addition,
  (ii) matrix multiplication, and
  (iii) scalar multiplication by real numbers;

that is, show that the sum of two matrices in $\mathcal{M}$ is a matrix in $\mathcal{M}$, and so on.

(b) Let the matrix

$$\begin{bmatrix} a & b \\ -b & a \end{bmatrix}$$

correspond to the complex number $a + bi$. What complex numbers correspond to the following matrices?

$$\begin{bmatrix} 1 & 3 \\ -3 & 1 \end{bmatrix}; \quad \begin{bmatrix} 5 & 1 \\ -1 & 5 \end{bmatrix}; \quad \begin{bmatrix} 0 & 1 \\ -1 & 0 \end{bmatrix}; \quad \begin{bmatrix} 1 & 0 \\ 0 & 1 \end{bmatrix}.$$

(c) What matrices in $\mathcal{M}$ correspond to real numbers?

(d) Using the correspondence of part (b) between the collection $\mathcal{M}$ and $\mathscr{C}$ show that $\mathcal{M}$ is a matrix model of the complex numbers. That is, show that matrix addition agrees with addition of the corresponding complex numbers and that matrix multiplication agrees with multiplication of the corresponding complex numbers.

5. Let $A \in \mathscr{F}_{mn}$, $B \in \mathscr{F}_{np}$, $C \in \mathscr{F}_{pq}$. Show that the element in the ith row, jth column of $AB(C)$ is $\sum_{k=1}^{n} \sum_{r=1}^{p} a_{ik} b_{kr} c_{rj}$. What is the element in the ith row, jth column of $A(BC)$?

Let us observe, by means of an example, that it is possible to find matrices $A, B \in \mathscr{F}_{nn}$ such that $AB \neq BA$.

Example 2. Let

$$A = \begin{bmatrix} 0 & 1 \\ 0 & 0 \end{bmatrix}, \quad B = \begin{bmatrix} 1 & 0 \\ 0 & 0 \end{bmatrix}.$$

Then $AB = 0$, while $BA = A \neq 0$. This example illustrates that matrix multiplication, unlike the ordinary multiplication of numbers, is not commutative. This does not exclude the possibility that $AB = BA$ for some pairs A, B.

2.3 Matrix Multiplication

The fundamental properties of matrix multiplication are given in the following theorem.

Theorem 1. *If A, B, C are matrices, then*

(i) $A(BC) = (AB)C$,

(ii) $A(B + C) = AB + AC$,

(iii) $(A + B)C = AC + BC$,

(iv) $\alpha(AB) = (\alpha A)B = A(\alpha B)$ *for all α in F,*

whenever the indicated operations are defined.

Proof. We illustrate by proving (ii); the other parts are similar. (Note that part (i) is contained in Exercise 5.)

Let $A \in \mathscr{F}_{mn}$, and let $B, C \in \mathscr{F}_{np}$. Let $D = B + C$, $F = AD$, $G = AB$, $H = AC$, and $M = G + H$. Then $A(B + C) = AD = F$ and $AB + AC = G + H = M$. We must prove that $F = M$. We have

$$f_{ij} = \sum_{k=1}^{n} a_{ik} d_{kj} = \sum_{k=1}^{n} a_{ik}(b_{kj} + c_{kj})$$

$$= \sum_{k=1}^{n} (a_{ik} b_{kj} + a_{ik} c_{kj})$$

$$= \sum_{k=1}^{n} a_{ik} b_{kj} + \sum_{k=1}^{n} a_{ik} c_{kj}$$

$$= g_{ij} + h_{ij} = m_{ij} \qquad (i = 1, \ldots, m; j = 1, \ldots, p)$$

and thus $F = M$. ∎

We observe that, if $A, B, C \in \mathscr{F}_{nn}$, then all operations in Theorem 1 are defined.

● EXERCISES

6. Find two matrices $A, B \in \mathscr{F}_{33}$ such that $AB \neq BA$.

7. Show that the matrix

$$A = \begin{bmatrix} 3 & -1 & 1 \\ 2 & 0 & 1 \\ 1 & -1 & 2 \end{bmatrix}$$

satisfies the equation $A^3 - 5A^2 + 8A - 41 = 0$. (Here A^2 means AA and A^3 means $A^2A = (AA)A$.)

8. Prove parts (iii) and (iv) of Theorem 1.

2.4 Nonsingular Matrices and Inverses

The identity element for multiplication in $\mathscr{F}$ is the number 1. That is, for every $\alpha \in \mathscr{F}$, $\alpha 1 = 1\alpha = \alpha$. For multiplication in $\mathscr{F}_{nn}$, we have the following analog.

Definition 1. *The matrix I in $\mathscr{F}_{nn}$ with entries e_{ij}, where*

$$e_{ij} = \begin{cases} 0 & \text{if } i \neq j \\ 1 & \text{if } i = j \end{cases}$$

is called the $n \times n$ identity matrix.

Where it is necessary to distinguish, we shall denote the $n \times n$ identity matrix by I_n. An identity matrix has the following important properties:

● **EXERCISES**

1. Prove that $AI = IA = A$ for every $A \in \mathscr{F}_{nn}$.
2. Prove that if $A \in \mathscr{F}_{mn}$, then $I_m A = A$ and $AI_n = A$.

The reader will recall that in $\mathscr{F}$ every $\alpha \neq 0$ has a multiplicative inverse α^{-1}; that is, an element of $\mathscr{F}$ with the property that $\alpha\alpha^{-1} = \alpha^{-1}\alpha = 1$. In $\mathscr{F}_{nn}$, this fact has no precise analog, as is shown by the following example.

Example 1. Let

$$A = \begin{bmatrix} 1 & 0 \\ 0 & 0 \end{bmatrix}.$$

Suppose there exists a matrix $X \in \mathscr{F}_{22}$ such that $AX = I_2$. From the definition of multiplication,

$$AX = \begin{bmatrix} x_{11} & x_{12} \\ 0 & 0 \end{bmatrix}.$$

In order to have $AX = I_2$, we must, therefore, have

$$\begin{bmatrix} x_{11} & x_{12} \\ 0 & 0 \end{bmatrix} = \begin{bmatrix} 1 & 0 \\ 0 & 1 \end{bmatrix},$$

which is impossible. Thus, there is no $X \in \mathscr{F}_{22}$ such that $AX = I_2$.

2.4 Nonsingular Matrices and Inverses

On the other hand, if

$$A = \begin{bmatrix} 2 & 0 \\ 0 & 3 \end{bmatrix},$$

then the matrix

$$B = \begin{bmatrix} \frac{1}{2} & 0 \\ 0 & \frac{1}{3} \end{bmatrix}$$

satisfies $AB = BA = I_2$, as the reader may easily verify. To discuss a somewhat more complicated situation, we consider the following example.

Example 2. Let

$$A = \begin{bmatrix} 3 & 4 \\ 2 & 3 \end{bmatrix}.$$

We wish to find a matrix

$$X = \begin{bmatrix} x_{11} & x_{12} \\ x_{21} & x_{22} \end{bmatrix}$$

such that $AX = I$.

From the definition of multiplication, we see that

$$AX = \begin{bmatrix} 3 & 4 \\ 2 & 3 \end{bmatrix} \begin{bmatrix} x_{11} & x_{12} \\ x_{21} & x_{22} \end{bmatrix} = \begin{bmatrix} 3x_{11} + 4x_{21} & 3x_{12} + 4x_{22} \\ 2x_{11} + 3x_{21} & 2x_{12} + 3x_{22} \end{bmatrix}.$$

Thus, $AX = I$ is equivalent to the systems of linear equations

$$\begin{aligned} 3x_{11} + 4x_{21} &= 1 \\ 2x_{11} + 3x_{21} &= 0 \end{aligned} \tag{2.4a}$$

$$\begin{aligned} 3x_{12} + 4x_{22} &= 0 \\ 2x_{12} + 3x_{22} &= 1 \end{aligned} \tag{2.4b}$$

Solving (2.4a), we obtain $x_{11} = 3$, $x_{21} = -2$. Then solving (2.4b), we obtain $x_{12} = -4$, $x_{22} = 3$. Thus, the desired matrix X is given by

$$X = \begin{bmatrix} 3 & -4 \\ -2 & 3 \end{bmatrix}.$$

EXERCISES

3. Solve the pairs of Eq. 2.4a and 2.4b above.
4. (a) By direct computation, verify that, with A and X as in Example 2 above, $AX = I$ and $XA = I$.
   (b) Use the fact that $XA = AX = I$ to solve the system of equations

   $3y_1 + 4y_2 = 7$
   $2y_1 + 3y_2 = 5,$

   and also the more general system

   $3y_1 + 4y_2 = z_1$
   $2y_1 + 3y_2 = z_2,$

   where z_1 and z_2 are any given real numbers.

In view of the fact that for some matrices A there exists a matrix X such that $AX = 1$, we make the following definition.

Definition 2. *Let $A \in \mathscr{F}_{nn}$. If there exists a matrix $X \in \mathscr{F}_{nn}$ such that $AX = I$ and $XA = I$, then X is said to be an inverse of A.*

Thus, in Example 2 above, we have shown that the matrix

$$\begin{bmatrix} 3 & -4 \\ -2 & 3 \end{bmatrix}$$

is an inverse of the matrix

$$\begin{bmatrix} 3 & 4 \\ 2 & 3 \end{bmatrix}.$$

The reader should note that the argument in Example 2 showed that the inverse was unique. In fact, this is true in general.

Theorem 1. *Let $X, \hat{X} \in \mathscr{F}_{nn}$ be inverses of $A \in \mathscr{F}_{nn}$. Then $X = \hat{X}$.*

Proof. By the definition of inverse, we have $XA = I$ and $A\hat{X} = I$. Hence, $X = XI = X(A\hat{X}) = (XA)\hat{X} = I\hat{X} = \hat{X}.$ ∎

This result says that if a matrix A has an inverse, then it has only one inverse, and we speak of the inverse of A.

Definition 3. *A matrix $A \in \mathscr{F}_{nn}$ which has an inverse is said to be nonsingular, and the (unique) inverse of A is denoted by A^{-1}.*

2.4 Nonsingular Matrices and Inverses

Definition 4. *A matrix $A \in \mathscr{F}_{nn}$ which does not have an inverse is said to be singular.*

In Chapter 3 (more precisely, in Exercise 6, Section 3.8), we will see that if $AX = I$, then also $XA = I$. Thus, to find the inverse of a matrix A, it is sufficient to find a matrix X such that $AX = I$. Note that this is what we did in Example 2, and that we verified in Exercise 4 that $XA = I$.

Theorem 2. *If A and B are nonsingular matrices in $\mathscr{F}_{nn}$, then AB is a nonsingular matrix in $\mathscr{F}_{nn}$, and*

$$(AB)^{-1} = B^{-1}A^{-1}.$$

Proof. To prove the result, we need only verify that $(B^{-1}A^{-1})(AB) = I$ and $(AB)(B^{-1}A^{-1}) = I$. But

$$(B^{-1}A^{-1})(AB) = B^{-1}(A^{-1}A)B = B^{-1}IB = B^{-1}B = I,$$
$$(AB)(B^{-1}A^{-1}) = A(BB^{-1})A^{-1} = AIA^{-1} = AA^{-1} = I. \blacksquare$$

It is easy to prove the following corollary by induction.

Corollary to Theorem 2. *Let $A_1, A_2, \ldots, A_k$ be nonsingular matrices in $\mathscr{F}_{nn}$. Then $A_1 A_2 \cdots A_n$ is nonsingular and*

$$(A_1 A_2 \cdots A_k)^{-1} = A_k^{-1} A_{k-1}^{-1} \cdots A_1^{-1}.$$

● **EXERCISE**

5. Prove the above corollary.

We have shown that the set of all nonsingular matrices in $\mathscr{F}_{nn}$ is closed under multiplication, and is associative under multiplication (Theorem 1(i), Section 2.3). Also, there is an identity matrix I such that $AI = IA = A$ for every A in $\mathscr{F}_{nn}$, and for every nonsingular matrix A in $\mathscr{F}_{nn}$ there is a nonsingular matrix A^{-1} such that $AA^{-1} = A^{-1}A = I$. The reader should compare these with the properties (i), (iii), (iv) of Theorem 1, Section 2.2. A system, such as the set of all nonsingular matrices in $\mathscr{F}_{nn}$, with these properties is called a (multiplicative) *group*. Since it is not necessarily true that $AB = BA$ for two nonsingular matrices A, B in $\mathscr{F}_{nn}$, this group is *noncommutative*.

In Example 2, we observed that it is possible to find nonzero matrices A, B such that $AB = 0$. However, if one of the matrices A, B is nonsingular, this cannot happen.

Theorem 3. Let $A \in \mathscr{F}_{nn}$ be nonsingular and let $B \in \mathscr{F}_{np}$ be such that $AB = 0 \in \mathscr{F}_{np}$. Then $B = 0$.

Proof. Since $AB = 0$ and A is nonsingular,

$$B = IB = (A^{-1}A)B = A^{-1}(AB) = A^{-1}0 = 0. \blacksquare$$

As an application of the ideas of this section, we consider a system of n algebraic equations in n unknowns

$$A\mathbf{x} = \mathbf{b}, \tag{2.5}$$

where $A \in \mathscr{F}_{nn}$ and $\mathbf{b} \in \mathscr{F}_{n1}$ are given. By a solution, we mean a vector $\mathbf{u} \in \mathscr{F}_{n1}$ such that $A\mathbf{u} = \mathbf{b}$. If A is nonsingular and $\mathbf{u}$ is a solution, multiplication of both sides of the equation $A\mathbf{u} = \mathbf{b}$ on the left shows that $\mathbf{u} = A^{-1}\mathbf{b}$, that is, $\mathbf{u} = A^{-1}\mathbf{b}$ is the only possible solution of (2.5). Direct verification shows that $\mathbf{u} = A^{-1}\mathbf{b}$ does satisfy Eq. 2.5. Thus, we have shown that if A is nonsingular, then the system of n equations in n unknowns $A\mathbf{x} = \mathbf{b}$ has the **unique** solution $\mathbf{u} = A^{-1}\mathbf{b}$. In Chapter 3, we shall consider linear systems of algebraic equations when the number of equations is not necessarily the same as the number of unknowns.

● **EXERCISES**

6. Find the inverses of the following matrices by solving the systems of algebraic equations $AX = I$.

(a) $\begin{bmatrix} 1 & 0 \\ 0 & 2 \end{bmatrix}$.

(b) $\begin{bmatrix} 1 & 0 & 0 \\ 0 & 2 & 0 \\ 0 & 0 & 3 \end{bmatrix}$.

(c) $\begin{bmatrix} 1 & 2 & 3 \\ 0 & 4 & 5 \\ 0 & 0 & 6 \end{bmatrix}$.

(d) $\begin{bmatrix} 4 & 3 \\ 1 & 5 \end{bmatrix}$.

7. A square $n \times n$ matrix whose only nonzero entries occur in the positions a_{ii}, $i = 1, \ldots, n$, is called a **diagonal matrix**, (that is, $a_{ij} = 0$ if $i \neq j$). In general, $a_{11}, a_{22}, a_{33}, \ldots, a_{nn}$ is called the **main diagonal**.
 (a) Prove that the product of two diagonal matrices is a diagonal matrix.
 (b) Prove that if A is diagonal with nonzero entries on the main diagonal, then A is nonsingular. Describe A^{-1}.

8. Square matrices whose only nonzero entries occur on or above the main diagonal are called **upper triangular matrices** (that is, $a_{ij} = 0$ if $i > j$). Prove

2.4 Nonsingular Matrices and Inverses

that the product of two upper triangular matrices T_1, T_2 is an upper triangular matrix.

9. (a) Show that an upper triangular matrix whose main diagonal has no zero entries is nonsingular. [*Hint:* Follow the argument used in Example 2].
   (b) Show that an upper triangular matrix with a zero entry on the main diagonal is singular. [*Hint:* Follow the argument used in Example 1.]

10. A **strictly upper triangular matrix** is a triangular matrix whose diagonal entries are zero (that is, $a_{ij} = 0$ if $i \geq j$). Prove that the product of a strictly upper triangular matrix S and an upper triangular matrix T is strictly upper triangular. (One must check ST and TS.)

11. Let S be any $n \times n$ strictly upper triangular matrix. Prove that $S^n = 0$.

12. Show in two different ways that a strictly upper triangular matrix is singular. [*Hint:* One way is to use Exercise 9; another is to use Theorem 3 and Exercise 11.]

13. Show that the $n \times n$ matrix

$$A = \begin{bmatrix} 1 & 1 & 0 & \cdots & 0 \\ 0 & 0 & 1 & \cdot & \vdots \\ \vdots & \vdots & & \cdot & \cdot \\ \vdots & \vdots & & \cdot & 0 \\ 0 & 0 & \cdots & 0 & 1 \\ \varepsilon & 0 & \cdots & 0 & 0 \end{bmatrix}$$

is nonsingular if $\varepsilon \neq 0$.

14. Let S be an $n \times n$ strictly upper triangular matrix. Show that
$$(I - S)^{-1} = I + S + S^2 + \cdots + S^{n-1}.$$
[*Hint:* See Exercise 11.]

15. Let $A, B \in \mathscr{F}_{nn}$ and suppose $AB = BA$.
    (a) Show that $(A + B)^2 = A^2 + 2AB + B^2$.
    (b) Prove that for every integer $k \geq 1$,
    $$(A + B)^k = A^k + \binom{k}{1} A^{k-1} B + \binom{k}{2} A^{k-2} B^2 + \cdots + B^k,$$
    where $\binom{k}{j}$ is the binomial coefficient $k!/j!(k-j)!$

16. Find two square matrices, C and D, such that $(C + D)^2 \neq C^2 + 2CD + D^2$. Why does equality not hold in general?

17. Find all 2×2 matrices X such that $X^2 = I = \begin{bmatrix} 1 & 0 \\ 0 & 1 \end{bmatrix}$.

18. If A is nonsingular and $AB = AC$, show that $B = C$.

19. If $A \in \mathscr{F}_{nn}$ has a column of zeros or a row of zeros, show that A must be singular.

Matrix Operations

20. In each of the following, write the linear system of algebraic equations in the matrix form $A\mathbf{x} = \mathbf{b}$, that is, in the form

$$\begin{bmatrix} a_{11} & a_{12} \\ a_{21} & a_{22} \end{bmatrix} \begin{bmatrix} x_1 \\ x_2 \end{bmatrix} = \begin{bmatrix} b_{11} \\ b_{12} \end{bmatrix},$$

where the a_{ij}'s and the b_{ij}'s are given constants. Then multiply both sides of the matrix equation on the left by A^{-1} (that is, $A^{-1}A\mathbf{x} = I\mathbf{x} = A^{-1}\mathbf{b}$) and simplify to obtain the solution of the given system.

(a) $3x_1 + 4x_2 = 0$
$2x_1 + 3x_2 = 5;$ $\quad A^{-1} = \begin{bmatrix} 3 & -4 \\ 2 & 3 \end{bmatrix}.$

(b) $3x_1 + 2x_2 = -3$
$7x_1 + 5x_2 = 1;$ $\quad A^{-1} = \begin{bmatrix} 5 & -2 \\ -7 & 3 \end{bmatrix}.$

(c) $7x_1 + 3x_2 = 12$
$9x_1 + 4x_2 = 7$ $\quad A^{-1} = \begin{bmatrix} 4 & -3 \\ -9 & 7 \end{bmatrix}.$

(d) $2x_1 - 5x_2 = -3$
$1x_1 - 2x_2 = 4;$ $\quad A^{-1} = \begin{bmatrix} -2 & 5 \\ -1 & 2 \end{bmatrix}.$

21. In Exercise 7, Section 2.3, it was shown that the matrix

$$A = \begin{bmatrix} 3 & -1 & 1 \\ 2 & 0 & 1 \\ 1 & -1 & 2 \end{bmatrix}$$

satisfies the equation $A^3 - 5A^2 + 8A - 4I = 0$. Deduce that A is nonsingular and that $A^{-1} = \frac{1}{4}(A^2 - 5A + 8I)$.

Chapter 3 | LINEAR SYSTEMS OF ALGEBRAIC EQUATIONS

Some of the problems formulated in Chapter 1 lead to linear systems of algebraic equations. In Chapter 2 we have developed a notation which will be convenient for the systematic study of such systems. In this chapter, we study the solution of linear systems of algebraic equations. In addition to providing the means of solving some of the problems formulated in Chapter 1, this will also lay the foundation for a study of abstract vector spaces in Chapter 5.

3.1 Three Examples

In this chapter we shall study linear systems of m equations in n unknowns. To illustrate some of the main ideas, we begin with some examples.

Example 1. Solve the system of equations

$$\begin{aligned} x_1 + x_2 + x_3 &= 0 \\ x_2 + x_3 &= -1 \\ x_1 + x_2 \phantom{{}+x_3} &= 1. \end{aligned} \tag{3.1}$$

We may write this in matrix-vector form (see Section 2.3) as

$$A\mathbf{x} = \mathbf{b},$$

36 Linear Systems of Algebraic Equations

where

$$A = \begin{bmatrix} 1 & 1 & 1 \\ 0 & 1 & 1 \\ 1 & 1 & 0 \end{bmatrix}, \quad \mathbf{x} = \begin{bmatrix} x_1 \\ x_2 \\ x_3 \end{bmatrix}, \quad \mathbf{b} = \begin{bmatrix} 0 \\ -1 \\ 1 \end{bmatrix}.$$

By a **solution** of (3.1), we mean a vector

$$\mathbf{u} = \begin{bmatrix} u_1 \\ u_2 \\ u_3 \end{bmatrix} \in \mathscr{F}_{31}$$

such that $A\mathbf{u} = \mathbf{b}$, or written out in terms of entries such that

$$\begin{aligned} u_1 + u_2 + u_3 &= 0 \\ u_2 + u_3 &= -1 \\ u_1 + u_2 &= 1. \end{aligned} \qquad (3.2)$$

To solve the system (3.1), we proceed as follows. If $\mathbf{u}$ is a solution of (3.1), that is, if (3.2) is satisfied, then we may subtract the second equation in (3.2) from the first to obtain $u_1 = 1$. Similarly, we may subtract the third equation in (3.2) from the first to obtain $u_3 = -1$, and therefore $u_2 = 0$. Thus, the only possible solution of (3.1) is

$$\mathbf{u} = \begin{bmatrix} 1 \\ 0 \\ -1 \end{bmatrix}.$$

On the other hand, defining $\mathbf{u}$ to be this vector, we see by direct substitution that $A\mathbf{u} = \mathbf{b}$. Thus,

$$\mathbf{u} = \begin{bmatrix} 1 \\ 0 \\ -1 \end{bmatrix}$$

is the unique solution of (3.1).

● **EXERCISE**

1. Verify that

$$\mathbf{u} = \begin{bmatrix} 1 \\ 0 \\ -1 \end{bmatrix}$$

is a solution of (3.1).

3.1 Three Examples

The reader should not suppose that every linear system of three equations in three unknowns has a unique solution. Consider the following example.

Example 2. Solve the system of equations

$$\begin{aligned} x_1 + 2x_2 + x_3 &= 0 \\ x_2 + x_3 &= -1 \\ x_1 + x_2 &= 1. \end{aligned} \qquad (3.3)$$

Again, we may write this as $A\mathbf{x} = \mathbf{b}$, where

$$A = \begin{bmatrix} 1 & 2 & 1 \\ 0 & 1 & 1 \\ 1 & 1 & 0 \end{bmatrix}, \quad \mathbf{x} = \begin{bmatrix} x_1 \\ x_2 \\ x_3 \end{bmatrix}, \quad \mathbf{b} = \begin{bmatrix} 0 \\ -1 \\ 1 \end{bmatrix}.$$

If $\mathbf{u}$ is a solution of (3.3), then $A\mathbf{u} = \mathbf{b}$, or

$$\begin{aligned} u_1 + 2u_2 + u_3 &= 0 \\ u_2 + u_3 &= -1 \\ u_1 + u_2 &= 1. \end{aligned} \qquad (3.4)$$

From the last two equations in (3.4), we obtain $u_3 = -1 - u_2$, $u_1 = 1 - u_2$. Therefore, letting $u_2 = \alpha$, we must have $u_1 = 1 - \alpha$, $u_3 = -1 - \alpha$. We verify by direct substitution that the vector

$$\mathbf{u} = \begin{bmatrix} 1 - \alpha \\ \alpha \\ -1 - \alpha \end{bmatrix}$$

is a solution of (3.3) for **every** $\alpha \in \mathscr{F}$. Thus, the system (3.3) has infinitely many solutions.

● EXERCISE

2. Verify that

$$\mathbf{u} = \begin{bmatrix} 1 - \alpha \\ \alpha \\ -1 - \alpha \end{bmatrix}$$

is a solution of (3.3).

The reader should not suppose that every linear system of three equations in three unknowns has a solution. Consider the following example.

Example 3. Solve the system of equations

$$\begin{aligned} x_1 + 2x_2 + x_3 &= 1 \\ x_2 + x_3 &= -1 \\ x_1 + x_2 &= 1. \end{aligned} \qquad (3.5)$$

Again, we may write this as $A\mathbf{x} = \mathbf{b}$, where

$$A = \begin{bmatrix} 1 & 2 & 1 \\ 0 & 1 & 1 \\ 1 & 1 & 0 \end{bmatrix}, \quad \mathbf{x} = \begin{bmatrix} x_1 \\ x_2 \\ x_3 \end{bmatrix}, \quad \mathbf{b} = \begin{bmatrix} 1 \\ -1 \\ 1 \end{bmatrix}.$$

If $\mathbf{u}$ is a solution of (3.5), then $A\mathbf{u} = \mathbf{b}$, or

$$\begin{aligned} u_1 + 2u_2 + u_3 &= 1 \\ u_2 + u_3 &= -1 \\ u_1 + u_2 &= 1. \end{aligned}$$

Adding the last two equations, we obtain $u_1 + 2u_2 + u_3 = 0$. But we are given that $u_1 + 2u_2 + u_3 = 1$, and these equalities cannot both hold. Thus, there is no solution of (3.5).

These examples are very special cases of a general theory, which we shall develop in the remainder of this chapter.

● **EXERCISES**

3. Indicate for each of the following systems of equations whether there is no solution, a unique solution, or an infinite number of solutions.

(a) $\quad x_1 + x_2 = 0$
$\quad\quad 2x_1 - x_2 = 0.$

(b) $\quad x_1 + x_2 = 0.$
$\quad\quad 2x_1 + x_2 = 0.$

(c) $\quad x_1 + x_2 = 0$
$\quad\quad 2x_1 + 2x_2 = 0.$

(d) $\quad x_1 + x_2 = 0$
$\quad\quad 2x_1 + 2x_2 = 2.$

4. Can a linear system of algebraic equations have exactly two solutions? [*Hint:* Show that if $\mathbf{u}$ and $\mathbf{v}$ are different solutions, then $\alpha\mathbf{u} + (1 - \alpha)\mathbf{v}$ is also a solution for every real value of α.]

3.2 Solutions and Equivalent Systems

We now consider the linear system of m algebraic equations in n unknowns

$$A\mathbf{x} = \mathbf{b} \qquad (3.6)$$

where $A \in \mathscr{F}_{mn}$, $\mathbf{b} \in \mathscr{F}_{m1}$, and where $\mathbf{x}$ is an unknown n-dimensional column vector.

3.2 Solutions and Equivalent Systems

Definition 1. *A solution of the system* $A\mathbf{x} = \mathbf{b}$ *is a vector* $\mathbf{u} \in \mathscr{F}_{n1}$ *such that* $A\mathbf{u} = \mathbf{b}$. *The set of all solutions of* $A\mathbf{x} = \mathbf{b}$ *constitutes the solution set.*

Definition 2. *The systems* $A\mathbf{x} = \mathbf{b}$ *and* $C\mathbf{x} = \mathbf{d}$ *are said to be equivalent if and only if* (i) $A, C \in \mathscr{F}_{mn}$, (ii) $\mathbf{b}, \mathbf{d} \in \mathscr{F}_{m1}$, *and* (iii) *the two systems have the same solution set.*

We observe that all the information about the system (3.6) is contained in the matrix A and the column vector $\mathbf{b}$. It is convenient to define the **augmented matrix**

$$[A\,|\,\mathbf{b}] = \begin{bmatrix} a_{11} & \cdots & a_{1n} & b_1 \\ \vdots & & \vdots & \vdots \\ a_{m1} & \cdots & a_{mn} & b_m \end{bmatrix},$$

where $[A\,|\,\mathbf{b}] \in \mathscr{F}_{m,\,n+1}$.

A general procedure for solving a system such as (3.6) is to reduce it to an equivalent system of a simpler form. An example will illustrate the procedure.

Example 1. Consider the system

$$\begin{aligned} x_1 + x_2 + 5x_3 &= 11 \\ 2x_1 + x_2 + 7x_3 &= 15 \\ 2x_1 + 4x_3 &= 8. \end{aligned} \qquad (3.7)$$

A systematic procedure to solve this system is the following:

STEP 1. Subtract twice the first equation from the second, to obtain

$$\begin{aligned} x_1 + x_2 + 5x_3 &= 11 \\ -x_2 - 3x_3 &= -7 \\ 2x_1 + 4x_3 &= 8. \end{aligned}$$

STEP 2. Subtract twice the first equation from the third, to obtain

$$\begin{aligned} x_1 + x_2 + 5x_3 &= 11 \\ -x_2 - 3x_3 &= -7 \\ -2x_2 - 6x_3 &= -14. \end{aligned}$$

STEP 3. Multiply the second equation by -1, to obtain

$$\begin{aligned} x_1 + x_2 + 5x_3 &= 11 \\ x_2 + 3x_3 &= 7 \\ -2x_2 - 6x_3 &= -14. \end{aligned}$$

STEP 4. Subtract the second equation from the first, to obtain

$$x_1 \quad\quad + 2x_3 = 4$$
$$x_2 + 3x_3 = 7$$
$$-2x_2 - 6x_3 = -14.$$

STEP 5. Add twice the second equation to the third, to obtain

$$x_1 \quad\quad + 2x_3 = 4$$
$$x_2 + 3x_3 = 7 \tag{3.8}$$
$$0 = 0.$$

Assuming that (3.8) is equivalent to (3.7) (as will be shown later), we see that if x_3 is any element of $\mathscr{F}$, say $x_3 = \alpha$, then we can read off from the last system that $x_1 = 4 - 2\alpha$, $x_2 = 7 - 3\alpha$, $x_3 = \alpha$. Thus,

$$\mathbf{u} = \begin{bmatrix} 4 - 2\alpha \\ 7 - 3\alpha \\ \alpha \end{bmatrix} \tag{3.9}$$

is a solution for every $\alpha \in \mathscr{F}$.

● **EXERCISES**

1. Show by direct substitution that the vector given by (3.9) is a solution of the system (3.7).

2. Show that (3.7) and (3.8) are equivalent. [*Hint:* Suppose that **u** is a solution of (3.7). Proceed as in the examples in Section 3.1 to show that **u** satisfies (3.8), and conversely.]

We now rework Example 1 from a different point of view. We write down the augmented matrix of (3.7)

$$S_0 = \begin{bmatrix} 1 & 1 & 5 & | & 11 \\ 2 & 1 & 7 & | & 15 \\ 2 & 0 & 4 & | & 8 \end{bmatrix}.$$

The augmented matrix obtained after Step 1 above is

$$S_1 = \begin{bmatrix} 1 & 1 & 5 & | & 11 \\ 0 & -1 & -3 & | & -7 \\ 2 & 0 & 4 & | & 8 \end{bmatrix},$$

and we observe that S_1 is obtained from S_0 by subtracting twice the first row of S_0 from the second row of S_0. Similarly, the augmented matrix obtained after Step 2 above is

$$S_2 = \begin{bmatrix} 1 & 1 & 5 & | & 11 \\ 0 & -1 & -3 & | & -7 \\ 0 & -2 & -6 & | & -14 \end{bmatrix},$$

which is obtained from S_1 by subtracting twice the first row of S_1 from the third row of S_1. Analogously, we obtain respectively the matrices

$$S_3 = \begin{bmatrix} 1 & 1 & 5 & | & 11 \\ 0 & 1 & 3 & | & 7 \\ 0 & -2 & -6 & | & -14 \end{bmatrix},$$

$$S_4 = \begin{bmatrix} 1 & 0 & 2 & | & 4 \\ 0 & 1 & 3 & | & 7 \\ 0 & -2 & -6 & | & -14 \end{bmatrix},$$

$$S_5 = \begin{bmatrix} 1 & 0 & 2 & | & 4 \\ 0 & 1 & 3 & | & 7 \\ 0 & 0 & 0 & | & 0 \end{bmatrix}$$

from Steps 3, 4, and 5.

● **EXERCISES**

3. What operations on the rows give S_3 from S_2, S_4 from S_3, and S_5 from S_4?
4. What operations on the rows give S_4 from S_5, S_3 from S_4, S_2 from S_3, S_1 from S_2, and S_0 from S_1?

The operations on the rows of the augmented matrix correspond exactly to the operations on equations, and it is much more convenient to work directly with the matrices. We will therefore discuss the general theory of row operations and then apply it to linear systems of algebraic equations.

3.3 Elementary Row Operations and Elementary Matrices

We begin with the following definitions.

Definition 1. *The elementary row operations on a matrix $A \in \mathscr{F}_{mn}$ are of the following types*:

TYPE I. *Interchange two distinct rows (symbolically $a_{r*} \to a_{s*}$ and $a_{s*} \to a_{r*}$, $r \neq s$).* (Recall that a_{r*} denotes the rth row of A; see Section 2.3.)

TYPE II. *Multiply a row by a nonzero scalar λ (symbolically $a_{r*} \to \lambda a_{r*}$, $\lambda \neq 0$).*

TYPE III. *Add a scalar multiple of one row to a different row (symbolically $a_{r*} \to a_{r*} + \lambda a_{s*}$; $r \neq s$).*

In Example 1, Section 3.2, beginning with the matrix S_0 and applying the elementary row operation of Type III: $a_{2*} \to a_{2*} - 2a_{1*}$, we obtain the matrix S_1. Similarly, the elementary row operation of Type III: $a_{3*} \to a_{3*} - 2a_{1*}$ yields S_2.

● **EXERCISES**

1. In each of the following examples, state which elementary row operation applied to the matrix A yields the matrix B.

(a) $A = \begin{bmatrix} 1 & 0 & 0 \\ 0 & 2 & 0 \\ 0 & 0 & 3 \end{bmatrix}$, $B = \begin{bmatrix} 0 & 0 & 3 \\ 0 & 2 & 0 \\ 1 & 0 & 0 \end{bmatrix}$.

(b) $A = \begin{bmatrix} 1 & 0 & 1 \\ 0 & 1 & 2 \end{bmatrix}$, $B = \begin{bmatrix} 1 & 1 & 3 \\ 0 & 1 & 2 \end{bmatrix}$.

(c) $A = \begin{bmatrix} 1 & -1 \\ 0 & 2 \end{bmatrix}$, $B = \begin{bmatrix} \pi & -\pi \\ 0 & 2 \end{bmatrix}$.

2. In Exercise 1 above state which elementary row operation applied to B yields A.

3. For each of the matrices A in Exercise 1, perform the following sequence of elementary row operations and determine the resulting matrix C.
 (a) Interchange the first and last rows.
 (b) Multiply the second row by 2.
 (c) Add 3 times the last row to first row.

Let I_m be the $m \times m$ identity matrix. If we perform on I_m the elementary row operation of Type I: $a_{r*} \to a_{s*}$ and $a_{s*} \to a_{r*}$ ($r \neq s$), we obtain a matrix called an **elementary matrix of Type I** denoted by

$$E_1 = \begin{bmatrix} 1 & & & & & & & & \\ & \ddots & & & & & & & \\ & & 1 & & & & & & \\ & & & 0 & \cdots & 1 & & & \leftarrow r \\ & & & & 1 & & & & \\ & & & \vdots & & \ddots & & \vdots & \\ & & & & & & 1 & & \\ & & & 1 & \cdots & 0 & & & \leftarrow s \\ & & & & & & & 1 & \\ & & & & & & & & \ddots \\ & & & & & & & & & 1 \end{bmatrix} \quad (3.10)$$

$$\uparrow \qquad \uparrow$$
$$r \qquad s$$

3.3 Elementary Row Operations and Elementary Matrices

Thus, a matrix E is an elementary matrix of Type I if all diagonal elements are 1 except that $e_{rr} = e_{ss} = 0$ and the off diagonal elements are zero except that $e_{rs} = e_{sr} = 1$.

If we perform on I_m the elementary row operation of Type II: $a_{r*} \to \lambda a_{r*}$, $\lambda \neq 0$, we obtain a matrix called an **elementary matrix of Type II** denoted by

$$E_{\text{II}} = \begin{bmatrix} 1 & 0 & \cdots & & & 0 \\ & \ddots & & & & \\ 0 & & 1 & & & \vdots \\ & & & \lambda & & \\ \vdots & & & & 1 & 0 \\ & & & & & \ddots \\ 0 & \cdots & & & 0 & 1 \end{bmatrix} \leftarrow r \qquad (3.11)$$
$$\phantom{E_{\text{II}} = }\uparrow_{\;r}$$

Finally, if we perform on I_m the elementary row operation of Type III: $a_{r*} \to a_{r*} + \lambda a_{s*}$, $r \neq s$, we obtain a matrix called an **elementary matrix of Type III** denoted by

$$E_{\text{III}} = \begin{bmatrix} 1 & & & & & \\ & \ddots & & & & \\ & & 1 & 0 \ldots 0 & \lambda & \\ & & & & 0 & \\ & & & \ddots & \vdots & \\ & & & & 0 & \\ & & & & 1 & \\ & & & & & \ddots \\ & & & & & & 1 \end{bmatrix} \begin{matrix} \leftarrow r \\ \\ \\ \\ \\ \to s \\ \\ \end{matrix} \qquad (3.12)$$
$$\phantom{E_{\text{III}}=}\uparrow_{\;r}\;\;\uparrow_{\;s}$$

We now observe that every elementary row operation of Type I on a matrix $A \in \mathscr{F}_{mn}$ can be accomplished by premultiplying A by an elementary matrix of Type I and conversely for every elementary matrix E_{I} the operation $A \to E_{\text{I}} A$ is an elementary row operation of Type I. For, if

$$A = \begin{bmatrix} a_{11} & \cdots & a_{1n} \\ \vdots & & \vdots \\ a_{r1} & \cdots & a_{rn} \\ \vdots & & \vdots \\ a_{s1} & \cdots & a_{sn} \\ \vdots & & \vdots \\ a_{m1} & \cdots & a_{mn} \end{bmatrix}$$

and if $B = E_I$ is the matrix of (3.10), then $C = BA = E_I A$ can be computed by using formula (2.2). Thus, if $i \neq r$ and $i \neq s$,

$$c_{i*} = b_{i1} a_{1*} + b_{i2} a_{2*} + \cdots + b_{ii} a_{i*} + \cdots + b_{im} a_{m*}$$
$$= 0 \cdot a_{1*} + 0 a_{2*} + \cdots + 0 a_{(i-1)*} + 1 a_{i*} + 0 a_{(i+1)*} + \cdots + 0 a_{m*}$$
$$= a_{i*}.$$

On the other hand,

$$c_{r*} = b_{r1} a_{1*} + \cdots + b_{rr} a_{r*} + \cdots + b_{rs} a_{s*} + \cdots + b_{rm} a_{m*}$$
$$= 0 a_{1*} + \cdots + 0 a_{r*} + \cdots + 0 a_{(s-1)*} + 1 a_{s*} + 0 a_{(s+1)*} + \cdots + 0 a_{m*}$$
$$= a_{s*}.$$

Similarly, $c_{s*} = a_{r*}$. Hence,

$$C = E_I A = \begin{bmatrix} a_{11} & \cdots & a_{1n} \\ \vdots & & \vdots \\ a_{s1} & \cdots & a_{sn} \\ \vdots & & \vdots \\ a_{r1} & \cdots & a_{rn} \\ \vdots & & \vdots \\ a_{m1} & \cdots & a_{mn} \end{bmatrix}.$$

If E_{II} is an elementary matrix of Type II, the operation $A \to E_{II} A$ is an elementary row operation of Type II and conversely. For

$$E_{II} A = \begin{bmatrix} a_{11} & \cdots & a_{1n} \\ \vdots & & \vdots \\ \lambda a_{r1} & \cdots & \lambda a r_n \\ \vdots & & \vdots \\ a_{s1} & \cdots & a_{sn} \\ \vdots & & \vdots \\ a_{m1} & \cdots & a_{mn} \end{bmatrix}.$$

Finally, if E_{III} is an elementary matrix of Type III, the operation $A \to E_{III} A$ is an elementary row operation of Type III and conversely. For

3.3 Elementary Row Operations and Elementary Matrices

$$E_{\text{III}} A = \begin{bmatrix} a_{11} & \cdots & a_{1n} \\ \vdots & & \vdots \\ a_{r1} + \lambda a_{s1} & \cdots & a_{rn} + \lambda a_{sn} \\ \vdots & & \vdots \\ a_{s1} & \cdots & a_{sn} \\ \vdots & & \vdots \\ a_{m1} & \cdots & a_{mn} \end{bmatrix}.$$

Thus, premultiplication by elementary matrices corresponds completely to elementary row operations.

● EXERCISES

4. In each part of Exercise 1, find the elementary matrix P such that $B = PA$.

5. In each part of Exercise 2, find the elementary matrix Q such that $A = QB$.

6. In each part of Exercise 3, determine the matrix P as a product of elementary matrices such that $C = PA$.

7. Find two different elementary row operations which can be applied to the matrix A to yield the matrix B, if

$$A = \begin{bmatrix} 1 & 2 \\ 2 & 2 \\ 2 & 2 \end{bmatrix}, \quad B = \begin{bmatrix} 1 & 2 \\ 2 & 2 \\ 4 & 4 \end{bmatrix}.$$

It is important to establish that elementary matrices are nonsingular. In fact, we have the following result.

Lemma 1. *The inverse of any elementary matrix is an elementary matrix of the same type.*

Proof. The reader can easily observe that if E_{I} is a fixed elementary matrix of Type I, such as is given by (3.10), then $E_{\text{I}}^2 = I_m$. Thus, $E_{\text{I}}^{-1} = E_{\text{I}}$. Similarly, the inverse of the elementary matrix E_{II} of Type II, given by (3.11), is

$$E_{\text{II}}^{-1} = \begin{bmatrix} 1 & & & \cdots & & & 0 \\ & \ddots & & & & & \\ & & 1 & & & & \vdots \\ \vdots & & & 1/\lambda & & & \\ & & & & 1 & & 0 \\ & & & & & \ddots & \\ 0 & & & \cdots & & 0 & 1 \end{bmatrix} \leftarrow r,$$
$$\uparrow$$
$$r$$

which is an elementary matrix of Type II.

46   *Linear Systems of Algebraic Equations*

Finally, the inverse of the elementary matrix E_{III} of Type III given by (3.12) is

$$E_{\text{III}}^{-1} = \begin{bmatrix} 1 & & & & & & & \\ & \ddots & & & & & & \\ & & 1 & 0 & \cdots & 0 & -\lambda & \\ & & & & & & 0 & \\ & & & \ddots & & & \vdots & \\ & & & & & & 0 & \\ & & & & & & 1 & \\ & & & & & & & \ddots \\ & & & & & & & & 1 \end{bmatrix} \begin{matrix} \leftarrow r \\ \\ \\ \\ \leftarrow s \\ \\ \end{matrix}$$

$$\phantom{E_{\text{III}}^{-1} = \begin{bmatrix}}\uparrow\uparrow$$
$$\phantom{E_{\text{III}}^{-1} = \begin{bmatrix}xx}rs$$

which is an elementary matrix of Type III. ∎

● **EXERCISE**

8. Show that the matrix called E_{III}^{-1} is actually the inverse of the elementary matrix E_{III} given in (3.12).

3.4 Row Equivalence and Equivalent Systems of Equations

In order to relate the theory of elementary row operations to the theory of linear systems of algebraic equations, we introduce another concept.

Definition 1. *Let A, $B \in \mathscr{F}_{mn}$. Then B is said to be row equivalent to A (notation $A \overset{R}{\sim} B$) if and only if B is obtained from A by a finite sequence of elementary row operations.*

By the correspondence between elementary row operations and premultiplication by elementary matrices, $A \overset{R}{\sim} B$, if and only if there exist elementary matrices $E_1, E_2, \ldots, E_s$ (each of which may be of Type I, II, or III) such that

$$B = E_s E_{s-1} \cdots E_1 A. \tag{3.13}$$

It follows from (3.13) and Lemma 1, Section 3.3, that

(i) $A \overset{R}{\sim} A$ for every $A \in \mathscr{F}_{mn}$.
(ii) If $A \overset{R}{\sim} B$, then $B \overset{R}{\sim} A$.
(iii) If $A \overset{R}{\sim} B$, and $B \overset{R}{\sim} C$, then $A \overset{R}{\sim} C$.

3.4 Row Equivalence and Equivalent Systems of Equations

● EXERCISES

1. Prove (i), (ii), (iii).

2. Let $A \overset{R}{\sim} B$, where $A, B \in \mathscr{F}_{mn}$. Show that there is a nonsingular matrix $P \in \mathscr{F}_{mm}$ such that $B = PA$ and $A = P^{-1}B$. [*Hint:* P is a product of elementary matrices.]

3. For each of the following pairs of matrices A and B, find a nonsingular matrix P such that $B = PA$.

(a) $A = \begin{bmatrix} 1 & 2 & 3 \\ 2 & 4 & 6 \\ 1 & 2 & 3 \end{bmatrix}$, $B = \begin{bmatrix} 1 & 2 & 3 \\ 0 & 0 & 0 \\ 0 & 0 & 0 \end{bmatrix}$.

(b) $A = \begin{bmatrix} 1 & 1 & 1 \\ 1 & 0 & 1 \\ 0 & 0 & 0 \end{bmatrix}$, $B = \begin{bmatrix} 1 & 1 & 1 \\ 2 & 0 & 2 \\ 2 & 1 & 2 \end{bmatrix}$.

(c) $A = \begin{bmatrix} 1 & 0 \\ 0 & 0 \end{bmatrix}$, $B = \begin{bmatrix} 0 & 0 \\ 1 & 0 \end{bmatrix}$.

(d) $A = \begin{bmatrix} 1 & 1 & 1 \\ 0 & 1 & 0 \\ 0 & 1 & 0 \end{bmatrix}$, $B = \begin{bmatrix} 1 & 0 & 1 \\ 0 & 1 & 0 \\ 0 & 0 & 0 \end{bmatrix}$.

4. For each of the following matrices A, $\mathbf{b}$, $G = [A|\mathbf{b}]$ and P verify that $PG = [PA|P\mathbf{b}]$.

(a) $A = \begin{bmatrix} 1 & 2 \\ -2 & 4 \\ -1 & 7 \end{bmatrix}$, $\mathbf{b} = \begin{bmatrix} 3 \\ 6 \\ -3 \end{bmatrix}$, $P = \begin{bmatrix} 1 & 0 & 1 \\ 0 & 1 & 2 \\ 0 & 0 & -1 \end{bmatrix}$.

(b) $A = \begin{bmatrix} -1 & 0 & 5 \\ 4 & -i & 3 \end{bmatrix}$, $\mathbf{b} = \begin{bmatrix} i \\ -i \end{bmatrix}$, $P = \begin{bmatrix} 1 & i \\ -i & 1 \end{bmatrix}$.

We will show later, in Section 3.8, that the converse of Exercise 2 is also true; that is, if $B = PA$, where P is nonsingular, then $A \overset{R}{\sim} B$.

We are now in a position to relate the concept of row equivalence of matrices to the concept of equivalent systems of algebraic equations. [See Definition 2, Section 3.2.]

Theorem 1. *Let $[A|\mathbf{b}]$ and $[C|\mathbf{d}]$ be row-equivalent matrices in $\mathscr{F}_{m, n+1}$. Then the systems $A\mathbf{x} = \mathbf{b}$ and $C\mathbf{x} = \mathbf{d}$ are equivalent.*

Proof. By Exercise 2, there exists a nonsingular matrix $P \in \mathscr{F}_{mm}$ such that

$$[C|\mathbf{d}] = P[A|\mathbf{b}] \text{ and } [A|\mathbf{b}] = P^{-1}[C|\mathbf{d}].$$

By the definitions of matrix multiplication and augmented matrices, it follows that $C = PA$ and $\mathbf{d} = P\mathbf{b}$ (compare with Exercise 4 above). Also,

$$A = P^{-1}C \quad \text{and} \quad \mathbf{b} = P^{-1}\mathbf{d}.$$

Let $\mathbf{u} \in \mathscr{F}_{n1}$ be any solution of $A\mathbf{x} = \mathbf{b}$. Then $A\mathbf{u} = \mathbf{b}$. Hence,

$$C\mathbf{u} = PA\mathbf{u} = P\mathbf{b} = \mathbf{d}.$$

Thus, $\mathbf{u}$ is also a solution of $C\mathbf{x} = \mathbf{d}$. Conversely, let $\mathbf{v} \in \mathscr{F}_{n1}$ be any solution of $C\mathbf{x} = \mathbf{d}$, so that $C\mathbf{v} = \mathbf{d}$. Then

$$A\mathbf{v} = P^{-1}C\mathbf{v} = P^{-1}\mathbf{d} = \mathbf{b},$$

and $\mathbf{v}$ is also a solution of $A\mathbf{x} = \mathbf{b}$. Thus, the systems $A\mathbf{x} = \mathbf{b}$ and $C\mathbf{x} = \mathbf{d}$ have the same solution set and are equivalent. ∎

Corollary to Theorem 1. *Let A and C be row-equivalent matrices in $\mathscr{F}_{mn}$. Then the systems $A\mathbf{x} = \mathbf{0}$ and $C\mathbf{x} = \mathbf{0}$ are equivalent.*

Proof. There exists a nonsingular matrix $P \in \mathscr{F}_{mm}$ such that $C = PA$ and $A = P^{-1}C$. It follows easily as in the proof of Theorem 1 above that $A\mathbf{u} = \mathbf{0}$ if and only if $C\mathbf{u} = \mathbf{0}$. Thus, the systems $A\mathbf{x} = \mathbf{0}$ and $C\mathbf{x} = \mathbf{0}$ have the same solution set, and are equivalent. ∎

● **EXERCISES**

    5. (a) Show that the matrices S_0 and S_1 in Section 3.2 are row equivalent.
       (b) Show that $S_1 \stackrel{R}{\sim} S_2$, where S_1 and S_2 are defined in Section 3.2.
       (c) Similarly, show that $S_2 \stackrel{R}{\sim} S_3$, $S_3 \stackrel{R}{\sim} S_4$, $S_4 \stackrel{R}{\sim} S_5$, where S_2, S_3, S_4, S_5 are defined in Section 3.2.

    6. Show that the systems (3.7) and (3.8) are equivalent by finding the nonsingular matrix $P \in \mathscr{F}_{33}$ such that

$$\begin{bmatrix} 1 & 0 & 2 & | & 4 \\ 0 & 1 & 3 & | & 7 \\ 0 & 0 & 0 & | & 0 \end{bmatrix} = P \begin{bmatrix} 1 & 1 & 5 & | & 11 \\ 2 & 1 & 7 & | & 15 \\ 2 & 0 & 4 & | & 8 \end{bmatrix}.$$

[*Hint:* Use Exercise 5 and Theorem 1.]

    7. For what values of α, if any, are the matrices

$$\begin{bmatrix} 1 & 1 & 1 \\ 0 & \alpha & 1 \\ 0 & 0 & 1 \end{bmatrix}, \quad \begin{bmatrix} 1 & 0 & 0 \\ 0 & 2 & 0 \\ 0 & 0 & 3 \end{bmatrix}$$

row equivalent?

3.5 Row Echelon Form

We are now ready to give a systematic procedure for solving general linear systems of algebraic equations. We have seen that all the information about the system is contained in the augmented matrix. We will, therefore, define a simple standard form of a matrix and show that every matrix in $\mathscr{F}_{mn}$ can be reduced to such a standard form by a sequence of elementary row operations. This standard form will have the property that if the augmented matrix of a linear system of algebraic equations is in this standard form, we can read off the solutions by inspection. This has already been done in one specific case in Example 1, Section 3.2; the matrix S_5 (Section 3.2) is the standard form in question. Thus, given any linear system of algebraic equations, we may reduce its augmented matrix to this standard form, read off the solutions, and then deduce from Theorem 1, Section 3.4, that these are the solutions of the original system.

Definition 1. *The leading entry of a nonzero row a_{i*} of matrix A is the first nonzero entry in that row.*

Thus, if $a_{ij} = 0$ for $j = 1, 2, \ldots, l-1$ and if $a_{il} \neq 0$, then a_{il} is the leading entry of a_{i*}. Observe also that a zero row has no leading entry.

Definition 2. *A matrix $A \in \mathscr{F}_{mn}$ is said to be in row echelon form* (REF for short) *if and only if each of the following properties holds:*

(i) *All the nonzero rows are above all the zero rows.*
(ii) *The leading entry of every nonzero row is 1.*
(iii) *The leading entries move to the right as we descend through the nonzero rows of A.*
(iv) *Any column containing a leading entry has all other entries zero.*

A brief comment on this definition is in order. Suppose A is in REF. If A has r nonzero rows $r \leq m$, then $a_{i*} \neq 0$ for $i = 1, \ldots, r$, but $a_{i*} = 0$ for $i = r+1, \ldots, m$. Further, if $1 \leq i < k \leq r$ and if $a_{il} = 1$ is the leading entry of the ith row a_{i*} and if $a_{kp} = 1$ is the leading entry of the kth row a_{k*}, then $l < p$. Finally, $a_{ql} = 0$ for all $q \neq i$.

Example 1. The matrix $A \in \mathscr{F}_{45}$:

$$A = \begin{bmatrix} 1 & x & 0 & y & 0 \\ 0 & 0 & 1 & z & 0 \\ 0 & 0 & 0 & 0 & 1 \\ 0 & 0 & 0 & 0 & 0 \end{bmatrix},$$

50   Linear Systems of Algebraic Equations

where x, y, z are any elements of $\mathscr{F}$, is in REF. However, the matrix

$$\begin{bmatrix} 1 & x & 0 & y & 0 \\ 0 & 0 & 1 & z & 0 \\ 0 & 1 & 0 & 0 & 0 \\ 0 & 0 & 0 & 0 & 0 \end{bmatrix}$$

is not in REF.

The reader may also note that matrix S_5 of Example 1, Section 3.2, is in REF. Moreover, we note that in that example, the matrix S_0 was reduced to the form S_5 by a sequence of elementary row operations. Thus, $S_5 \stackrel{R}{\sim} S_0$. This is a special case of the following general result.

Theorem 1. *Every matrix $A \in \mathscr{F}_{mn}$ is row equivalent to a matrix A_R in REF.*

Discussion of Proof. Instead of writing down a formal proof, we shall outline a method which reduces any given matrix A to a matrix A_R in REF by a sequence of elementary row operations. Such a recipe is called an **algorithm**. In fact, a computer can be programmed to carry out this task.

STEP 1. If $A = 0$, then $A = A_R$ is already in REF and there is no more to be done. Thus, let $A \neq 0$. Then A has a first nonzero column, say a_{*p}. Let a_{ip} be the first nonzero element in that column. Interchange rows a_{i*} and a_{1*}. (If i happens to be 1 this operation is, of course, unnecessary.) Next, divide the new row a_{1*} by a_{1p}. This results in a matrix of the form

$$\begin{bmatrix} 0 & \cdots & 0 & 1 & x & x & x & x & x & x \\ 0 & \cdots & 0 & x & x & x & x & x & x & x \\ \vdots & & \vdots & \vdots & \vdots & & & & & \vdots \\ 0 & \cdots & 0 & x & & & \cdots & & & x \end{bmatrix},$$

with pth col. indicated,

where the x's are possibly nonzero elements of $\mathscr{F}$. We next make all entries, except a_{1p}, of the column a_{*p} zero by performing the elementary row operations of Type III:

$$a_{i*} \to a_{i*} - a_{ip} a_{1*} \quad (i = 2, \ldots, m).$$

This yields a matrix of the form

$$\begin{bmatrix} 0 & \cdots & 0 & 1 & x & x & \cdots & x \\ 0 & \cdots & 0 & 0 & x & x & \cdots & x \\ \vdots & & \vdots & \vdots & \vdots & \vdots & & \vdots \\ 0 & \cdots & 0 & 0 & x & x & \cdots & x \end{bmatrix}.$$

3.5 Row Echelon Form

STEP 2. Look at the matrix $A_1 \in \mathscr{F}_{m-1,n}$ consisting of the last $m-1$ rows of the matrix obtained at the end of Step 1. Apply Step 1 to the matrix A_1. This yields a matrix of the form:

$$\begin{bmatrix} 0 & \cdots & 0 & 1 & x & x & & x & & \cdots & x \\ 0 & \cdots & 0 & 0 & 0 & & \cdots & 0 & 1 & x & \cdots & x \\ & & & & & & & 0 & 0 & x & \cdots & x \\ \vdots & & \vdots & \vdots & \vdots & & & \vdots & \vdots & & & \\ 0 & \cdots & 0 & 0 & 0 & & \cdots & 0 & 0 & x & \cdots & x \end{bmatrix},$$

with pth col. and qth col. indicated,

where $q \geq p + 1$. Next, we perform the elementary row operation of Type III:

$$a_{1*} \to a_{1*} - a_{1q} a_{2*}$$

in order to eliminate the element a_{1q}. This yields a matrix of the form:

$$\begin{bmatrix} 0 & \cdots & 0 & 1 & x & x & \cdots & x & 0 & x & \cdots & x \\ 0 & \cdots & 0 & 0 & 0 & & \cdots & 0 & 1 & x & \cdots & x \\ 0 & & & & & & & 0 & 0 & x & \cdots & x \\ \vdots & & & & \cdots & & & & \vdots & \vdots & x & \cdots & x \\ & & & & & & & & & & \vdots & & \\ 0 & & & & \cdots & & & & 0 & 0 & x & \cdots & x \end{bmatrix}.$$

with pth col. and qth col. indicated.

Observe that the matrix consisting of the first two rows of the above matrices is in REF. Moreover, the matrix consisting of the last $m-2$ rows of the above matrix has its first q columns zero.

STEP 3. Look at the matrix $A_2 \in \mathscr{F}_{m-2,n}$ consisting of the last $m-2$ rows of the matrix obtained at the end of Step 2. Apply Step 1 to A_2. Then, if a_{3r}, $r \geq q + 1$, is the leading element of the new a_{3*}, eliminate a_{1r} and a_{2r} as at the end of Step 2.

Continue in this way until after **at most** m **steps** either we are applying Step 1 to a zero matrix or we have used up all m rows. This final matrix is in REF and we call it A_R. ∎

As a matter of fact, it can be shown that corresponding to a given matrix $A \in \mathscr{F}_{mn}$ there is precisely one matrix in REF which is row equivalent to A.

Linear Systems of Algebraic Equations

We omit the proof of this more difficult result. However, in view of this theorem we shall call A_R **the row echelon form of** A.

Example 2. Compute A_R if

$$A = \begin{bmatrix} 0 & 0 & 3 & -1 \\ 0 & -1 & 4 & 7 \\ 0 & -1 & 7 & 6 \end{bmatrix}.$$

STEP 1. $A \neq 0$. The first nonzero column is a_{*2} and $a_{22} \neq 0$ is the nonzero entry of a_{*2}. Thus, interchange a_{1*} and a_{2*} and

$$A \overset{R}{\sim} \begin{bmatrix} 0 & -1 & 4 & 7 \\ 0 & 0 & 3 & -1 \\ 0 & -1 & 7 & 6 \end{bmatrix}.$$

Now, divide the new a_{1*} by -1. Thus,

$$A \overset{R}{\sim} \begin{bmatrix} 0 & 1 & -4 & -7 \\ 0 & 0 & 3 & -1 \\ 0 & -1 & 7 & 6 \end{bmatrix}.$$

Finally, add a_{1*} to a_{3*} resulting in

$$A \overset{R}{\sim} \begin{bmatrix} 0 & 1 & -4 & -7 \\ 0 & 0 & 3 & -1 \\ 0 & 0 & 3 & -1 \end{bmatrix}.$$

STEP 2. Apply Step 1 to the matrix A_1 consisting of the shaded part of A below

$$A \overset{R}{\sim} \begin{bmatrix} 0 & 1 & -4 & -7 \\ 0 & 0 & 3 & -1 \\ 0 & 0 & 3 & -1 \end{bmatrix}.$$

The first nonzero column of A_1 is a_{*3}. Thus, we obtain

$$A \overset{R}{\sim} \begin{bmatrix} 0 & 1 & -4 & -7 \\ 0 & 0 & 1 & -1/3 \\ 0 & 0 & 3 & -1 \end{bmatrix} \overset{R}{\sim} \begin{bmatrix} 0 & 1 & -4 & -7 \\ 0 & 0 & 1 & -1/3 \\ 0 & 0 & 0 & 0 \end{bmatrix}$$

$$A_1 \overset{R}{\sim} \begin{bmatrix} 0 & 1 & 0 & -25/3 \\ 0 & 0 & 1 & -1/3 \\ 0 & 0 & 0 & 0 \end{bmatrix}.$$

STEP 3. Look at the matrix A_2 consisting of the more heavily shaded part of A_1.

$$A \overset{R}{\sim} \begin{bmatrix} 0 & 1 & 0 & -25/3 \\ 0 & 0 & 1 & -1/3 \\ 0 & 0 & 0 & 0 \end{bmatrix}.$$

Since A_2 is the zero matrix, we are done, and

$$A \sim A_R = \begin{bmatrix} 0 & 1 & 0 & -25/3 \\ 0 & 0 & 1 & -1/3 \\ 0 & 0 & 0 & 0 \end{bmatrix}.$$

● **EXERCISES**

1. Find the REF of each of the following matrices:

(a) $A = \begin{bmatrix} 1 & -3 & 0 & 2 \\ 1 & -3 & 1 & 1 \\ 0 & -3 & 0 & 1 \end{bmatrix}.$ (b) $A = \begin{bmatrix} 1 & 2 & 3 & 4 \\ 2 & 3 & 4 & 5 \\ 3 & 5 & 7 & 9 \end{bmatrix}.$

(c) $A = \begin{bmatrix} 4 & 2 \\ 9 & 5 \end{bmatrix}.$ (d) $A = \begin{bmatrix} -1 & 1 & -1 \\ 1 & -1 & 1 \\ -1 & 1 & 1 \end{bmatrix}.$

(e) $A = \begin{bmatrix} 1 & 2 & 3 \\ 2 & 4 & 6 \\ -3 & -6 & -9 \end{bmatrix}.$ (f) $A = \begin{bmatrix} 1 & 0 & 1 \\ 1 & 1 & 0 \\ -1 & 1 & 1 \end{bmatrix}.$

(g) $A = \begin{bmatrix} 3 & 2 & -3 & 1 \\ 2 & -1 & 1 & 0 \\ 1 & 1 & 1 & 2 \end{bmatrix}.$

2. Find the REF of each of the following matrices:

(a) $A = \begin{bmatrix} 1 \\ 2 \\ 3 \\ 4 \end{bmatrix}.$ (b) $A = \begin{bmatrix} 1 & 2 \\ -1 & 0 \\ 1 & 3 \\ 6 & 4 \end{bmatrix}.$

(c) $A = \begin{bmatrix} 1 & 2 & 1 \\ 3 & 1 & 4 \\ 7 & -2 & 6 \\ 0 & 0 & 1 \end{bmatrix}.$
(d) $A = \begin{bmatrix} 1 & 4 \\ 9 & 4 \\ 3 & 4 \end{bmatrix}.$

As we have seen in Exercise 2, Section 3.4, if two matrices A and B in $\mathscr{F}_{mn}$ are row equivalent, then there exists a nonsingular matrix $P \in \mathscr{F}_{mm}$ such that $B = PA$. This fact, together with Theorem 1, yields the following corollary.

Corollary to Theorem 1. *If A_R is the row echelon form of $A \in \mathscr{F}_{mn}$, then*

$$A_R = PA,$$

for some nonsingular matrix $P \in \mathscr{F}_{mm}$.

● **EXERCISES**

3. (a) Find the row echelon A_R of the matrix

$$A = \begin{bmatrix} 1 & 2 & 3 & 1 & 0 & 0 \\ 0 & 1 & -4 & 0 & 1 & 0 \\ 0 & 0 & 1 & 0 & 0 & 1 \end{bmatrix}.$$

(b) Let B be the 3×3 matrix consisting of the last three columns of A_R, and let C be the 3×3 matrix consisting of the first three columns of A. Show that $B = C^{-1}$.
(c) Can you generalize this procedure for finding when a matrix is nonsingular and for calculating its inverse?

A concept of great importance is the **row rank of a matrix.** At this time we introduce a temporary definition based on the row echelon form and we call this the **rank** of the matrix. The definition given here depends on the fact that the REF is unique, which is difficult to prove. In Section 5.6 we give the usual definition of row rank and show that it is equivalent to the definition given here.

Definition 3. *The rank of a matrix $A \in \mathscr{F}_{mn}$ is the number of nonzero rows in the row echelon form A_R of A.*

The reader should note that if $A \in \mathscr{F}_{mn}$, then the rank of A is obviously not greater than m, the number of rows of A. Further, since each row contains a leading entry and since there is at most one leading entry in each column, it follows that the rank is not greater than n, the number of columns of A. Hence, if $A \in \mathscr{F}_{mn}$, then

$$\text{rank } A \leq \min(m, n). \tag{3.14}$$

● **EXERCISES**

4. Find the rank of each of the matrices in Exercises 1 and 2.
5. Find the rank of the identity matrix $I_n \in \mathscr{F}_{nn}$.
6. (a) Show that if $A \in \mathscr{F}_{nn}$ and A has rank n, then $A_R = I_n$.
   (b) Deduce from part (a) that the matrix P such that $A_R = PA$ is actually A^{-1}.
   (c) Show that $A \in \mathscr{F}_{nn}$ is nonsingular if and only if A has rank n.
7. Let A be a $n \times n$ matrix in row echelon form with rank $r < n$. Let $\mathbf{b}$ be an $n \times 1$ column vector with $b_i \neq 0$ for some $i > r$. Show that the row rank of $[A|\mathbf{b}]$ is $r + 1$, where $[A|\mathbf{b}]$ denotes the $n \times (n + 1)$ matrix whose first n columns are identical to the n columns of A and whose $n + 1$ column is the vector $\mathbf{b}$.
8. Let A be an arbitrary matrix and $A_R = PA$, where P is a product of elementary matrices. Prove $[A|\mathbf{b}] \overset{R}{\sim} [A_R|P\mathbf{b}]$.
9. Prove that row rank $[A|B] \geq$ row rank A, where $[A|B]$ is the matrix obtained by juxtaposing B to the right of A.

3.6 Linear Homogeneous Systems of Algebraic Equations

In this and the following sections we shall study the solutions of the linear system of algebraic equations

$$A\mathbf{x} = \mathbf{b},$$

where $A \in \mathscr{F}_{mn}$, $\mathbf{b} \in \mathscr{F}_{m1}$.

We first consider the case where $\mathbf{b} = \mathbf{0}$, so that the system is $A\mathbf{x} = \mathbf{0}$.

Definition 1. *A system of the form $A\mathbf{x} = \mathbf{0}$ is said to be a linear homogeneous system.*

It is obvious that the vector $\mathbf{u} = \mathbf{0} \in \mathscr{F}_{n1}$ is always a solution of $A\mathbf{x} = \mathbf{0}$. This solution is called the trivial solution, as it is generally of little interest. The interesting question is whether there are any other solutions.

If A_R is the row echelon form of A, then $A_R \overset{R}{\sim} A$. By the corollary to Theorem 1, Section 3.4, the system $A\mathbf{x} = \mathbf{0}$ is equivalent to the system $A_R\mathbf{x} = \mathbf{0}$. As we shall see, the solutions of $A_R\mathbf{x} = \mathbf{0}$ are easily obtained. Thus, the first step in finding the solutions of $A\mathbf{x} = \mathbf{0}$ is always to find the REF A_R of A. To see how the solutions of $A_R\mathbf{x} = \mathbf{0}$ are obtained, we begin with the following simple example.

Example 1. Solve the system $A\mathbf{x} = \mathbf{0}$, where

$$A = \begin{bmatrix} 1 & -2 & 0 & 3 & 0 \\ 0 & 0 & 1 & -1 & 0 \\ 0 & 0 & 0 & 0 & 1 \end{bmatrix} \in \mathscr{F}_{35}.$$

We observe that A is in REF. The corresponding system of equations is

$$x_1 - 2x_2 + 3x_4 = 0$$
$$x_3 - x_4 = 0$$
$$x_5 = 0,$$

which we may write as

$$x_1 = 2x_2 - 3x_4$$
$$x_3 = x_4$$
$$x_5 = 0.$$

Observe that once x_2 and x_4 are prescribed, then x_1, x_3, x_5 are determined, and that for each choice of values for x_2, x_4 we obtained a solution for x_1, x_3, x_5. Thus, if we put $x_2 = \alpha_1$, $x_4 = \alpha_2$, we obtain a solution

$$\mathbf{u} = \begin{bmatrix} 2\alpha_1 - 3\alpha_2 \\ \alpha_1 \\ \alpha_2 \\ \alpha_2 \\ 0 \end{bmatrix} = \alpha_1 \begin{bmatrix} 2 \\ 1 \\ 0 \\ 0 \\ 0 \end{bmatrix} + \alpha_2 \begin{bmatrix} -3 \\ 0 \\ 1 \\ 1 \\ 0 \end{bmatrix}, \qquad (3.15)$$

and every solution is of this form with some suitable choice of α_1, α_2. We point out that the variables which can be assigned arbitrarily are those with subscripts corresponding to columns **without** leading entries in $A = A_R$. The variables whose subscripts correspond to columns with leading entries in $A = A_R$ are then determined.

● **EXERCISE**

1. Verify that **u** given by (3.15) is a solution of the system $A\mathbf{x} = \mathbf{0}$ in Example 1 for any choice of α_1, α_2.

The situation in the general case is analogous to that in the above example. We separate the components of **x** into two classes:

CLASS 1. Components whose subscripts correspond to those columns of A_R which do not contain the leading entry of any row.

CLASS 2. Components whose subscripts correspond to columns of A_R which do contain the leading entry of some row.

If A_R has rank r, there are r components of Class 2 and $n - r$ components of Class 1.

3.6 Linear Homogeneous Systems of Algebraic Equations

Because of the definition of the REF, if x_k is of Class 2, the kth column of A_R has only one nonzero entry, and this nonzero entry is 1. This means that in the system of equations $A_R \mathbf{x} = \mathbf{0}$, x_k occurs exactly once. The equation which contains x_k has the form x_k minus a sum of terms of the form $c_p x_p$ equals zero, where all x_p appearing in the equation are of Class 1. Thus, we can assign arbitrary values to all $(n - r)$ components of Class 1, and the r components of Class 2 are then determined. This method yields all the solutions of $A_R \mathbf{x} = \mathbf{0}$.

Example 2. Solve the system $A\mathbf{x} = \mathbf{0}$, where A is in REF,

$$A = \begin{bmatrix} 0 & 1 & 4 & 0 & 2 & 0 \\ 0 & 0 & 0 & 1 & -2 & 0 \\ 0 & 0 & 0 & 0 & 0 & 1 \\ 0 & 0 & 0 & 0 & 0 & 0 \end{bmatrix} \in \mathscr{F}_{46}.$$

The components in Class 1 are x_1, x_3, x_5 and the components in Class 2 are x_2, x_4, x_6. The equations for the components in Class 2 may be written as

$$\begin{aligned} x_2 &= -4x_3 - 2x_5 \\ x_4 &= 2x_5 \\ x_6 &= 0. \end{aligned}$$

Thus, if we choose $x_1 = \alpha_1$, $x_3 = \alpha_2$, $x_5 = \alpha_3$, we obtain the solution vector $\mathbf{u}$ of $A\mathbf{x} = \mathbf{0}$ of the form

$$\mathbf{u} = \begin{bmatrix} \alpha_1 \\ -4\alpha_2 - 2\alpha_3 \\ \alpha_2 \\ 2\alpha_3 \\ \alpha_3 \\ 0 \end{bmatrix} = \alpha_1 \begin{bmatrix} 1 \\ 0 \\ 0 \\ 0 \\ 0 \\ 0 \end{bmatrix} + \alpha_2 \begin{bmatrix} 0 \\ -4 \\ 1 \\ 0 \\ 0 \\ 0 \end{bmatrix} + \alpha_3 \begin{bmatrix} 0 \\ -2 \\ 0 \\ 2 \\ 1 \\ 0 \end{bmatrix},$$

and every solution has this form.

Example 3. Solve the system $A\mathbf{x} = \mathbf{0}$, where

$$A = \begin{bmatrix} 1 & -1 & 0 \\ -1 & 0 & 1 \\ 0 & 1 & -1 \end{bmatrix} \in \mathscr{F}_{33}.$$

58 Linear Systems of Algebraic Equations

To solve, we first put the matrix A into row echelon form. We have

$$A = \begin{bmatrix} 1 & -1 & 0 \\ -1 & 0 & 1 \\ 0 & 1 & -1 \end{bmatrix} \overset{R}{\sim} \begin{bmatrix} 1 & -1 & 0 \\ 0 & -1 & 1 \\ 0 & 1 & -1 \end{bmatrix} \overset{R}{\sim} \begin{bmatrix} 1 & -1 & 0 \\ 0 & 1 & -1 \\ 0 & 1 & -1 \end{bmatrix}$$

$$\overset{R}{\sim} \begin{bmatrix} 1 & -1 & 0 \\ 0 & 1 & -1 \\ 0 & 0 & 0 \end{bmatrix} \overset{R}{\sim} \begin{bmatrix} 1 & 0 & -1 \\ 0 & 1 & -1 \\ 0 & 0 & 0 \end{bmatrix} = A_R.$$

The only component in Class 1 is x_3. Therefore, we can choose $x_3 = \alpha_1$ arbitrarily and then $x_2 = \alpha_1$, $x_1 = \alpha_1$. Hence, the solution set of $A_R \mathbf{x} = \mathbf{0}$ consists of all vectors $\mathbf{u}$ of the form

$$\mathbf{u} = \begin{bmatrix} \alpha_1 \\ \alpha_1 \\ \alpha_1 \end{bmatrix} = \alpha_1 \begin{bmatrix} 1 \\ 1 \\ 1 \end{bmatrix}.$$

Since $A \overset{R}{\sim} A_R$, the solution set of $A\mathbf{x} = \mathbf{0}$ is the same.

In Example 2 above, we saw that the solution set consisted of all vectors of the form

$$\mathbf{u} = \alpha_1 \mathbf{u}_1 + \alpha_2 \mathbf{u}_2 + \alpha_3 \mathbf{u}_3,$$

where $\alpha_1, \alpha_2, \alpha_3 \in \mathcal{F}$ and where

$$\mathbf{u}_1 = \begin{bmatrix} 1 \\ 0 \\ 0 \\ 0 \\ 0 \\ 0 \end{bmatrix}, \quad \mathbf{u}_2 = \begin{bmatrix} 0 \\ -4 \\ 1 \\ 0 \\ 0 \\ 0 \end{bmatrix}, \quad \mathbf{u}_3 = \begin{bmatrix} 0 \\ -2 \\ 0 \\ 2 \\ 1 \\ 0 \end{bmatrix}.$$

In particular, $\mathbf{u}_1$, $\mathbf{u}_2$, and $\mathbf{u}_3$ are themselves solutions. (Put $\alpha_1 = 1$, $\alpha_2 = 0$, $\alpha_3 = 0$, to obtain $\mathbf{u}_1$, etc.) We may observe that $\mathbf{u}_1$ is the solution obtained when the first component of Class 1, namely x_1, is chosen to be 1, and all other components of Class 1 are chosen to be 0. Similarly, choosing the second component of Class 1, namely x_3, equal to 1 and all other components of Class 1 to be 0, we obtain $\mathbf{u}_2$. Finally, choosing x_5, the remaining component of Class 1 equal to 1 and the other components of Class 1 to be 0, we obtain $\mathbf{u}_3$.

A vector of the form $\mathbf{u} = \alpha_1 \mathbf{u}_1 + \alpha_2 \mathbf{u}_2 + \alpha_3 \mathbf{u}_3$, where $\alpha_1, \alpha_2, \alpha_3 \in \mathcal{F}$ is called a **linear combination** of the vectors $\mathbf{u}_1, \mathbf{u}_2, \mathbf{u}_3$ with coefficients in $\mathcal{F}$. For example, if

3.6 Linear Homogeneous Systems of Algebraic Equations

$$\mathbf{u}_1 = \begin{bmatrix} 2 \\ i \\ -1 \\ 0 \end{bmatrix}, \quad \mathbf{u}_2 = \begin{bmatrix} 3 \\ 0 \\ 1 \\ i \end{bmatrix} \in C_{41},$$

then the vector

$$\mathbf{u} = 2\mathbf{u}_2 - \mathbf{u}_1 = 2\begin{bmatrix} 2 \\ i \\ -1 \\ 0 \end{bmatrix} - \begin{bmatrix} 3 \\ 0 \\ 1 \\ i \end{bmatrix} = \begin{bmatrix} 1 \\ 2i \\ -3 \\ -i \end{bmatrix}$$

is a linear combination of $\mathbf{u}_1$ and $\mathbf{u}_2$.

The general theory of linear combinations will be developed in Chapter 5. As we have seen in Example 2, and may deduce in general from the remarks immediately preceding it, we can make the following statement about solutions of the system $A\mathbf{x} = \mathbf{0}$.

Theorem 1. *Let $A \in \mathscr{F}_{mn}$ have rank r. Then the solution set of the system $A\mathbf{x} = \mathbf{0}$ consists of all linear combinations of $(n - r)$ vectors $\mathbf{u}_1, \mathbf{u}_2, \ldots, \mathbf{u}_{n-r}$ with coefficients in $\mathscr{F}$. The vectors $\mathbf{u}_1, \mathbf{u}_2, \ldots, \mathbf{u}_{n-r}$ are determined from the row echelon form of A as follows: $\mathbf{u}_j$ has the jth component of Class 1 equal to one and all other components of Class 1 equal to zero, for $j = 1, \ldots, n - r$.*

Another way of stating this result is to say that if $\mathbf{u}$ is any solution of $A\mathbf{x} = \mathbf{0}$, there exist $(n - r)$ constants $\alpha_1, \alpha_2, \ldots, \alpha_{n-r} \in \mathscr{F}$ such that $\mathbf{u} = \alpha_1 \mathbf{u}_1 + \alpha_2 \mathbf{u}_2 + \cdots + \alpha_{n-r} \mathbf{u}_{n-r}$, and every vector of this form is a solution. In this solution $\mathbf{u}$, the kth component of Class 1 is exactly α_k.

Corollary 1. *If r, the rank of A, is equal to n, then $A\mathbf{x} = \mathbf{0}$ has only the trivial solution, and conversely if $A\mathbf{x} = \mathbf{0}$ has only the trivial solution, then $r = n$.*

Proof. In this case there are no components of Class 1; all components are of Class 2. That is, every one of the n columns of A_R, the REF of A has exactly one 1 in it, and all other entries are zero. Thus, the system of equations $A_R \mathbf{x} = \mathbf{0}$ is $x_1 = 0, x_2 = 0, \ldots, x_n = 0$.

Conversely, if $A\mathbf{x} = \mathbf{0}$ has only the trivial solution, then there can be no nonzero vectors $\mathbf{u}_j$ as in Theorem 1, and thus $n = r$. ∎

Corollary 2. *Let $A \in \mathscr{F}_{mn}$. If $m < n$, the system $A\mathbf{x} = \mathbf{0}$ has at least one nontrivial solution.*

Linear Systems of Algebraic Equations

Proof. By Eq. 3.14, the rank r of A satisfies $r \leq \min(m, n) = m < n$. Hence, by Theorem 1, there exist nontrivial solutions $\mathbf{u}_1, \mathbf{u}_2, \ldots, \mathbf{u}_{n-r}$. ∎

• **EXERCISES**

2. Find the solution set of each of the following systems $A\mathbf{x} = \mathbf{0}$, where A is specified below. In each case determine the rank of A and write any solution $\mathbf{u}$ as a linear combination of suitably chosen vectors.

(a) $A = \begin{bmatrix} 1 & -3 & 0 & 2 \\ 1 & -3 & 1 & 1 \\ 0 & -3 & 0 & 1 \end{bmatrix}$
(b) $A = \begin{bmatrix} 1 & 2 & 3 & 4 \\ 2 & 3 & 4 & 5 \\ 3 & 5 & 5 & 7 \end{bmatrix}$.

(c) $A = \begin{bmatrix} 4 & 2 \\ 9 & 5 \end{bmatrix}$.
(d) $A = \begin{bmatrix} -1 & 1 & -1 \\ 1 & -1 & 1 \\ -1 & 1 & 1 \end{bmatrix}$.

(e) $A = \begin{bmatrix} 1 & 0 & 1 \\ 1 & 1 & 0 \\ -1 & 1 & 1 \end{bmatrix}$.
(f) $A = \begin{bmatrix} 3 & 2 & -3 & 1 \\ 2 & -1 & 1 & 0 \\ 1 & 1 & 1 & 2 \end{bmatrix}$.

3. Consider the system of equations

$$x_1 + x_2 + x_3 + x_4 = 0$$
$$2x_1 + x_2 + 2x_3 + x_4 = 0.$$

(a) Find the solution $\mathbf{u}$ with $u_3 = 2$, $u_4 = -1$.
(b) Find the solution $\mathbf{v}$ with $v_3 = v_4 = 0$.

3.7 General Linear Systems of Algebraic Equations

We will now study the system of equations

$$A\mathbf{x} = \mathbf{b}, \tag{3.16}$$

where $A \in \mathscr{F}_{mn}$, $\mathbf{b} \in \mathscr{F}_{m1}$, and where $\mathbf{b}$ is not necessarily the zero vector. If $\mathbf{b} = \mathbf{0}$, then (3.16) always has at least the zero solution. If $\mathbf{b} \neq \mathbf{0}$, then the examples in Section 3.1 show that (3.16) may or may not have a solution. This suggests the following definition.

Definition 1. *The system $A\mathbf{x} = \mathbf{b}$ is called* consistent *if and only if it has a solution. If no solution exists, it is called* inconsistent.

3.7 General Linear Systems of Algebraic Equations

To solve (3.16), we consider the augmented matrix $[A\,|\,\mathbf{b}]$. Let A_R be the REF of A, so that by the corollary to Theorem 1, Section 3.5, $A_R = PA$ for a suitably chosen nonsingular $P \in \mathscr{F}_{mm}$. Let $\mathbf{b}_R = P\mathbf{b}$. Then

$$[A\,|\,\mathbf{b}] \overset{R}{\sim} [PA\,|\,P\mathbf{b}] = [A_R\,|\,\mathbf{b}_R].$$

Thus, by Theorem 1, Section 3.4, the system (3.16) is equivalent to the system

$$A_R \mathbf{x} = \mathbf{b}_R. \qquad (3.17)$$

We may, therefore, assume at the start that the original system (3.16) has its coefficient matrix A in REF.

In order to lead up to the main result we suppose the system (3.16) is consistent, with solution $\mathbf{u}$. Let A have rank r. This means, since A is in REF, that the last $(m - r)$ rows of A have only zeros, and therefore the last $(m - r)$ equations of (3.16) have the form

$$0 \cdot u_1 + 0 \cdot u_2 + \cdots + 0 \cdot u_n = b_i \qquad (i = r+1, \ldots, m).$$

Hence, we have $b_i = 0$ $(i = r+1, \ldots, m)$ and we have established one half of the following theorem.

Theorem 1. *Let $A \in \mathscr{F}_{mn}$ be in REF and have rank r. Then the system $A\mathbf{x} = \mathbf{b}$ is consistent if and only if $b_{r+1} = b_{r+2} = \cdots = b_m = 0$. In particular, if A has rank m, then the system is consistent.*

Proof. It remains to be shown that if $b_i = 0$, for $i = r+1, \ldots, m$, then the system $A\mathbf{x} = \mathbf{b}$ (where A is in REF) is consistent. We accomplish this by exhibiting a particular solution $\mathbf{u}$ as follows. Separate the components of $\mathbf{u}$ into those of Class 1 and those of Class 2 as in Section 3.6. If u_k is of Class 1, we put $u_k = 0$. If u_k is of Class 2, then the kth column of A contains a leading entry; suppose that this leading entry is $a_{ik} = 1$. Then we put $u_k = b_i$. We claim that the vector $\mathbf{u}$ thus defined is a solution of $A\mathbf{x} = \mathbf{b}$. By the definition of the REF, the ith equation in (3.16) contains only one component of Class 2, for $i = 1, \ldots, r$. It, therefore, has the form x_k plus a sum of terms of the form $c_p x_p$ equaling b_i, where all x_p appearing in the equation are of Class 1. This equation is obviously satisfied if we take $u_k = b_i$, $u_p = 0$. ∎

A form of Theorem 1 that does not assume that A is in REF is the following.

Theorem 2. *Let $A \in \mathscr{F}_{mn}$. Then the system of equations $A\mathbf{x} = \mathbf{b}$ is consistent if and only if rank $[A\,|\,\mathbf{b}] = $ rank A.*

● EXERCISES

1. Prove Theorem 2. [*Hint:* Show that rank $[A|\mathbf{b}] = $ rank A if and only if the last $(m-r)$ components of $\mathbf{b}_R$ are zero, where $\mathbf{b}_R$ is as in (3.17). Then apply Theorem 1.]

2. Consider the systems of equations $A\mathbf{x} = \mathbf{b}$, where A and $\mathbf{b}$ are given below.
   (i) Determine which of the systems are consistent.
   (ii) Find a particular solution of the consistent ones.

(a) $\begin{bmatrix} 3 & 2 & -3 \\ 2 & -1 & 1 \\ 1 & 1 & 1 \end{bmatrix}; \begin{bmatrix} 6 \\ 2 \\ 1 \end{bmatrix}.$
(b) $\begin{bmatrix} 1 & 2 \\ -1 & -2 \\ 3/2 & 3 \end{bmatrix}; \begin{bmatrix} 2 \\ -2 \\ 3/2 \end{bmatrix}.$

(c) $\begin{bmatrix} 4 & 8 & 12 \\ 2 & 2 & -2 \\ 5 & -5 & 5 \end{bmatrix}; \begin{bmatrix} 16 \\ -4 \\ 15 \end{bmatrix}.$
(d) $\begin{bmatrix} -1 & 1 & -1 \\ 1 & -1 & 1 \\ -1 & 1 & 1 \end{bmatrix}; \begin{bmatrix} 1 \\ 1 \\ 1 \end{bmatrix}.$

(e) $\begin{bmatrix} 4 & -12 & 32 & 4 \\ 1 & -1 & -1 & 1 \\ 1 & 10 & 0 & 1 \end{bmatrix}; \begin{bmatrix} 8 \\ 2 \\ -15 \end{bmatrix}.$

(f) $\begin{bmatrix} 4 & -3 & -2 & -1 \\ 0 & 3 & -2 & 1 \\ 1 & 1 & 3 & 5 \\ 1 & 1 & 1 & 3 \end{bmatrix}; \begin{bmatrix} 1 \\ 2 \\ 3 \\ 4 \end{bmatrix}.$

(g) $\begin{bmatrix} 3/2 & 1 & -1/2 & -1 \\ 0 & 2 & 2 & -2 \\ 1 & 1 & 0 & -1 \\ 4 & 5 & 1 & -5 \end{bmatrix}; \begin{bmatrix} 2 \\ -2 \\ 2 \\ -2 \end{bmatrix}.$

(h) $\begin{bmatrix} 1 & 2 & 0 \\ 3 & 6 & -1 \\ -1 & -2 & 3 \end{bmatrix}; \begin{bmatrix} 1 \\ 3 \\ 0 \end{bmatrix}.$
(i) $\begin{bmatrix} 1 & 2 & 0 \\ 3 & 6 & -1 \\ -1 & -2 & 3 \end{bmatrix}; \begin{bmatrix} 2 \\ 5 \\ 1 \end{bmatrix}.$

3. Show that if A is nonsingular then $A\mathbf{x} = \mathbf{b}$ is consistent, where $A \in \mathscr{F}_{nn}$ and $\mathbf{b} \in \mathscr{F}_{n1}$.

4. Let A be a given $n \times n$ matrix and suppose that $A\mathbf{x} = \mathbf{b}$ is consistent for every $\mathbf{b} \in \mathscr{F}_{n1}$. Show that A is nonsingular. [*Hint:* Let $\mathbf{w}_i$ be a solution to $A\mathbf{x} = \mathbf{e}_i$, where the entries of $\mathbf{e}_i$ are all zero except for the *i*th row which contains 1. Then consider the matrix $[\mathbf{w}_1|\mathbf{w}_2|\cdots|\mathbf{w}_n]$.]

The proof of Theorem 1 exhibits only one solution of a consistent system $A\mathbf{x} = \mathbf{b}$, where A has rank r. However, it suggests that other solutions may be obtained by a different choice of the components of Class 1. It even

3.7 General Linear Systems of Algebraic Equations

suggests that an arbitrary choice of the $(n - r)$ components of Class 1 will yield a solution vector. These facts can be seen more clearly as a consequence of the following result, which is basic to the theory of linear equations of all types (see, for example, Section 6.5 for a similar result about linear differential equations).

Theorem 3. *Let $A\mathbf{x} = \mathbf{b}$ be a consistent system, and let $\mathbf{u}$ be a particular solution. Then the solution set of $A\mathbf{x} = \mathbf{b}$ consists of all vectors of the form $\mathbf{w} = \mathbf{u} + \mathbf{v}$, where $\mathbf{v}$ is a solution of the corresponding homogeneous system $A\mathbf{x} = \mathbf{0}$.*

Proof. Let $\mathbf{v}$ be a solution of $A\mathbf{x} = \mathbf{0}$, and let $\mathbf{w} = \mathbf{u} + \mathbf{v}$. Then

$$A\mathbf{w} = A(\mathbf{u} + \mathbf{v}) = A\mathbf{u} + A\mathbf{v} = \mathbf{b} + \mathbf{0} = \mathbf{b},$$

whence $\mathbf{w}$ is a solution of $A\mathbf{x} = \mathbf{b}$. Conversely, let $\mathbf{w}$ be a solution of $A\mathbf{x} = \mathbf{b}$, and let $\mathbf{v} = \mathbf{w} - \mathbf{u}$. Then

$$A\mathbf{v} = A(\mathbf{w} - \mathbf{u}) = A\mathbf{w} - A\mathbf{u} = \mathbf{b} - \mathbf{b} = \mathbf{0},$$

so that $\mathbf{w} = \mathbf{u} + \mathbf{v}$, where $\mathbf{v}$ is a solution of $A\mathbf{x} = \mathbf{0}$. ∎

In practice, given the system $A\mathbf{x} = \mathbf{b}$ we first test for consistency by putting A into REF. If the system is consistent, we find a particular solution as indicated in Theorem 1. We then find the solution set of the system $A\mathbf{x} = \mathbf{0}$, as indicated in Section 3.6, and add, as suggested by Theorem 3. Observe that if A has rank r, there will be $(n - r)$ arbitrary constants in the solution. We illustrate by the following example.

Example 1. Solve the system $A\mathbf{x} = \mathbf{b}$ over $\mathscr{R}$, where

$$A = \begin{bmatrix} 1 & 3 & 5 & -1 \\ -1 & -2 & -5 & 4 \\ 0 & 1 & 1 & -1 \\ 1 & 4 & 6 & -2 \end{bmatrix}, \quad \mathbf{b} = \begin{bmatrix} 1 \\ 2 \\ 4 \\ 5 \end{bmatrix}.$$

We have

$$[A \mid \mathbf{b}] = \begin{bmatrix} 1 & 3 & 5 & -1 & \bigm| & 1 \\ -1 & -2 & -5 & 4 & \bigm| & 2 \\ 0 & 1 & 1 & -1 & \bigm| & 4 \\ 1 & 4 & 6 & -2 & \bigm| & 5 \end{bmatrix}.$$

As is easily verified

$$A_R = \begin{bmatrix} 1 & 0 & 0 & 10 \\ 0 & 1 & 0 & 3 \\ 0 & 0 & 1 & -4 \\ 0 & 0 & 0 & 0 \end{bmatrix}$$

and

$$[A_R | b_R] = \begin{bmatrix} 1 & 0 & 0 & 10 & | & -13 \\ 0 & 1 & 0 & 3 & | & 3 \\ 0 & 0 & 1 & -4 & | & 1 \\ 0 & 0 & 0 & 0 & | & 0 \end{bmatrix}.$$

Since rank A_R = rank $[A_R | b_R]$, $Ax = b$ is consistent. Now,

$$b_R = \begin{bmatrix} -13 \\ 3 \\ 1 \\ 0 \end{bmatrix}$$

so if we let $u_1 = -13$, $u_2 = 3$, $u_3 = 1$, and $u_4 = 0$ (as in the proof of Theorem 1) then

$$u = \begin{bmatrix} -13 \\ 3 \\ 1 \\ 0 \end{bmatrix}$$

is a solution of $A_R x = b_R$ and, hence, of $Ax = b$. But

$$\begin{bmatrix} -10\alpha \\ -3\alpha \\ 4\alpha \\ \alpha \end{bmatrix} = \alpha \begin{bmatrix} -10 \\ -3 \\ 4 \\ 1 \end{bmatrix}$$

is a solution of $A_R x = 0$ (and of $Ax = 0$) for every α. Therefore, by Theorem 3, the solutions of the nonhomogeneous system are all vectors of the form $\mathbf{w} = \mathbf{u} + \alpha \mathbf{v}_1$, where

$$\mathbf{v}_1 = \begin{bmatrix} -10 \\ -3 \\ 4 \\ 1 \end{bmatrix}$$

and where $\alpha \in \mathcal{R}$ is an arbitrary constant.

● **EXERCISE**

5. Find all solutions over the reals of each of the consistent systems $A\mathbf{x} = \mathbf{b}$ in Exercise 2.

To complete the theory of nonhomogeneous systems, we state the following results.

Theorem. 4 *Suppose the system* $A\mathbf{x} = \mathbf{b}$, *where* $A \in \mathcal{F}_{mn}$, $\mathbf{b} \in \mathcal{F}_{m1}$ *is consistent. Then it has a unique solution if and only if rank* $A = n$.

● **EXERCISE**

6. Prove Theorem 4. [*Hint:* Apply Corollary 1, Theorem 1, Section 3.6.]

Corollary to Theorem 4. *Consider the system* $A\mathbf{x} = \mathbf{b}$, *where* $A \in \mathcal{F}_{mn}$, $\mathbf{b} \in \mathcal{F}_{m1}$ *and* $m \leq n$. *There is a unique solution if and only if rank* $A = n$, *and hence* $m = n$.

Proof. $n = \text{rank } A \leq m \leq n$; thus, $n = m = \text{rank } A$. ∎

● **EXERCISE**

7. Discuss why the above corollary is false if $m > n$. Illustrate by a specific example.

3.8 The Inverse of a Nonsingular Matrix

As an application of the theory of the preceding section, we first prove a result for square matrices which will make it possible to give a systematic procedure for computing the inverse of a nonsingular matrix.

Theorem 1. Let $A \in \mathscr{F}_{nn}$. Then the following statements are equivalent.

(i) A is nonsingular.
(ii) Rank $A = n$.
(iii) $A_R = I_n$, where A_R is the REF of A.

Proof. If A is nonsingular, then it follows from Theorem 3, Section 2.4, that the system $A\mathbf{x} = \mathbf{0}$ has $\mathbf{x} = \mathbf{0}$ as its only solution. By Corollary 1, Theorem 1, Section 3.6., rank $A = n$. If the rank of A is n, there are no zero rows; hence, each row of the REF of A has a leading entry, and there are n rows. Thus, by the definition of the REF, $A_R = I_n$. Finally, if $A_R = I_n$, there exists a nonsingular matrix $P \in \mathscr{F}_{nn}$ such that $A_R = PA = I_n$. Hence, A is nonsingular and $P = A^{-1}$. ∎

Corollary 1 to Theorem 1. *A matrix $A \in \mathscr{F}_{nn}$ is nonsingular if and only if it is a product of elementary matrices.*

● **EXERCISE**

1. Prove Corollary 1, Theorem 1.

Incidentally, we can now prove the converse of Exercise 2, Section 3.4, namely that if $B = PA$, where $A, B \in \mathscr{F}_{mn}$ and $P \in \mathscr{F}_{mm}$ is nonsingular, then $A \overset{R}{\sim} B$. Because, by the above corollary, $P = E_1 \cdots E_k$, where $E_1, \ldots, E_k$ are elementary matrices. Thus, $B = E_1 \cdots E_k A$ and B is obtained from A by a sequence of elementary row operations, which establishes that $A \overset{R}{\sim} B$.

Corollary 2 to Theorem 1. *The system $A\mathbf{x} = \mathbf{0}$, where $A \in \mathscr{F}_{nn}$, has a nontrivial solution if and only if rank $A < n$.*

● **EXERCISE**

2. Prove Corollary 2, Theorem 1.

Theorem 1 suggests the following procedure for finding the inverse of a nonsingular matrix $A \in \mathscr{F}_{nn}$:

STEP 1. Consider the matrix $[A \mid I_n] \in \mathscr{F}_{n, 2n}$.

STEP 2. Reduce this matrix to row echelon form.

By Theorem 1, if P is a nonsingular matrix such that $PA = A_R = I_n$, we have

$$P[A \mid I_n] = [PA \mid P] = [I_n \mid P]$$

3.8 The Inverse of a Nonsingular Matrix

as this row echelon form. But, since $P = A^{-1}$, we read off A^{-1} as the last n columns of the matrix $[I_n | P]$. If A_R does not turn out to be the identity matrix, then A is, of course, singular; see also Exercise 3, Section 3.5 and Exercise 4, Section 3.7.

Example 1. Find the inverse, if it exists, of the matrix

$$A = \begin{bmatrix} 1 & 0 & 0 & 1 \\ 0 & 1 & 1 & 0 \\ 0 & 1 & -1 & 0 \\ -1 & 0 & 0 & 1 \end{bmatrix}.$$

$$[A | I_4] = \begin{bmatrix} 1 & 0 & 0 & 1 & | & 1 & 0 & 0 & 0 \\ 0 & 1 & 1 & 0 & | & 0 & 1 & 0 & 0 \\ 0 & 1 & -1 & 0 & | & 0 & 0 & 1 & 0 \\ -1 & 0 & 0 & 1 & | & 0 & 0 & 0 & 1 \end{bmatrix}$$

$$\underset{\sim}{R} \begin{bmatrix} 1 & 0 & 0 & 1 & | & 1 & 0 & 0 & 0 \\ 0 & 1 & 1 & 0 & | & 0 & 1 & 0 & 0 \\ 0 & 1 & -1 & 0 & | & 0 & 0 & 1 & 0 \\ 0 & 0 & 0 & 2 & | & 1 & 0 & 0 & 1 \end{bmatrix}$$

$$\underset{\sim}{R} \begin{bmatrix} 1 & 0 & 0 & 1 & | & 1 & 0 & 0 & 0 \\ 0 & 1 & 1 & 0 & | & 0 & 1 & 0 & 0 \\ 0 & 0 & -2 & 0 & | & 0 & -1 & 1 & 0 \\ 0 & 0 & 0 & 2 & | & 1 & 0 & 0 & 1 \end{bmatrix}$$

$$\underset{\sim}{R} \begin{bmatrix} 1 & 0 & 0 & 1 & | & 1 & 0 & 0 & 0 \\ 0 & 1 & 1 & 0 & | & 0 & 1 & 0 & 0 \\ 0 & 0 & 1 & 0 & | & 0 & \tfrac{1}{2} & -\tfrac{1}{2} & 0 \\ 0 & 0 & 0 & 2 & | & 1 & 0 & 0 & 1 \end{bmatrix}$$

$$\underset{\sim}{R} \begin{bmatrix} 1 & 0 & 0 & 1 & | & 1 & 0 & 0 & 0 \\ 0 & 1 & 0 & 0 & | & 0 & \tfrac{1}{2} & \tfrac{1}{2} & 0 \\ 0 & 0 & 1 & 0 & | & 0 & \tfrac{1}{2} & -\tfrac{1}{2} & 0 \\ 0 & 0 & 0 & 2 & | & 1 & 0 & 0 & 1 \end{bmatrix}$$

$$\underset{\sim}{R} \begin{bmatrix} 1 & 0 & 0 & 1 & | & 1 & 0 & 0 & 0 \\ 0 & 1 & 0 & 0 & | & 0 & \tfrac{1}{2} & \tfrac{1}{2} & 0 \\ 0 & 0 & 1 & 0 & | & 0 & \tfrac{1}{2} & -\tfrac{1}{2} & 0 \\ 0 & 0 & 0 & 1 & | & \tfrac{1}{2} & 0 & 0 & \tfrac{1}{2} \end{bmatrix}$$

$$\underset{\sim}{R} \begin{bmatrix} 1 & 0 & 0 & 0 & | & \tfrac{1}{2} & 0 & 0 & -\tfrac{1}{2} \\ 0 & 1 & 0 & 0 & | & 0 & \tfrac{1}{2} & \tfrac{1}{2} & 0 \\ 0 & 0 & 1 & 0 & | & 0 & \tfrac{1}{2} & -\tfrac{1}{2} & 0 \\ 0 & 0 & 0 & 1 & | & \tfrac{1}{2} & 0 & 0 & \tfrac{1}{2} \end{bmatrix}.$$

Therefore,

$$A^{-1} = \tfrac{1}{2}\begin{bmatrix} 1 & 0 & 0 & -1 \\ 0 & 1 & 1 & 0 \\ 0 & 1 & -1 & 0 \\ 1 & 0 & 0 & 1 \end{bmatrix}.$$

● **EXERCISES**

3. Let $A \in \mathscr{F}_{nn}$ with $a_{i*} = 0$ for some i. Prove that A is singular.

4. Let $B \in \mathscr{F}_{nn}$ with $b_{i*} = b_{j*}$ for some $i \neq j$. Prove that B is singular.

5. Let A and B be square matrices. Prove that if AB is nonsingular, then both A and B are nonsingular. [*Hint:* Consider nontrivial solutions of $B\mathbf{x} = \mathbf{0}$ and $AB\mathbf{x} = \mathbf{0}$ to show that B is nonsingular. A, then, is the product of two nonsingular matrices.]

6. Let A and B be square matrices. If $AB = I$, show that $BA = I$. [*Hint:* From Exercise 5, A and B are nonsingular. Therefore, $A = E_1 E_2 \cdots E_n$, a product of elementary matrices. By multiplying by elementary matrices, $(E_1 \cdots E_n)B = I$ can be changed to $B(E_1 \cdots E_n) = I$.]

7. Find the inverses, if they exist, of the following matrices.

(a) $\begin{bmatrix} 3 & 2 & -1 \\ -1 & 2 & 3 \\ -3 & 1 & 3 \end{bmatrix}.$ (b) $\begin{bmatrix} 2 & 1 \\ 1 & 2 \end{bmatrix}.$ (c) $\begin{bmatrix} 1 & 1 & 1 \\ -1 & 2 & -1 \\ 0 & 0 & 3 \end{bmatrix}.$

(d) $\begin{bmatrix} 1 & 2 \\ 3 & 4 \end{bmatrix}.$ (e) $\begin{bmatrix} 2 & 0 & 5 & 0 \\ 0 & 7 & 0 & 2 \\ 3 & 0 & 7 & 0 \\ 0 & 0 & 0 & 6 \end{bmatrix}.$

(f) $\begin{bmatrix} 3 & 4 & 0 & 0 \\ 3 & 3 & 0 & 0 \\ 0 & 0 & 7 & 0 \\ 0 & 0 & 0 & 6 \end{bmatrix}.$ (g) $\begin{bmatrix} 1 & 2 & 1 & 2 & 3 \\ 2 & 3 & 1 & 0 & 1 \\ 2 & 2 & 1 & 0 & 0 \\ 1 & 1 & 1 & 1 & 1 \\ 0 & -2 & 0 & 2 & -2 \end{bmatrix}.$

(h) $\begin{bmatrix} 2 & 5 & -1 \\ 4 & -1 & 2 \\ 6 & 4 & 1 \end{bmatrix}.$ (i) $\begin{bmatrix} 1 & -1 & 2 \\ 3 & 2 & 4 \\ 0 & 1 & -2 \end{bmatrix}.$

8. Let

$$A = \begin{bmatrix} a & b \\ c & d \end{bmatrix},$$

where a, b, c, d are complex numbers. Show, using elementary row operations, that A is nonsingular if and only if $ad - bc \neq 0$.

3.8 The Inverse of a Nonsingular Matrix

● **MISCELLANEOUS EXERCISES**

1. (a) If $A, B \in \mathscr{F}_{nn}$ and if $X = AB - BA$, show that

$$x_{11} + x_{22} + \cdots + x_{nn} = 0.$$

(b) If $x \in \mathscr{F}_{22}$ and if $x_{11} + x_{22} = 0$, show that there exist matrices $A, B \in \mathscr{F}_{22}$ such that $X = AB - BA$.

2. Solve completely each of the following systems of equations (that is, determine if a solution exists and, if so, describe all solutions):

(a) $5x_1 + 2x_2 - x_3 = 0$
     $3x_1 + 5x_2 + 3x_3 = 0$
     $x_1 + 8x_2 + 7x_3 = 0.$

(b) $2x_1 - x_2 + 3x_3 = 5$
     $3x_1 + 2x_2 - 2x_3 = 1$
     $7x_1 + 4x_3 = 11.$

(c) $3x_1 - x_2 = 7$
     $2x_1 + x_2 = 1.$

(d) $x_1 + 2x_2 = -1$
     $3x_1 + 6x_2 = -3.$

3. Solve completely:

(a) $5x_1 - 4x_2 + 7x_3 = -4$
     $9x_1 - 5x_2 + 5x_3 = 8$
     $4x_1 + 3x_2 + 3x_3 = 2.$

(b) $x_1 - x_2 + x_3 = 2$
     $2x_1 + 3x_2 - x_3 = -1$
     $x_1 + 2x_2 - 3x_3 = 3.$

(c) $3x_1 - 2x_3 + 5x_4 = 2$
     $-2x_1 - x_2 + 4x_3 = 0$
     $x_1 + 2x_2 - 3x_3 - 5x_4 = 1$
     $ 3x_2 + 5x_3 + 2x_4 = -3.$

(d) $x_1 + 2x_2 - x_3 = 5$
     $3x_1 - x_2 + x_3 = 5$
     $2x_1 + 4x_2 - 2x_3 = 10.$

(e) $2x_1 - 2x_2 + x_3 = -2$
     $x_1 + 2x_2 = 7$
     $3x_1 + x_3 = 5.$

4. Let

$$A = \begin{bmatrix} 1 & 2 & 3 \\ 2 & 4 & 2 \\ 0 & 1 & 0 \end{bmatrix}.$$

Show that for every vector $\mathbf{b} \in \mathscr{F}_{31}$, the system $A\mathbf{x} = \mathbf{b}$ has a unique solution.

5. (a) Let

$$A = \begin{bmatrix} 1 & 2 \\ 2 & \lambda \end{bmatrix}, \quad \mathbf{b}_1 = \begin{bmatrix} 0 \\ 0 \end{bmatrix}, \quad \mathbf{b}_2 = \begin{bmatrix} 3 \\ 6 \end{bmatrix}, \quad \mathbf{b}_3 = \begin{bmatrix} 1 \\ 3 \end{bmatrix}.$$

For each of the systems $A\mathbf{x} = \mathbf{b}_1$, $A\mathbf{x} = \mathbf{b}_2$, and $A\mathbf{x} = \mathbf{b}_3$, in turn, determine for which values of λ the system has (i) no solution, (ii) a unique solution, or (iii) more than one solution.

(b) Repeat part (a), for

$$A = \begin{bmatrix} 1 & 3 & 2 \\ 2 & 2 & 0 \\ 0 & 1 & \lambda \end{bmatrix}, \quad \mathbf{b}_1 = \begin{bmatrix} 0 \\ 0 \\ 0 \end{bmatrix}, \quad \mathbf{b}_2 = \begin{bmatrix} 2 \\ 1 \\ 4 \end{bmatrix}, \quad \mathbf{b}_3 = \begin{bmatrix} 6 \\ 4 \\ 0 \end{bmatrix}.$$

Chapter 4 | DETERMINANTS

In dealing with square matrices it is convenient to associate with every matrix $A \in \mathscr{F}_{nn}$ an element of $\mathscr{F}$, called the determinant of A, and denoted by det A. It is the purpose of this chapter to define the determinant and to study its properties. We will then discuss Cramer's rule for writing down the solution of a system of n linear algebraic equations in n unknowns in terms of determinants. Finally, we will establish the useful criterion that a square matrix is nonsingular if and only if its determinant is not zero.

There are in the literature several different ways of defining a determinant. Ours is inductive and is, we believe, the most expedient one. The interested reader will find two distinctly different approaches in C. W. Curtis, *Linear Algebra—An Introductory Approach*, (Allyn and Bacon, Boston, 1968), and H. Schneider and G. P. Barker, *Matrices and Linear Algebra* (Holt, Rinehart, & Winston, New York, 1968).

4.1 Definition and Notation

We define the determinant of a square matrix inductively as follows: *If $A \in \mathscr{F}_{11}$, we define its determinant to be* det $A = A$. *If $n > 1$ and $A \in \mathscr{F}_{nn}$, we denote by A_{ij} the matrix in $\mathscr{F}_{n-1, n-1}$ obtained from A by deleting the ith* row and jth column; that is,

$$A_{ij} = \begin{bmatrix} a_{11} & \cdots & a_{1j} & \cdots & a_{1n} \\ \vdots & & \vdots & & \vdots \\ a_{i1} & \cdots & a_{ij} & \cdots & a_{in} \\ \vdots & & \vdots & & \vdots \\ a_{n1} & \cdots & a_{ij} & \cdots & a_{nn} \end{bmatrix}.$$

Assuming that the determinant of an $(n-1) \times (n-1)$ matrix has already been defined, we define

$$\det A = \sum_{j=1}^{n} a_{1j} |A_{1j}|, \tag{4.1}$$

where

$$|A_{1j}| = (-1)^{1+j} \det A_{1j} \qquad (j = 1, \ldots, n).$$

The quantity $|A_{1j}|$ is called **the cofactor** of a_{1j} in A and formula (4.1) is called the expansion of the determinant of A by means of cofactors of the first row. The reader will note that to perform the expansion we form the sum of products of each element of the first row with its cofactor.

Example 1. Compute $\det A$ if $A \in \mathscr{F}_{22}$. Here

$$A = \begin{bmatrix} a_{11} & a_{12} \\ a_{21} & a_{22} \end{bmatrix}$$

and using formula (4.1), we obtain

$$\det A = a_{11}|A_{11}| + a_{12}|A_{12}|$$
$$= a_{11}(-1)^{1+1}a_{22} + a_{12}(-1)^{1+2}a_{21} = a_{11}a_{22} - a_{12}a_{21}.$$

Example 2. Compute $\det A$ if $A \in \mathscr{F}_{33}$. Here

$$A = \begin{bmatrix} a_{11} & a_{12} & a_{13} \\ a_{21} & a_{22} & a_{23} \\ a_{31} & a_{32} & a_{33} \end{bmatrix}$$

and using formula (4.1) and Example 1, we obtain

$$\det A = a_{11}|A_{11}| + a_{12}|A_{12}| + a_{13}|A_{13}|$$

$$= a_{11}(-1)^{1+1} \det \begin{bmatrix} a_{22} & a_{23} \\ a_{32} & a_{33} \end{bmatrix} + a_{12}(-1)^{1+2} \det \begin{bmatrix} a_{21} & a_{23} \\ a_{31} & a_{33} \end{bmatrix}$$

$$+ a_{13}(-1)^{1+3} \det \begin{bmatrix} a_{21} & a_{22} \\ a_{31} & a_{32} \end{bmatrix}$$

$$= a_{11}(a_{22}a_{33} - a_{23}a_{32}) - a_{12}(a_{21}a_{33} - a_{23}a_{31})$$

$$+ a_{13}(a_{21}a_{32} - a_{22}a_{31})$$

$$= a_{11}a_{22}a_{33} - a_{11}a_{23}a_{32} - a_{12}a_{21}a_{33} + a_{12}a_{23}a_{31}$$

$$+ a_{13}a_{21}a_{32} - a_{13}a_{22}a_{31}.$$

● **EXERCISE**

1. Calculate det A for each of the following matrices A.

   (a) $A = \begin{bmatrix} 1 & -1 \\ 3 & i \end{bmatrix}$.

   (b) $A = \begin{bmatrix} -1 & 2 & 3 \\ 3 & 2 & 1 \\ -2 & 4 & 6 \end{bmatrix}$.

   (c) $A = \begin{bmatrix} 1 & -2 & 5 & 0 \\ 0 & 2 & i & 3 \\ 0 & 0 & 3 & -1 \\ 0 & 0 & 0 & 4 \end{bmatrix}$.

   (d) $A = \begin{bmatrix} 2 & 0 & 1 \\ 5 & -2 & 3 \\ 1 & 1 & 1 \end{bmatrix}$.

   (e) $A = \begin{bmatrix} 1 & 0 & 3 \\ 2 & a & 4 \\ 0 & 6 & 0 \end{bmatrix}$, where a is a constant.

   (f) $A = \begin{bmatrix} a_{11} & a_{12} & a_{13} & a_{14} \\ a_{21} & a_{22} & a_{23} & a_{24} \\ a_{31} & a_{32} & a_{33} & a_{34} \\ a_{41} & a_{42} & a_{43} & a_{44} \end{bmatrix}$.

74 Determinants

It is important to be able to compute the determinant of the $n \times n$ zero and identity matrices.

Theorem 1. $\det 0_n = 0$

$\det I_n = 1,$

where 0_n is the $n \times n$ zero matrix and I_n is the $n \times n$ identity matrix.

Proof. Since each element of the first row of 0_n is zero, it is obvious from the definition that $\det 0_n = 0$. To calculate $\det I_n$, we proceed by induction. Clearly, $\det I_1 = 1$. We assume the result for all $m < n$. Then, by (4.1) $\det I_n = 1 \,|\, I_{11}|$, and I_{11} is the $(n-1) \times (n-1)$ identity matrix. Thus, by the induction assumption $|I_{11}| = 1$, and $\det I_n = 1$. ∎

● EXERCISES

2. Evaluate the determinant of the matrix

$$A = \begin{bmatrix} 1 & 0 & 0 & 0 \\ 0 & 2 & 0 & 0 \\ 0 & 0 & 3 & 0 \\ 0 & 0 & 0 & 4 \end{bmatrix}.$$

3. Prove by induction that

$$\det \begin{bmatrix} d_1 & 0 & \cdots & & 0 \\ 0 & d_2 & 0 & \cdots & 0 \\ \vdots & & \ddots & & \vdots \\ & & & \ddots & 0 \\ 0 & \cdots & & 0 & d_n \end{bmatrix} = d_1 d_2 \cdots d_n;$$

that is, the determinant of a diagonal matrix is the product of the diagonal elements.

4. Prove by induction that if A is a square matrix and some row of A is zero, then $\det A = 0$. (You will probably want to consider the cases of $a_{1*} = 0$ and $a_{2*} = 0$ for $i > 1$ separately.)

5. Show, by exhibiting an example, that it is not true in general that

$\det A + \det B = \det(A + B)$.

The reader may wonder whether the determinant of a matrix must be evaluated by expanding by cofactors of the first row as is done in Formula (4.1). As a matter of fact, *we can evaluate the determinant by expanding by cofactors of any row or any column obtaining the same value.* We will establish this important fact in Section 4.3. The relevant formulas are

$$\det A = \sum_{j=1}^{n} a_{ij} |A_{ij}| \tag{4.2}$$

for any fixed i, $1 \leq i \leq n$, and, similarly,

$$\det A = \sum_{i=1}^{n} a_{ij} |A_{ij}| \tag{4.3}$$

for any fixed j, $1 \leq j \leq n$, where in (4.2) and (4.3)

$$|A_{ij}| = (-1)^{i+j} \det A_{ij} \qquad (i, j = 1, \ldots, n). \tag{4.4}$$

● **EXERCISES**

6. For each of the matrices A in Exercise 1, evaluate $\det A$ by expanding by cofactors of every row and every column.
[*Remark.* Here we assume, of course, the validity of formulas (4.2) and (4.3); note that if a row or column of a matrix has several elements zero, the expansion by cofactors of such a row or column becomes particularly simple.]

7. In each of the following matrices, there is at least one row or column for which expansion by cofactors is particularly easy. Evaluate $\det A$ for each matrix A.

(a) $A = \begin{bmatrix} 0 & 1 & 1 \\ 0 & 1 & 2 \\ 1 & 0 & 1 \end{bmatrix}$. (b) $A = \begin{bmatrix} 1 & 2 & 2 \\ 0 & 1 & 1 \\ 1 & 0 & 1 \end{bmatrix}$.

(c) $A = \begin{bmatrix} 1 & 0 & 0 & 2 \\ 0 & 2 & 1 & 1 \\ 1 & 3 & 0 & 4 \\ 2 & 0 & 0 & 2 \end{bmatrix}$. (d) $A = \begin{bmatrix} 1 & 6 & 2 & -1 \\ -2 & 2 & 0 & 0 \\ 4 & 4 & 0 & 0 \\ 2 & -3 & 1 & 5 \end{bmatrix}$.

(e) $A = \begin{bmatrix} 1 & 5 & 7 & 5 \\ -2 & 0 & 2 & 1 \\ 1 & 4 & -1 & 0 \\ 0 & 2 & 0 & 1 \end{bmatrix}$. (f) $A = \begin{bmatrix} 0.0 & 0.5 & 0.0 \\ -0.4 & -0.7 & 4.0 \\ 0.2 & 0.3 & 3.0 \end{bmatrix}$.

4.2 Some Properties of Determinants

We shall first consider the effect of certain operations concerning the columns of a square matrix on the value of its determinant. These operations are called **elementary column operations**, and they correspond exactly to elementary row operations discussed in Chapter 3. They are as follows.

(i) Interchange of two columns (Type I).
(ii) Multiplication of a column by a nonzero element $\lambda \in \mathscr{F}$ (Type II).
(iii) Addition of a multiple of one column to another column (Type III).

We shall again use the notation a_{*k} for the kth column of A, so that $A = [a_{*1}, a_{*2}, \ldots, a_{*n}]$.

Lemma 1. *Let $A \in \mathscr{F}_{nn}$. Define*

$$B = [a_{*1}, \ldots, a_{*(k-1)}, b_{*k}, a_{*(k+1)}, \ldots, a_{*n}]$$
$$C = [a_{*1}, \ldots, a_{*(k-1)}, c_{*k}, a_{*(k+1)}, \ldots, a_{*n}]$$

for some fixed k, $1 \leq k \leq n$. If $a_{*k} = \beta b_{*k} + \gamma c_{*k}$ for some $\beta, \gamma \in \mathscr{F}$, then

$$\det A = \beta \det B + \gamma \det C. \tag{4.5}$$

Proof. We proceed by induction on n. Obviously, Lemma 1 is true for $n = 1$. Suppose the result has been established for any $m \times m$ matrix, where $m < n$. By definition

$$\det A = \sum_{j=1}^{n} a_{1j} |A_{1j}|.$$

We observe that $a_{1k} = \beta b_{1k} + \gamma c_{1k}$ and $A_{1k} = B_{1k} = C_{1k}$; thus, $a_{1k}|A_{1k}| = \beta b_{1k}|B_{1k}| + \gamma c_{1k}|C_{1k}|$. If $j \neq k$, then $a_{1j} = b_{1j} = c_{1j}$ and for each fixed j the matrix A_{1j} in $\mathscr{F}_{(n-1)(n-1)}$ is identical to each of the matrices B_{1j} and C_{1j} except for one column which can be obtained by adding β times the same column of B_{1j} to γ times the same column of C_{1j}. Thus, by the induction assumption $\det A_{1j} = \beta \det B_{1j} + \gamma \det C_{1j}$ and so $|A_{1j}| = \beta |B_{1j}| + \gamma |C_{1j}|$. It now follows that

$$a_{1j}|A_{1j}| = \beta b_{1j}|B_{1j}| + \gamma c_{1j} C_{1j}|$$

for $j \neq k$. Since the formula has already been established for $j = k$, one has

$$\det A = \sum_{j=1}^{n} a_{1j}|A_{1j}| = \sum_{j=1}^{n} (\beta b_{1j}|B_{1j}| + \gamma c_{1j}|C_{1j}|)$$
$$= \beta \det B + \gamma \det C. \quad \blacksquare$$

4.2 Some Properties of Determinants

By choosing $\gamma = 0$ in (4.5), we obtain the following result.

Lemma 2. *If $A \in \mathscr{F}_{nn}$, then*

$$\det [a_{*1}, \ldots, a_{*(k-1)}, \beta a_{*k}, a_{*(k+1)}, \ldots, a_{*n}] = \beta \det A. \tag{4.6}$$

By choosing $\beta = 0$ in (4.6) we obtain the following result.

Lemma 3. *If $A \in \mathscr{F}_{nn}$ and $a_{*k} = 0$ for some fixed k, $1 \leq k \leq n$, then $\det A = 0$.*

In order to obtain additional properties of determinants, we establish the following preliminary result.

Lemma 4. *If $A \in \mathscr{F}_{nn}$ and if two adjacent columns of A are equal, then $\det A = 0$.*

Proof. The proof is by induction on n. The lemma is obviously true for $n = 2$, as may be seen from Example 1, Section 4.1. For $n > 2$, we assume that the result is true for all $m \times m$ matrices with $m < n$. Then for $A \in \mathscr{F}_{nn}$, let $a_{*k} = a_{*(k+1)}$, where $1 \leq k \leq n - 1$. We expand $\det A$ by the formula (4.1). It is obvious that $A_{1k} = A_{1(k+1)}$ and, therefore,

$$|A_{1k}| = (-1)^{1+k} \det A_{1k} = -(-1)^{1+(k+1)} \det A_{1k}$$
$$= -(-1)^{k+2} \det A_{1(k+1)} = -|A_{1(k+1)}|.$$

Thus, $a_{1k}|A_{1k}| + a_{1(k+1)}|A_{1k(+1)}| = 0$. If $j \neq k$ or $k + 1$, then A_{1j} has two identical columns and by the induction hypothesis, $|A_{1j}| = 0$. Thus, $\det A = \sum_{j=1}^{n} a_{1j}|A_{1j}| = a_{1k}|A_{1k}| + a_{1(k+1)}|A_{1(k+1)}| = 0$. ∎

The effect of interchanging two adjacent columns is given by the following result.

Lemma 5. *If $A \in \mathscr{F}_{nn}$, then for any fixed k, $1 \leq k \leq n - 1$,*

$$\det [a_{*1}, \ldots, a_{*(k-1)}, a_{*k}, a_{*(k+1)}, \ldots, a_{*n}]$$
$$= -\det [a_{*1}, \ldots, a_{*(k-1)}, a_{*(k+1)}, a_{*k}, \ldots, a_{*n}].$$

Proof. Consider the matrix

$$B = [a_{*1}, \ldots, a_{*(k-1)}, a_{*k} + a_{*(k+1)}, a_{*k} + a_{*(k+1)}, \ldots, a_{*n}]$$

78 *Determinants*

By Lemma 4, det $B = 0$ On the other hand, by Lemma 1,

$$\begin{aligned}
0 = \det B &= \det [a_{*1}, \ldots, a_{*(k-1)}, a_{*k}, a_{*k} + a_{*(k+1)}, \ldots, a_{*n}] \\
&\quad + \det [a_{*1}, \ldots, a_{*(k-1)}, a_{*(k+1)}, a_{*k} + a_{*(k+1)}, \ldots, a_{*n}] \\
&= \det [a_{*1}, \ldots, a_{*(k-1)}, a_{*k}, a_{*k}, \ldots, a_{*n}] \\
&\quad + \det [a_{*1}, \ldots, a_{*(k-1)}, a_{*k}, a_{*(k+1)}, \ldots, a_{*n}] \\
&\quad + \det [a_{*1}, \ldots, a_{*(k-1)}, a_{*(k+1)}, a_{*k}, \ldots, a_{*n}] \\
&\quad + \det [a_{*1}, \ldots, a_{*(k-1)}, a_{*(k+1)}, a_{*(k+1)}, \ldots, a_{*n}] \\
&= \det [a_{*1}, \ldots, a_{*(k-1)}, a_{*k}, a_{*(k+1)}, \ldots, a_{*n}] \\
&\quad + \det [a_{*1}, \ldots, a_{*(k-1)}, a_{*(k+1)}, a_{*k}, \ldots, a_{*n}],
\end{aligned}$$

by Lemma 4, and the result follows. ∎

As easy consequences, we have the following useful results.

Lemma 6. *If $A \in \mathscr{F}_{nn}$ and if $a_{*j} = a_{*k}$ for any j, k, where $1 \leq j$, $k \leq n$, with $j \neq k$, then det $A = 0$.*

Proof. Suppose $j < k$. Using Lemma 5 repeatedly, we have

$$\begin{aligned}
\det A &= \det [a_{*1}, \ldots, a_{*(j-1)}, a_{*j}, a_{*(j+1)}, a_{*(j+2)}, \ldots, a_{*n}] \\
&= -\det [a_{*1}, \ldots, a_{*(j-1)}, a_{*(j+1)}, a_{*j}, a_{*(j+2)}, \ldots, a_{*n}] \\
&= \det [a_{*1}, \ldots, a_{*(j-1)}, a_{*(j+1)}, a_{*(j+2)}, a_{*j}, \ldots, a_{*n}].
\end{aligned}$$

We continue, moving a_{*j} to the right until a_{*j} is adjacent to a_{*k}. The result of this is that

$$\det A = \pm \det [a_{*1}, \ldots, a_{*(j-1)}, a_{*(j+1)}, \ldots, a_{*j}, a_{*k}, \ldots, a_{*n}]$$

which is zero by Lemma 4. ∎

Lemma 7. *If $A \in \mathscr{F}_{nn}$, and if $\gamma \in \mathscr{F}$, and if $j \neq k$, then*

$$\det [a_{*1}, \ldots, a_{*j}, \ldots, a_{*(k-1)}, a_{*k} + \gamma a_{*j}, a_{*(k+1)}, \ldots, a_{*n}] = \det A.$$

Proof. Applying successively Lemmas 1 and 6, we have

$$\begin{aligned}
\det &[a_{*1}, \ldots, a_{*j}, \ldots, a_{*(k-1)}, a_{*k} + \gamma a_{*j}, a_{*(k+1)}, \ldots, a_{*n}] \\
&= \det A + \gamma \det [a_{*1}, \ldots, a_{*j}, \ldots, a_{*(k-1)}, a_{*j}, a_{*(k+1)}, \ldots, a_{*n}] \\
&= \det A. \quad \blacksquare
\end{aligned}$$

4.2 Some Properties of Determinants

Lemma 8. *If $A \in \mathscr{F}_{nn}$, and if $j \neq k$, then*

$$\det [a_{*1}, \ldots, a_{*k}, \ldots, a_{*j}, \ldots, a_{*n}]$$
$$= -\det [a_{*1}, \ldots, a_{*j}, \ldots, a_{*k}, \ldots, a_{*n}].$$

Proof. The proof resembles the proof of Lemma 5. Using Lemmas 6 and 1, we have

$$0 = \det [a_{*1}, \ldots, a_{*j} + a_{*k}, \ldots, a_{*j} + a_{*k}, \ldots, a_{*n}]$$
$$= \det [a_{*1}, \ldots, a_{*j}, \ldots, a_{*j}, \ldots, a_{*n}]$$
$$+ \det [a_{*1}, \ldots, a_{*j}, \ldots, a_{*k}, \ldots, a_{*n}]$$
$$+ \det [a_{*1}, \ldots, a_{*k}, \ldots, a_{*j}, \ldots, a_{*n}]$$
$$+ \det [a_{*1}, \ldots, a_{*k}, \ldots, a_{*k}, \ldots, a_{*n}]$$
$$= \det [a_{*1}, \ldots, a_{*j}, \ldots, a_{*k}, \ldots, a_{*n}]$$
$$+ \det [a_{*1}, \ldots, a_{*k}, \ldots, a_{*j}, \ldots, a_{*n}],$$

which yields the result. ∎

We now single out those of the above properties for computation of determinants which correspond to the elementary column operations introduced at the beginning of this section.

Theorem 1. *If $A \in \mathscr{F}_{nn}$, and if $\lambda \in \mathscr{F}$, then*

(i) *interchange of two columns of A changes the sign of* det A (Lemma 8);
(ii) *addition of λ times any column of A to any other column of A leaves the value of* det A *unchanged* (Lemma 7);
(iii) *multiplication of any column of A by λ multiplies* det A *by λ* (Lemma 2).

In Chapter 3 we showed that an elementary row operation on a matrix A corresponded to premultiplication of A by an elementary matrix of the appropriate type. The reader can easily verify that an elementary column operation on a matrix A corresponds to postmultiplication of A by an elementary matrix.

● EXERCISES

1. In each of the following, find an elementary matrix E such that $B = AE$.

(a) $A = \begin{bmatrix} 1 & 1 & 1 \\ 1 & 2 & 3 \\ 1 & 3 & 9 \end{bmatrix}$, $B = \begin{bmatrix} 1 & 1 & 1 \\ 1 & 3 & 2 \\ 1 & 9 & 3 \end{bmatrix}$.

80 Determinants

(b) $A = \begin{bmatrix} 1 & 1 & 1 \\ 1 & 2 & 3 \\ 1 & 3 & 9 \end{bmatrix}$, $B = \begin{bmatrix} 1 & -3 & 1 \\ 1 & -6 & 3 \\ 1 & -9 & 9 \end{bmatrix}$.

(c) $A = \begin{bmatrix} 1 & 1 & 1 \\ 1 & 2 & 3 \\ 1 & 3 & 9 \end{bmatrix}$, $B = \begin{bmatrix} -1 & 1 & 1 \\ -5 & 2 & 3 \\ -17 & 3 & 9 \end{bmatrix}$.

2. Find a sequence of elementary column operations that reduces A to B, and find a matrix Q such that $B = AQ$, where Q is the product of the corresponding elementary matrices, if

$$A = \begin{bmatrix} 2 & 1 & -2 \\ 0 & 1 & 1 \\ 14 & 5 & -16 \end{bmatrix}, \quad B = \begin{bmatrix} 1 & 0 & 0 \\ 0 & 1 & 0 \\ 7 & -2 & 0 \end{bmatrix}.$$

Using Theorem 1, we may now calculate the determinant of an elementary matrix.

Corollary 1 to Theorem 1.

(i) *If E is an elementary matrix of type* I, *then* $\det E = -1$.
(ii) *If E is an elementary matrix of Type* II, *then* $\det E = \lambda$, *where λ is as in Theorem* 1 (ii).
(iii) *If E is an elementary matrix of Type* III, *then* $\det E = 1$.

Proof. We shall prove (iii). Let E be the elementary matrix of Type III.

$$E = \begin{bmatrix} 1 & & & & & & \\ & \ddots & & & & & \\ & & 1 & \cdots & \lambda & & \\ & & & \ddots & \vdots & & \\ & & & & \ddots & & \\ & & & & 1 & & \\ & & & & & \ddots & \\ & & & & & & 1 \end{bmatrix} \begin{matrix} \\ \\ \leftarrow j\text{th row} \\ \\ \\ \leftarrow k\text{th row} \\ \\ \end{matrix}$$

$\quad\quad\quad\quad\quad\quad j\text{th} \quad k\text{th}$
$\quad\quad\quad\quad\quad\quad \uparrow \quad\quad \uparrow$
$\quad\quad\quad\quad\quad \text{column column}$

We observe that E can be obtained from the identity matrix by adding λ times the jth column to the kth column. Thus, by Theorem 1(iii), and

4.2 Some Properties of Determinants

Theorem 1, Section 4.1, we have $\det E = \det I = 1$. The proofs of the other two parts are similar. ∎

• EXERCISES

3. Let

$$A = \begin{bmatrix} 2 & 3 & 5 \\ 1 & -6 & -2 \\ 1 & -2 & 1 \end{bmatrix}.$$

(a) Evaluate $\det A$.
(b) Let B be obtained from A by interchanging the second and third columns of A. Evaluate $\det B$, and compare with Theorem 1(i).
(c) Let

$$C = \begin{bmatrix} 2 & 3 & 3 \\ 1 & -6 & -3 \\ 1 & -2 & 0 \end{bmatrix}.$$

Evaluate $\det C$, and compare with Theorem 1(ii).
(d) Let

$$D = \begin{bmatrix} -6 & 3 & 5 \\ -2 & -6 & -2 \\ -2 & -2 & 1 \end{bmatrix}.$$

Evaluate $\det D$, and compare with Theorem 1(iii).

4. Show by induction that if $A \in \mathscr{F}_{nn}$ and $B = \lambda A$ for some scalar λ, then $\det B = \lambda^n \det A$.

5. Prove parts (i) and (ii) of Corollary 1 to Theorem 1.

Corollary 2 to Theorem 1. *Let $A \in \mathscr{F}_{nn}$, and let E be an elementary matrix of any type. Then*

$$\det AE = \det A \det E. \tag{4.7}$$

Proof. We give the proof if E is an elementary matrix of Type I, and leave the other two cases to the reader. Let $B = AE$. By Theorem 1(i), $\det B = -\det A$, since B is obtained from A by interchanging two columns. On the other hand, by Corollary 1, $\det E = -1$. Therefore,

$$\det B = -\det A = \det A \det E. \blacksquare$$

Determinants

● **EXERCISE**

6. Prove Corollary 2 if E is an elementary matrix of Type II or III.

4.3 The Determinant of the Transposed Matrix

In order to establish the formulas (4.2) and (4.3), which enable us to expand a determinant by cofactors of any row or column, we begin by defining the **transpose of a matrix**.

Definition 1. *If $A \in \mathscr{F}_{mn}$, then the transpose of A, denoted by A^T, is the matrix in $\mathscr{F}_{nm}$ defined by $B = A^T$, where $b_{ij} = a_{ji}$ ($i = 1, \ldots, m; j = 1, \ldots, n$). Note that if $m = n$, then both A and A^T are elements of $\mathscr{F}_{nn}$. Note also that $(A^T)^T = A$.*

● **EXERCISES**

1. For each of the following matrices A, write down the matrix A^T.

(a) $A = \begin{bmatrix} 1 & 2 \\ -2 & 1 \end{bmatrix}$.   (b) $A = \begin{bmatrix} 0 & 1 \\ 0 & 0 \end{bmatrix}$.   (c) $A = \begin{bmatrix} 1 & 2 & 1 \\ 3 & 1 & 7 \\ 6 & 2 & 1 \end{bmatrix}$.

(d) $A = \begin{bmatrix} 1 & 1 & 0 \\ -1 & 2 & -4 \\ 0 & 4 & 3 \end{bmatrix}$.   (e) $A = \begin{bmatrix} -1 & 0 & 0 \\ 0 & -1 & 0 \\ 0 & 0 & -1 \end{bmatrix}$.

(f) $A = \begin{bmatrix} 3 & 2 & 4 \\ 2 & 7 & 1 \\ 4 & 1 & 6 \end{bmatrix}$.   (g) $A = \begin{bmatrix} 2 & -3 & 2 & 5 \\ 1 & -1 & 1 & 2 \\ 3 & 2 & 2 & 1 \\ 1 & 1 & -3 & 1 \end{bmatrix}$.

2. Prove that $(A^T)^T = A$, for $A \in \mathscr{F}_{mn}$.
3. Prove that $(AB)^T = B^T A^T$, whenever the product AB is defined.
4. Prove that $(A + B)^T = A^T + B^T$ for $A, B \in \mathscr{F}_{mn}$.
5. If $B = A^T$, show that $B_{ji} = (A_{ij})^T$.

As far as determinants are concerned, the important result is the following.

Theorem 1. *If $A \in \mathscr{F}_{nn}$, then*

$$\det A^T = \det A. \tag{4.8}$$

4.3 The Determinant of the Transposed Matrix

Proof. We proceed by induction. The theorem is obviously true for $n = 1$, and it is true for $n = 2$ since

$$\det A^T = \det \begin{bmatrix} a_{11} & a_{21} \\ a_{12} & a_{22} \end{bmatrix} = a_{11}a_{22} - a_{21}a_{12}$$
$$= \det A.$$

Suppose $n > 2$ and assume that the theorem is true for all $m \times m$ matrices with $m < n$. Observe that if $A \in \mathscr{F}_{nn}$ and $B = A^T$, then (compare Exercise 5) $B_{1i} = (A^T)_{1i} = (A_{i1})^T$. Since A_{i1} is an $(n-1) \times (n-1)$ matrix, we have $\det (A_{i1})^T = \det A_{i1}$ by the induction hypothesis. Therefore,

$$\det B = \sum_{i=1}^{n} b_{1i} |B_{1i}| = \sum_{i=1}^{n} a_{i1} |A_{i1}|, \tag{4.9}$$

since

$$|B_{1i}| = (-1)^{1+i} \det B_{1i} = (-1)^{1+i} \det (A_{i1})^T$$
$$= (-1)^{1+i} \det A_{i1} = |A_{i1}|.$$

Thus, our theorem (which is still to be proved) is equivalent to the statement that $\det A$ can be obtained by expansion by cofactors of the first column. The idea of the remainder of the proof is the following. For $i = 2$ we expand $\det A_{i1}$ by cofactors of the elements of its first row. This gives a double sum for $\det B$, which on interchanging the order of summation yields the expansion of $\det A$ by cofactors of elements of its first row. For $i \geq 2, j \geq 2$, we introduce the notation C_{ij} for the $(n-2) \times (n-2)$ matrix obtained from A by deleting the first and ith rows of A and the first and jth columns of A. Observe that C_{ij} is obtained from A_{1j} by deleting the ith row and first column of A, and C_{ij} is also obtained from A_{i1} by deleting the first row and jth column of A. In the matrix A_{i1}, a_{1j} is the element in the first row and $(j-1)$st column. Thus,

$$\det A_{i1} = \sum_{j=2}^{n} a_{1j}(-1)^{1+(j-1)} \det C_{ij} \quad (i \geq 2),$$

and (4.9) becomes

$$\det B = a_{11}|A_{11}| + \sum_{i=2}^{n} a_{i1} |A_{i1}|$$

84   Determinants

$$= a_{11}|A_{11}| + \sum_{i=2}^{n} a_{i1}(-1)^{i+1} \det A_{i1}$$

$$= a_{11}|A_{11}| + \sum_{i=2}^{n} a_{i1}(-1)^{i+1} \sum_{j=2}^{n} a_{1j}(-1)^{j} \det C_{ij}$$

$$= a_{11}|A_{11}| + \sum_{j=2}^{n} (-1)^{1+j} a_{1j} \sum_{i=2}^{n} a_{i1}(-1)^{i} \det C_{ij}.$$

But C_{ij} is obtained from A_{1j} by deleting the ith row and first column of A and a_{i1} is the element in the first column and $(i-1)$st row of A_{1j}. Therefore, by the induction hypothesis

$$\sum_{i=2}^{n} a_{i1}(-1)^{1+(i-1)} \det C_{ij} = \det A_{1j} \quad (i \geq 2).$$

Hence,

$$\det B = a_{11}|A_{11}| + \sum_{j=2}^{n} a_{1j}(-1)^{1+j} \det A_{1j} = \det A. \blacksquare$$

We shall now obtain an analog of Theorem 1, Section 4.2, for elementary row operations. This analog is a consequence of the following result.

Lemma 1. *If $A \in \mathscr{F}_{nn}$ and if E is an elementary matrix* (of any type) *in $\mathscr{F}_{nn}$, then*

$$\det EA = \det E \det A.$$

Proof. If E is an elementary matrix, then clearly E^T is also an elementary matrix. Therefore, by Theorem 1, Corollary 2 to Theorem 2, Section 4.2, and Exercise 3,

$$\det EA = \det (EA)^T = \det (A^T E^T)$$
$$= \det A^T \det E^T = \det A \det E$$
$$= \det E \det A. \blacksquare$$

It now follows immediately from Lemma 1 (since an elementary row operation can be affected by premultiplication by an elementary matrix), that the following analog of Theorem 1, Section 4.2, is valid.

4.3 The Determinant of the Transposed Matrix

Theorem 2. *If $A \in \mathscr{F}_{nn}$ and if $\lambda \in \mathscr{F}$, then*

(i) *interchange of two rows of A changes the sign of* det A;
(ii) *multiplication of any row of A by λ multiplies* det A *by λ*;
(iii) *addition of λ times any row to any other row of A leaves the value of* det A *unchanged.*

With the theory developed up to this point we can evaluate the determinant of any $n \times n$ matrix in a systematic way. We begin by showing that the determinant of a triangular matrix is the product of the diagonal elements of the matrix.

Theorem 3. *Let $A \in \mathscr{F}_{nn}$ be a triangular matrix; that is, suppose $a_{ij} = 0$ if $i > j$, so that*

$$A = \begin{bmatrix} a_{11} & a_{12} & a_{13} & \cdots & a_{1,n-1} & a_{1n} \\ 0 & a_{22} & a_{23} & \cdots & a_{2,n-1} & a_{2n} \\ 0 & 0 & a_{33} & \cdots & a_{3,n-1} & a_{3n} \\ \vdots & \vdots & & \ddots & \vdots & \vdots \\ & & & & a_{n-1,n-1} & \\ 0 & 0 & \cdots & & 0 & a_{nn} \end{bmatrix}.$$

Then

$$\det A = a_{11} a_{22} \cdots a_{nn}.$$

Proof. We proceed by induction. The result is obviously true for $n = 1$. If $n > 1$, we assume the result to be true for all $m \times m$ triangular matrices with $m < n$. Expansion of det A by cofactors of the first column gives det $A = a_{11} |A_{11}| = a_{11} \det A_{11}$. However, A_{11} is the $(n-1) \times (n-1)$ triangular matrix

$$A_{11} = \begin{bmatrix} a_{22} & a_{23} & \cdots & & a_{2n} \\ 0 & a_{33} & & & \vdots \\ \vdots & & \ddots & & \\ & & & & a_{n-1,n} \\ & \cdots & & 0 & a_{nn} \end{bmatrix}$$

and by the induction hypothesis, det $A_{11} = a_{22} a_{33} \cdots a_{nn}$. Thus, det $A = a_{11} a_{22} \cdots a_{nn}$. ∎

We have seen in Chapter 3 that any $n \times n$ matrix may be reduced to row echelon form by a sequence of elementary row operations. The row echelon

86 Determinants

form of a matrix is, of course, triangular. Therefore, an $n \times n$ matrix may be reduced to a triangular matrix by a sequence of elementary row operations. The reduction to a triangular form may be simpler than the reduction to row echelon form, and for the purpose of evaluating determinants it is just as good. Now, we can evaluate the determinant of a matrix A by finding a triangular matrix B which is row equivalent to A, using Theorem 2 to express det A in terms of det B, and then using Theorem 3 to evaluate det B.

Example 1. Evaluate det A, where

$$A = \begin{bmatrix} 1 & 2 & 3 \\ 3 & 1 & 2 \\ 2 & -4 & 3 \end{bmatrix}.$$

We have

$$\det A = \det \begin{bmatrix} 1 & 2 & 3 \\ 3 & 1 & 2 \\ 2 & -4 & 3 \end{bmatrix} = \det \begin{bmatrix} 1 & 2 & 3 \\ 0 & -5 & -7 \\ 2 & -4 & 3 \end{bmatrix}$$

$$= \det \begin{bmatrix} 1 & 2 & 3 \\ 0 & -5 & -7 \\ 0 & -8 & -3 \end{bmatrix} = -5 \det \begin{bmatrix} 1 & 2 & 3 \\ 0 & 1 & 7/5 \\ 0 & -8 & -3 \end{bmatrix}$$

$$= -5 \det \begin{bmatrix} 1 & 2 & 3 \\ 0 & 1 & 7/5 \\ 0 & 0 & 41/5 \end{bmatrix}$$

$$= (-5)(1)(1)(41/5) = -41.$$

Sometimes, it may be easier to simplify using both row and column operations.

Example 2. Evaluate

$$\det \begin{bmatrix} 4 & 4+i & -i \\ 2 & 3+i & -1 \\ 3 & 3+i & -i \end{bmatrix}.$$

We have

$$\det \begin{bmatrix} 4 & 4+i & -i \\ 2 & 3+i & -1 \\ 3 & 3+i & -i \end{bmatrix} = \det \begin{bmatrix} 4 & i & -i \\ 2 & 1+i & -1 \\ 3 & i & -i \end{bmatrix}$$

4.4 Expansions by Cofactors of Any Row or Column

$$= \det \begin{bmatrix} 4 & 0 & -i \\ 2 & i & -1 \\ 3 & 0 & -i \end{bmatrix} = 4 \det \begin{bmatrix} 1 & 0 & -i/4 \\ 2 & i & -1 \\ 3 & 0 & -i \end{bmatrix}$$

$$= 4 \det \begin{bmatrix} 1 & 0 & -i/4 \\ 0 & i & -1+i/2 \\ 3 & 0 & -i \end{bmatrix}$$

$$= 4 \det \begin{bmatrix} 1 & 0 & -i/4 \\ 0 & i & -1+i/2 \\ 0 & 0 & -i/4 \end{bmatrix}$$

$$= (4)(1)(i)\left(-\frac{i}{4}\right) = 1.$$

The reader should normally confine himself to row operations unless he sees an advantage to be gained by using column operations.

● **EXERCISE**

6. Evaluate det A if

(a) $A = \begin{bmatrix} 2 & 4 & 0 \\ 1 & 1 & 3 \\ 2 & 1 & -1 \end{bmatrix}$.

(b) $A = \begin{bmatrix} 2 & 2 & -1 & 9 \\ 11 & 10 & 3 & 22 \\ 7 & 6 & 5 & 4 \\ 4 & 2 & 1 & 17 \end{bmatrix}$.

(c) $A = \begin{bmatrix} 0 & -1 & 1 & 4 \\ 3 & 2 & -2 & 1 \\ 0 & 4 & 0 & 1 \\ 1 & 0 & -1 & 1 \end{bmatrix}$.

(d) $A = \begin{bmatrix} 1 & 3 & 2 \\ -1 & -1 & -1 \\ 2 & 3 & -2 \end{bmatrix}$.

(e) $A = \begin{bmatrix} 1 & 0 & 3 \\ 8 & 1 & -1 \\ 5 & 1 & -1 \end{bmatrix}$.

(f) $A = \begin{bmatrix} 3 & -1 & -4 & 2 \\ 2 & 3 & -2 & -4 \\ 2 & -1 & -3 & 2 \\ 1 & 2 & -1 & -3 \end{bmatrix}$.

(g) $A = \begin{bmatrix} 1 & 1 & 1 \\ a & b & c \\ a^2 & b^2 & c^2 \end{bmatrix}$.

4.4 Expansions by Cofactors of Any Row or Column

We are now ready to establish the formulas (4.2) and (4.3), which permit the expansion of det A either by cofactors of any row or by cofactors of any column.

Determinants

Theorem 1. Let $A \in \mathscr{F}_{nn}$. Let i be any fixed integer, $1 \leq i \leq n$. Then

$$\det A = \sum_{j=1}^{n} a_{ij} |A_{ij}|. \tag{4.2}$$

Let j be any fixed integer, $1 \leq j \leq n$. Then

$$\det A = \sum_{i=1}^{n} a_{ij} |A_{ij}|. \tag{4.3}$$

Proof. To prove formula (4.2), let B be the matrix

$$B = \begin{bmatrix} a_{i*} \\ a_{1*} \\ a_{2*} \\ \vdots \\ a_{(i-1)*} \\ a_{(i+1)*} \\ \vdots \\ a_{n*} \end{bmatrix}.$$

The matrix B is obtained from A by interchanging a_{i*} successively with $a_{(i-1)*}$, then with $a_{(i-2)*}$, until after $(i-1)$ interchanges a_{i*} is interchanged with a_{1*}. By Theorem 2, Section 4.3, $\det A = (-1)^{i-1} \det B$. Moreover, $b_{1j} = a_{ij}$ and $B_{1j} = A_{ij}$, where $j = 1, \ldots, n$. Therefore, by the definition

$$\det B = \sum_{j=1}^{n} b_{1j}(-1)^{1+j} \det B_{1j}$$

$$= \sum_{j=1}^{n} a_{ij}(-1)^{1+j} \det A_{ij}.$$

But

$$\det A = (-1)^{i-1} \det B = \sum_{j=1}^{n} (-1)^{i-1} a_{ij}(-1)^{1+j} \det A_{ij}$$

$$= \sum_{j=1}^{n} a_{ij}(-1)^{i+j} \det A_{ij} = \sum_{j=1}^{n} a_{ij} |A_{ij}|.$$

The reader will recall the definition of $|A_{ij}|$ in formula (4.4). This proves (4.2). A similar argument, or an application of Theorem 2, Section 4.3, establishes (4.3). ∎

● **EXERCISES**

1. Give two proofs of the formula (4.3).
2. Expand each of the determinants in Exercise 6, Section 4.3, by cofactors of every row and every column.

What happens if we try to expand a determinant using elements of one row and cofactors of a different row? This is called expansion by **alien cofactors** for obvious reasons.

Theorem 2. Let $A \in \mathscr{F}_{nn}$; then

$$\sum_{i=1}^{n} a_{ij}|A_{ip}| = \begin{cases} 0 & \text{if } j \neq p \\ \det A & \text{if } j = p \end{cases} \quad (1 \leq j, p \leq n) \tag{4.10}$$

and

$$\sum_{j=1}^{n} a_{ij}|A_{qj}| = \begin{cases} 0 & \text{if } i \neq q \\ \det A & \text{if } i = q \end{cases} \quad (1 \leq i, q \leq n). \tag{4.11}$$

Proof. To establish the first part, we observe that for $1 \leq j = p \leq n$ this is just the expansion of det A by the jth column (Theorem 1). If $j \neq p$, let B be the matrix which coincides with A except for the pth column, which is a_{*j} (the jth column of A). By Theorem 1 (using $b_{ip} = a_{ij}$ and $|B_{ip}| = |A_{ip}|$),

$$\sum_{i=1}^{n} b_{ip}|B_{ip}| = \sum_{i=1}^{n} a_{ij}|A_{ip}| = \det B \quad (1 \leq j \neq p \leq n).$$

But the jth and pth column of B are equal. Thus, det $B = 0$ and this proves the formula (4.10). The proof of the second part is similar. ∎

● **EXERCISE**

3. Prove the formula (4.11).

4.5 Inverses and Cramer's Rule

Theorem 2, Section 4.4, is useful for the computation of the inverse of a nonsingular matrix. In order to do this we introduce the adjugate of a matrix.

90 Determinants

Definition 1. Let $A \in \mathcal{F}_{nn}$. The matrix $B = \text{adj } A \in \mathcal{F}_{nn}$, called the *adjugate* of A, is defined by

$$b_{ij} = |A_{ji}| \qquad (i, j = 1, \ldots, n). \tag{4.12}$$

Theorem 1. If $A \in \mathcal{F}_{nn}$, then

$$A \text{ adj } A = \text{adj } A \cdot A = \det A \cdot I_n.$$

Hence, if $\det A \neq 0$, A is nonsingular and

$$A^{-1} = \frac{1}{\det A} \text{ adj } A. \tag{4.13}$$

Proof. Let $B = \text{adj } A$ and $C = A \text{ adj } A$. By Theorem 2, Section 4.4, the definition of the product of matrices, and the definition of adj A, we have

$$c_{iq} = \sum_{j=1}^{n} a_{ij} b_{jq} = \sum_{j=1}^{n} a_{ij} |A_{qj}| = \begin{cases} 0 & \text{if } i \neq q \\ \det A & \text{if } i = q. \end{cases}$$

Therefore, $A \text{ adj } A = \det A \cdot I_n$. Similarly, $\text{adj } A \cdot A = \det A \cdot I_n$. If $\det A \neq 0$, let $X = [1/(\det A)] \text{ adj } A$. Then $AX = XA = I$; thus, A is nonsingular and $X = A^{-1}$. ∎

● **EXERCISE**

1. Prove that $\text{adj } A \cdot A = \det A \cdot I_n$.

Example 1. Compute $\det A$, $\text{adj } A$, and A^{-1} (if it exists), where A is the matrix

$$A = \begin{bmatrix} 3 & 2 & -1 \\ -1 & 2 & 3 \\ -3 & 1 & 3 \end{bmatrix}.$$

We compute $|A_{11}| = 3$, $|A_{12}| = -6$, $|A_{13}| = 5$, $|A_{21}| = -7$, $|A_{22}| = 6$, $|A_{23}| = -9$, $|A_{31}| = +8$, $|A_{32}| = -8$, $|A_{33}| = 8$. Thus, from (4.12), we have

$$\text{adj } A = \begin{bmatrix} 3 & -7 & 8 \\ -6 & 6 & -8 \\ 5 & -9 & 8 \end{bmatrix}$$

and

$$\det A = a_{11}|A_{11}| + a_{12}|A_{12}| + a_{13}|A_{13}|$$
$$= 3 \cdot 3 + 2(-1) - 1 \cdot 5 = -8.$$

Therefore, A is nonsingular and, from (4.13)

$$A^{-1} = \begin{bmatrix} -\frac{3}{8} & \frac{7}{8} & -1 \\ \frac{3}{4} & -\frac{3}{4} & 1 \\ -\frac{5}{8} & \frac{9}{8} & -1 \end{bmatrix}.$$

• EXERCISES

2. Compute the adjugate, determinant, and inverse, if it exists, of each of the following matrices.

(a) $\begin{bmatrix} 1 & x & x^2 \\ 0 & 1 & 2x \\ 0 & 0 & 2 \end{bmatrix}$.

(b) $\begin{bmatrix} 1 & 0 & 1 \\ 1 & 0 & -1 \\ 1 & 1 & 1 \end{bmatrix}$.

(c) $\begin{bmatrix} x^2 & \sin x & e^x \\ 2x & \cos x & e^x \\ 2 & -\sin x & e^x \end{bmatrix}$.

(d) $\begin{bmatrix} e^x & e^{2x} & 3^{3x} \\ e^x & 2e^{2x} & 3e^{3x} \\ e^x & 4e^{2x} & 9e^{3x} \end{bmatrix}$.

(e) $\begin{bmatrix} \cos x & -\sin x \\ \sin x & \cos x \end{bmatrix}$.

(f) $\begin{bmatrix} e^x & e^{-x} \\ -e^x & e^{-x} \end{bmatrix}$.

(g) $\begin{bmatrix} \sin x & \cos x & e^x \\ \cos x & -\sin x & e^x \\ -\sin x & -\cos x & e^x \end{bmatrix}$.

(h) $\begin{bmatrix} \sin x & \cos x & x \\ \cos x & -\sin x & 1 \\ -\sin x & -\cos x & 0 \end{bmatrix}$.

(i) $\begin{bmatrix} \cos \theta & -r \sin \theta \\ \sin \theta & r \cos \theta \end{bmatrix}$.

(j) $\begin{bmatrix} \sin \phi \cos \theta & \rho \cos \phi \cos \theta & -\rho \sin \phi \sin \theta \\ \sin \phi \sin \theta & \rho \cos \phi \sin \theta & \rho \sin \phi \cos \theta \\ \cos \phi & -\rho \sin \phi & 0 \end{bmatrix}$.

3. If $ad - bc \neq 0$, find the inverse of the matrix $\begin{bmatrix} a & b \\ c & d \end{bmatrix}$.

92 Determinants

We can now give a criterion for testing whether a given matrix is nonsingular.

Theorem 2. *The matrix $A \in \mathscr{F}_{nn}$ is nonsingular if and only if*

$$\det A \neq 0.$$

Proof. In proving Theorem 1, we have already shown that if $\det A \neq 0$, then A is nonsingular. To prove the converse, if A is nonsingular, then A can be written as a product of elementary matrices, $A = E_1 E_2 \cdots E_s$. By Corollary 2, Theorem 1, Section 4.2, if $B = E_1 E_2 \cdots E_{s-1}$, then $\det A = \det BE_s = \det B \det E_s$. Hence, by an easy induction on the number of factors in the product, we obtain $\det A = \det E_1 \det E_2 \cdots \det E_s$. Since $\det E_k \neq 0$, where $k = 1, \ldots, s$ by Corollary 1 of Theorem 1, Section 4.2, $\det A \neq 0$. ∎

By a similar argument we can show how to compute the determinant of the product of two matrices.

Theorem 3. *If $A, B \in \mathscr{F}_{nn}$, then*

$$\det AB = \det A \cdot \det B.$$

Proof. The matrix B is either singular or nonsingular. If B is nonsingular, then, as in the proof of Theorem 2, we write

$$B = E_1 E_2 \cdots E_s,$$

and by Corollary 2 of Theorem 2, Section 4.2,

$$\det AB = \det AE_1 E_2 \cdots E_s$$
$$= \det (AE_1 E_2 \cdots E_{s-1})E_s$$
$$= \cdots = \det A \det E_1 \det E_2 \cdots \det E_s$$
$$= \det A \det B.$$

If B is singular, there exists a nonzero vector $\mathbf{u} \in \mathscr{F}_{n1}$ such that $B\mathbf{u} = \mathbf{0}$, by Theorem 1, Section 3.6. Therefore, $AB\mathbf{u} = \mathbf{0}$, and by the same theorem, AB is singular. By Theorem 2, $\det AB = 0$ and $\det B = 0$. Hence, $\det AB = 0 = \det A \det B$. ∎

4.5 Inverses and Cramer's Rule

● **EXERCISE**

4. Find all complex values of λ for which each of the following matrices is singular.

(a) $\begin{bmatrix} 1-\lambda & 1 \\ 2 & -\lambda \end{bmatrix}.$

(b) $\begin{bmatrix} 2-\lambda & 1 & 0 \\ 0 & 2-\lambda & 4 \\ 1 & 0 & -1-\lambda \end{bmatrix}.$

(c) $\begin{bmatrix} 1-\lambda & 0 & 3 \\ 1 & -1-\lambda & 0 \\ 0 & 2 & -\lambda \end{bmatrix}$

(d) $\begin{bmatrix} 1-\lambda & 2 \\ 4 & 3-\lambda \end{bmatrix}.$

(e) $\begin{bmatrix} 1-\lambda & 3 \\ 0 & 1-\lambda \end{bmatrix}.$

We may now use the theory developed in this section to give an explicit formula for the solution of a linear system of n algebraic equations in n unknowns with a nonsingular coefficient matrix. This formula is of limited use in practice, but has considerable theoretical importance.

Theorem 4. (Cramer's rule): *Let $A \in \mathscr{F}_{nn}$, $\mathbf{b} \in \mathscr{F}_{n1}$. If $\det A \neq 0$, then the system $A\mathbf{x} = \mathbf{b}$ has the unique solution $\mathbf{u}$ whose components are given by*

$$u_p = \frac{\det \Delta_p}{\det A} \quad (p = 1, \ldots, n),$$

where Δ_p is the matrix obtained from A by replacing the pth column a_{*p} of A by $\mathbf{b}$.

Proof. Since $\det A \neq 0$, A^{-1} exists by Theorem 2. Hence, $A\mathbf{x} = \mathbf{b}$ has a unique solution $\mathbf{u} = A^{-1}\mathbf{b}$, by Theorem 4, Section 3.7. Now,

$$A^{-1} = \frac{1}{\det A} \operatorname{adj} A,$$

and

$$\mathbf{u} = \frac{1}{\det A} \cdot \operatorname{adj} A \cdot \mathbf{b}.$$

Thus, the pth component of the solution $\mathbf{u}$ is given by

$$(\det A)u_p = \sum_{j=1}^{n} |A_{jp}| b_j.$$

94 **Determinants**

But the cofactor of b_j in Δ_p is $|A_{jp}|$; whence,

$$\det \Delta_p = \sum_{j=1}^{n} |A_{jp}| b_j = (\det A) u_p. \quad \blacksquare$$

● **EXERCISE**

5. Solve each of the following systems by Cramer's rule.

(a) $\quad \begin{aligned} x_1 + 3x_2 - x_3 &= 0 \\ x_1 + x_2 + x_3 &= 0 \\ x_1 \quad\quad - x_3 &= 1. \end{aligned}$

(b) $\quad \begin{aligned} x_1 - x_2 + 2x_3 &= 1 \\ -2x_1 + 3x_2 + x_3 &= 0 \\ x_1 - 3x_2 + 4x_3 &= 2. \end{aligned}$

(c) $\quad \begin{aligned} 3x_1 + 7x_2 + 4x_3 \quad\quad &= 1 \\ x_2 - 5x_3 + x_4 &= 2 \\ 2x_1 \quad\quad - 2x_3 - x_4 &= 3 \\ x_1 + x_2 \quad\quad + x_4 &= 4. \end{aligned}$

[*Note:* This exercise should convince you that there is an easier way to solve such problems. What is it? See Chapter 3 for some suggestions.]

4.6 Solution of Example 2, Chapter 1—A Plane Pin-Jointed Framework

In Example 2, Chapter 1, we arrived at the following system of linear algebraic equations (see Eqs. 1.4, 1.5, 1.6):

$$-T_1 \cos\theta_1 - T_2 \cos\theta_2 - \cdots - T_5 \cos\theta_5 = F_1$$
$$T_1 \sin\theta_1 + T_2 \sin\theta_2 + \cdots + T_5 \sin\theta_5 = F_2$$
$$-d_1 \cos\theta_i + d_2 \sin\theta_i = e_i \quad (i = 1, \ldots, 5)$$
$$k_i T_i = e_i \quad (i = 1, \ldots, 5).$$

(4.14)

The unknowns are T_i, e_i, where $i = 1, \ldots 5$, and d_1, d_2; the remaining qualities are given. We shall employ the so-called **equilibrium method** to solve the system (4.14). Define the vectors

$$\mathbf{t} = \begin{bmatrix} T_1 \\ T_2 \\ T_3 \\ T_4 \\ T_5 \end{bmatrix}, \quad \mathbf{e} = \begin{bmatrix} e_1 \\ e_2 \\ e_3 \\ e_4 \\ e_5 \end{bmatrix}, \quad \mathbf{f} = \begin{bmatrix} F_1 \\ F_2 \end{bmatrix}, \quad \mathbf{d} = \begin{bmatrix} d_1 \\ d_2 \end{bmatrix},$$

and the matrices

$$C = \begin{bmatrix} -\cos\theta_1 & -\cos\theta_2 & -\cos\theta_3 & -\cos\theta_4 & -\cos\theta_5 \\ \sin\theta_1 & \sin\theta_2 & \sin\theta_3 & \sin\theta_4 & \sin\theta_5 \end{bmatrix},$$

$$K = \begin{bmatrix} k_1 & 0 & \cdots & & 0 \\ 0 & k_2 & & & \\ \vdots & & k_3 & & \vdots \\ & & & k_4 & 0 \\ 0 & \cdots & & 0 & k_5 \end{bmatrix}.$$

Then the system (4.14) is conveniently written as follows. The first two equations are clearly expressible in the form

$$C\mathbf{t} = \mathbf{f}. \tag{4.15}$$

To write the next five equations in matrix vector form, we note that their coefficient matrix is precisely C^T, the transpose (see Section 4.3) of the matrix C. Note that since $C \in \mathcal{R}_{52}$, $C^T \in \mathcal{R}_{25}$. Then the five equations relating d_1, d_2 with $e_1, \ldots, e_5$ are

$$C^T\mathbf{d} = \mathbf{e}. \tag{4.16}$$

Finally, the last five equations are evidently represented by the matrix vector equation

$$K\mathbf{t} = \mathbf{e}. \tag{4.17}$$

Now, the equilibrium method consists of determining the vector $\mathbf{d}$ from the systems (4.15), (4.16), and (4.17). Since the vectors $\mathbf{e}$ and $\mathbf{t}$ are unknown, we shall eliminate them as follows. For physical reasons the quantities $k_i > 0$, $i = 1, 2, \ldots, 5$. Hence, K^{-1} exists and we have from (4.17)

$$\mathbf{t} = K^{-1}\mathbf{e}.$$

96 *Determinants*

Hence, (4.15) becomes

$$CK^{-1}\mathbf{e} = \mathbf{f}.$$

But from (4.16), $\mathbf{e} = C^T\mathbf{d}$ and, therefore, $\mathbf{d}$ can be determined as the solution of the system

$$CK^{-1}C^T\mathbf{d} = \mathbf{f}. \tag{4.18}$$

Consider the coefficient matrix in (4.18),

$$A = CK^{-1}C^T. \tag{4.19}$$

Since $K \in \mathcal{R}_{55}$ is a diagonal matrix, so is K^{-1} diagonal and $K^{-1} \in \mathcal{R}_{55}$. Now, $C \in \mathcal{R}_{25}$; hence, $CK^{-1} \in \mathcal{R}_{25}$ and $CK^{-1}C^T \in \mathcal{R}_{22}$. Let

$$A = \begin{bmatrix} a_{11} & a_{12} \\ a_{21} & a_{22} \end{bmatrix}. \tag{4.20}$$

Using Exercises 1 and 2 (Section 4.3) in (4.19), we have

$$A^T = (CK^{-1}C^T)^T = (C^T)^T(CK^{-1})^T$$
$$= C(K^{-1})^T C^T = CK^{-1}C^T = A.$$

Here, we have used the fact that since K^{-1} is a diagonal matrix, $(K^{-1})^T = K^{-1}$. A matrix, such as A, having the property that $A^T = A$ is called a **symmetric matrix**. For this reason we then must have $a_{12} = a_{21}$ in (4.20).

Notice that the matrix A is easy to calculate explicitly as follows.

$$K^{-1} = \begin{bmatrix} k_1^{-1} & 0 & \cdots & & 0 \\ 0 & k_2^{-1} & & & \vdots \\ \vdots & & \ddots & & 0 \\ 0 & \cdots & & 0 & k_5^{-1} \end{bmatrix}$$

Hence, by matrix multiplication

$$A = CK^{-1}C^T = \begin{bmatrix} \sum_{i=1}^{5}\left(\dfrac{\cos^2 \theta_i}{k_i}\right) & -\sum_{i=1}^{5}\left(\dfrac{\cos \theta_i \sin \theta_i}{k_i}\right) \\ -\sum_{i=1}^{5}\left(\dfrac{\cos \theta_i \sin \theta_i}{k_i}\right) & \sum_{i=1}^{5}\left(\dfrac{\sin^2 \theta_i}{k_i}\right) \end{bmatrix}$$

We may now solve the 2 × 2 system (4.18) for the quantities d_1, d_2 by any convenient method. The system (4.18) has a unique solution because in general A^{-1} exists, except for special and physically unrealistic values of the angles θ_i, $i = 1, \ldots, 5$ (for example, all $\theta_i = 0$ or π in which case we are not dealing with any sort of physically reasonable or interesting framework, or $\cos\theta_i/\cos\theta_j = \sin\theta_i/\sin\theta_j$, for all $i \neq j$). We then have

$$\mathbf{d} = A^{-1}\mathbf{f}, \tag{4.21}$$

which is all that is of interest physically. The elements of A^{-1} are called **flexibility coefficients**. However, having determined $\mathbf{d}$ we can find $\mathbf{e}$ from (4.16) as $\mathbf{e} = C^T\mathbf{d}$. (Note that this requires only matrix multiplication.) Then we find $\mathbf{t}$ from (4.17) as $\mathbf{t} = K^{-1}\mathbf{e} = K^{-1}C^T\mathbf{d}$.

The reader should realize that although we confined our attention to a very simple framework, the same method will work in much more complicated structures. We are always led to an equation like (4.18) provided we properly define the quantities C, K, $\mathbf{d}$, $\mathbf{f}$. Finally, we remark that analysis of pin-jointed frameworks are especially suitable for high-speed computers. We can easily set up a computer program to generate the entries of the matrix $A = CK^{-1}C^T$, which involves only matrix multiplication. Of course, the system (4.18) is also easily solved by machine. In fact, programs exist which provide the engineer with the matrices K, C for a particular framework. All he has to do is provide the computer with physical data for the framework.

4.7 Another Way of Evaluating Determinants

We recall that if $A \in \mathscr{F}_{22}$, then $\det A = a_{11}a_{22} - a_{12}a_{21}$. Observe that $\det A$ is a sum of two terms, each of which is, except for its sign, a product of two elements of A with exactly one element from each row and exactly one element from each column of A. Further, all such products appear in the sum. The reader should examine Example 2, Section 4.1, and observe that exactly the same holds if $A \in \mathscr{F}_{33}$. Accordingly, a product of elements of A with exactly one element from each row and exactly one element from each column will be called a **diagonal product** of A. For example, if $A \in \mathscr{F}_{44}$, one diagonal product is $a_{11}a_{22}a_{33}a_{44}$, another is $a_{13}a_{22}a_{44}a_{31}$. In general, for $A \in \mathscr{F}_{nn}$, each diagonal product is of the form $a_{1i_1}a_{2i_2}\cdots a_{ni_n}$, where $i_1, i_2, \ldots, i_n$ is a permutation (rearrangement) of $(1, 2, \ldots, n)$. Since we have n choices for i_1, then $(n-1)$ for i_2, and so on, there are precisely $n!$ diagonal products of A.

Theorem 1. *Let $A \in \mathscr{F}_{nn}$. Then det A is the sum of all diagonal products of A, each with coefficient ± 1. The coefficient of $a_{11}a_{22} \cdots a_{nn}$ is $+1$. Thus, det A is a polynomial in the elements of A.*

Proof. If $A = [a_{11}] \in \mathscr{F}_{11}$, then det $A = a_{11}$, and the theorem is obvious. So we proceed by induction. Suppose that $n > 1$, and that the theorem is true for all matrices in $\mathscr{F}_{mm}$, where $m < n$. Let $A \in \mathscr{F}_{nn}$. We expand det A by cofactors of the first row:

$$\det A = \sum_{j=1}^{n} a_{1j} |A_{1j}|.$$

By our induction assumption, $|A_{1j}| = \pm \det A_{ij}$, where $j = 1, \ldots, n$), is the sum of all diagonal products of A_{ij}, each with coefficient (± 1). If π is such a diagonal product, then $a_{1j}\pi$ is a diagonal product of A involving a_{1j}, and so $a_{1j}|A_{1j}|$ is the sum of all diagonal products of A involving a_{1j}, each with coefficient (± 1). It follows that det A is the sum of all diagonal products of A.

To determine the coefficient of $a_{11}a_{22} \cdots a_{nn}$ in det A, observe that $|A_{11}| = \det A_{11}$, and therefore, by our induction assumption, $(+1) a_{22} a_{33} \cdots a_{nn}$ is a term of $|A_{11}|$. Hence, it follows from det $A = a_{11}|A_{11}| + \sum_{j=2}^{n} a_{1j}|A_{1j}|$ that $(+1) a_{11}a_{22} \cdots a_{nn}$ is a term of det A, and this proves the second assertion of our theorem for A. The last assertion is now obvious. ∎

Theorem 1 does not explain how to determine whether the coefficient of a diagonal product in det A is $+1$ or -1 (except for the diagonal product $a_{11}a_{22} \cdots a_{nn}$). We can quite easily explain how to find the coefficient of each diagonal product, but we shall not give a proof. (For a further treatment of this point, see, for example, H. Schneider and G. Barker, *Matrices and Linear Algebra*, Holt, Rinehart and Winston, New York, 1968.) It is best to illustrate the general procedure by an example. Suppose $A \in \mathscr{F}_{22}$, and we wish to find the coefficient of $a_{13}a_{24}a_{32}a_{41}$. Write down (1, 2, 3, 4) in sequence in a row, and underneath write down (3, 2, 4, 1). Then join the two 1's, the two 2's, and so on. The joining lines need not be straight, but should be inside the "box" formed by the two rows (1, 2, 3, 4) and (3, 4, 2, 1), and three lines must not meet in a point (Figure 4.1). Count the number of intersections; in this case the number is four. Then the coefficient of $a_{13}a_{24}a_{32}a_{41}$ in det A is $(-1)^4 = +1$. In general, for $A \in \mathscr{F}_{nn}$, the sign of the diagonal product $a_{1i_1}a_{2i_2} \cdots a_{ni_n}$ is given by the following rule: Write down the row (1, 2, ..., n) and underneath it $(i_1, i_2, \ldots, i_n)$. Draw the n lines joining corresponding integers in the two rows, as explained above. Let p be the number of intersections in the diagram. Then the coefficient of $a_{1i_1}a_{2i_2} \cdots a_{ni_n}$ in det A is $(-1)^p$.

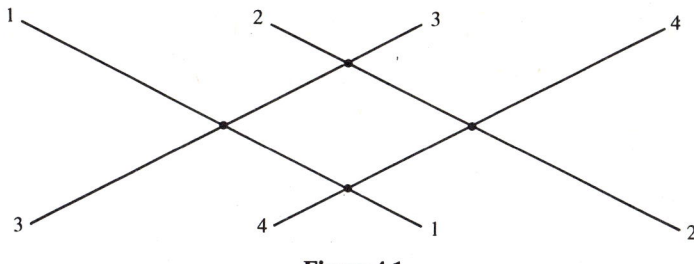

Figure 4.1

Example 1. Without actually expanding, find det A, where $\alpha_i \in \mathscr{F}$, $i = 1, \ldots, 4$, and

$$A = \begin{bmatrix} 1 & 1 & 1 & 1 \\ \alpha_1 & \alpha_2 & \alpha_3 & \alpha_4 \\ \alpha_1^2 & \alpha_2^2 & \alpha_3^2 & \alpha_4^2 \\ \alpha_1^3 & \alpha_2^3 & \alpha_3^3 & \alpha_4^3 \end{bmatrix}.$$

(This determinant is called a **Vandermonde determinant**.)

By Theorem 4.11, det A is a polynomial in the elements of A. Further, each diagonal product in det A is of total degree six. (This is a somewhat technical way of saying that if we put $\alpha_1 = \alpha_2 = \alpha_3 = \alpha_4 = \alpha$, then each diagonal product is $\pm \alpha^6$.) If $\alpha_1 = \alpha_2$, then $a_{*1} = a_{*2}$, whence, det $A = 0$. Hence, by a well-known theorem in algebra, it follows that $(\alpha_2 - \alpha_1)$ is a factor of det A. Similarly, $(\alpha_3 - \alpha_1)$, $(\alpha_3 - \alpha_2)$, $(\alpha_4 - \alpha_1)$, $(\alpha_4 - \alpha_4)$, and $(\alpha_4 - \alpha_3)$ are factors of det A. Hence, for some polynomial $p(\alpha, \beta, \gamma)$,

$$\det A = p(\alpha_1, \alpha_2, \alpha_3)(\alpha_2 - \alpha_1)(\alpha_3 - \alpha_1)(\alpha_3 - \alpha_2)(\alpha_4 - \alpha_1)(\alpha_4 - \alpha_2)$$
$$\times (\alpha_4 - \alpha_3). \tag{4.22}$$

But $(\alpha_2 - \alpha_1)(\alpha_3 - \alpha_1)(\alpha_3 - \alpha_2)(\alpha_4 - \alpha_1)(\alpha_4 - \alpha_2)(\alpha_4 - \alpha_3)$ is also of total degree six, hence, $p(\alpha, \beta, \gamma) = k$ (independent of α, β, γ). Now, by Theorem 1 the coefficient of $1\alpha_2 \alpha_3^2 \alpha_4^3$ in det A is $+1$, and the coefficient of the same term on the right-hand side of (4.22) is also $+1$. Hence, $k = 1$ and det $A = (\alpha_2 - \alpha_1)(\alpha_3 - \alpha_1)(\alpha_3 - \alpha_2)(\alpha_4 - \alpha_1)(\alpha_4 - \alpha_2)(\alpha_4 - \alpha_3)$.

● **EXERCISES**

1. For $A \in \mathscr{F}_{55}$, find the coefficients in det A of $a_{13}a_{24}a_{35}a_{42}a_{51}$ and $a_{23}a_{55}a_{14}a_{42}a_{31}$.

2. Evaluate the (Vandermonde) determinant of

$$A = \begin{bmatrix} 1 & 1 & \cdots & 1 \\ \alpha_1 & \alpha_2 & \cdots & \alpha_n \\ \alpha_1^2 & \alpha_2^2 & \cdots & \alpha_n^2 \\ \vdots & \vdots & & \vdots \\ \alpha_1^{n-1} & \alpha_2^{n-1} & \cdots & \alpha_n^{n-1} \end{bmatrix},$$

where $\alpha_i \in \mathscr{F}$, $i = 1, \ldots, n$.

● **MISCELLANEOUS EXERCISES**

1. Let A be an $n \times n$ matrix and consider the polynomial $\det(A - \lambda I)$ in the scalar λ.
  (a) Show that the coefficient of λ^n is $(-1)^n$.
  (b) Show that the coefficient of λ^{n-1} is $(-1)^{n-1} \sum_{i=0}^{n} a_{ii}$.
  (c) Show that the constant term is $\det A$.

2. We have seen that $\det AB = \det A \det B$. Is it true, in general, that $\det(A + B) = \det A + \det B$? Either prove the statement or give a counter-example.

3. Let

$$A = \begin{bmatrix} a & b \\ c & d \end{bmatrix}$$

be a 2×2 matrix.
  (a) Find the equation (in terms of a, b, c, d) which λ must satisfy in order to have $\det(A - \lambda I) = 0$.
  (b) If the equation in part (a) is $\alpha \lambda^2 + \beta \lambda + \gamma = 0$, for some α, β, γ, show that $\alpha A^2 + \beta A + \gamma I$ is the zero matrix.

4. Let P be a nonsingular $n \times n$ matrix, and let B be any $n \times n$ matrix. Show that $\det(B - \lambda I) = \det[P(B - \lambda I)P^{-1}] = \det[PBP^{-1} - \lambda I]$.

5. Let

$$P = \begin{bmatrix} 1 & 1 \\ 1 & -1 \end{bmatrix}, \quad A = \begin{bmatrix} 3 & -1 \\ -1 & 3 \end{bmatrix}.$$

  (a) Show that

$$PAP^{-1} = \begin{bmatrix} 2 & 0 \\ 0 & 4 \end{bmatrix}.$$

  (b) Using part (a) and Exercise 4, show that $\det(A - \lambda I) = 0$ if and only if $\lambda = 2$ or $\lambda = 4$.

6. Evaluate the determinant of each of the following matrices.

(a) $\begin{bmatrix} 0 & 1+i & 1+2i \\ 1-i & 0 & 2-3i \\ 1-2i & 2+3i & 0 \end{bmatrix}$. (b) $\begin{bmatrix} 2 & 2 & -2 \\ 1 & 2 & 3 \\ 2 & 3 & 4 \end{bmatrix}$.

(c) $\begin{bmatrix} 0 & 2 & 3 \\ -2 & 0 & 4 \\ -3 & -4 & 0 \end{bmatrix}$. (d) $\begin{bmatrix} 1 & 1 & 1 & 6 \\ 2 & 4 & 1 & 6 \\ 4 & 1 & 2 & 9 \\ 2 & 4 & 2 & 7 \end{bmatrix}$.

(e) $\begin{bmatrix} 0 & 0 & 0 & 0 & 0 & 1 \\ 0 & 0 & 0 & 0 & 2 & 1 \\ 0 & 0 & 0 & 3 & 2 & 1 \\ 0 & 0 & 4 & 3 & 2 & 1 \\ 0 & 5 & 4 & 3 & 2 & 1 \\ 6 & 5 & 4 & 3 & 2 & 1 \end{bmatrix}$.

(f) $\begin{bmatrix} 1 & 2 & 1 & 2 & 1 \\ 0 & 0 & 1 & 1 & 1 \\ 1 & 1 & 0 & 0 & 0 \\ 0 & 0 & 1 & 1 & 2 \\ 1 & 2 & 2 & 1 & 1 \end{bmatrix}$.

7. Solve each of the following systems using Cramer's rule.

(a) $2x_1 + x_2 + 5x_3 + x_4 = 5$
$3x_1 + 6x_2 - 2x_3 + x_4 = 8$
$x_1 + x_2 - 3x_3 - 4x_4 = -1$
$2x_1 + 2x_2 + 2x_3 - 3x_4 = 2.$

(b) $x_1 + x_2 + x_3 = 4$
$2x_1 + 5x_2 - 2x_3 = 3$
$x_1 + 7x_2 - 7x_3 = 5.$

(c) $x_1 - x_2 + x_3 + x_4 = 2$
$x_1 - x_2 + x_3 - x_4 = 4$
$x_1 + x_2 + x_3 + x_4 = 0$
$x_1 + x_2 - x_3 + x_4 = -4.$

8. For

$$A = \begin{bmatrix} a_{11} & a_{12} & a_{13} \\ a_{21} & a_{22} & a_{23} \\ a_{31} & a_{32} & a_{33} \end{bmatrix},$$

write out det A as a sum of terms $a_{1,\,i_1},\,a_{2,\,i_2},\,a_{3,\,i_3}$ with appropriate plus or minus signs, using the method of Section 4.7.

9. Write out det A as in Exercise 8, but using the definition of Section 4.1, and check for agreement of plus and minus signs with Exercise 8.

10. Compute adj A and A^{-1}, if it exists, for each of the matrices in Exercise 6.

11. Let

$$A = \begin{bmatrix} \cos\theta & -\sin\theta \\ \sin\theta & \cos\theta \end{bmatrix}, \quad B = \begin{bmatrix} e^{i\theta} & 0 \\ 0 & e^{-i\theta} \end{bmatrix}.$$

Using $e^{i\theta} = \cos\theta + i\sin\theta$, $e^{-i\theta} = \cos\theta - i\sin\theta$, show that $B = P^{-1}AP$ for $P = \begin{bmatrix} 1 & 1 \\ -i & i \end{bmatrix}$, and hence that $\det B = \det A$.

Chapter 5 | **VECTOR SPACES**

In this chapter we will study the theory of vector spaces. We begin with the special case of real Euclidean n-dimensional space and abstract its main properties, which leads us to an abstract vector space. This abstraction enables us to discuss many examples that will be useful in the development of the theory of linear differential equations in Chapter 6. In addition, the ideas here will give the reader a deeper understanding of the theory of linear algebraic equations presented in Chapter 3.

5.1 Real Euclidean n-Dimensional Space

We define **real Euclidean n space** to be the set $\mathscr{R}_{n1}$ of all column vectors with n real components; instead of writing $\mathscr{R}_{n1}$ we shall simply write $\mathscr{R}_n$. Recall from the definition of matrix addition and multiplication by scalars (Section 2.2) that if

$$\mathbf{x} = \begin{bmatrix} x_1 \\ \vdots \\ x_n \end{bmatrix}, \quad \mathbf{y} = \begin{bmatrix} y_1 \\ \vdots \\ y_n \end{bmatrix},$$

then $\mathbf{x} + \mathbf{y} = \mathbf{z}$ in $\mathscr{R}_n$ if $z_i = x_i + y_i$, where $i = 1, \ldots, n$, and $\alpha \mathbf{x} = \mathbf{w}$ in $\mathscr{R}_n$ for any real number α if $w_i = \alpha x_i$, where $i = 1, \ldots, n$.

The reader can easily verify that vector addition satisfies the following properties:

(A_1) $(\mathbf{u} + \mathbf{v}) + \mathbf{w} = \mathbf{u} + (\mathbf{v} + \mathbf{w})$ for all vectors $\mathbf{u}, \mathbf{v}, \mathbf{w} \in \mathscr{R}_n$.

(A_2) There is a unique vector in $\mathscr{R}_n$ denoted by $\mathbf{0}$ such that $\mathbf{u} + \mathbf{0} = \mathbf{u}$ for each vector $\mathbf{u} \in \mathscr{R}_n$ (of course, $\mathbf{0}$ is just the column vector with all entries 0).

(A_3) For every $\mathbf{u} \in \mathscr{R}_n$ there is a vector $\mathbf{v} \in \mathscr{R}_n$ such that $\mathbf{u} + \mathbf{v} = \mathbf{0}$; $\mathbf{v}$ is denoted by $-\mathbf{u}$ and is called the **additive inverse** (in fact, there is only one such $\mathbf{v}$, given by $v_i = -u_i$, $i = 1, \ldots, n$).

(A_4) $\mathbf{u} + \mathbf{v} = \mathbf{v} + \mathbf{u}$ for all vectors $\mathbf{u}, \mathbf{v} \in \mathscr{R}_n$.

● **EXERCISE**

1. Verify properties (A_1)–(A_4) for $\mathscr{R}_n$. [*Hint:* Observe that properties (A_1)–(A_4) hold for real numbers.]

Similarly, scalar multiplication satisfied the following properties:

(S_1) For all $\alpha \in \mathscr{R}$ and $\mathbf{u}, \mathbf{v} \in \mathscr{R}_n$, $\alpha(\mathbf{u} + \mathbf{v}) = \alpha\mathbf{u} + \alpha\mathbf{v}$.
(S_2) For all $\alpha, \beta \in \mathscr{R}$ and $\mathbf{u} \in \mathscr{R}_n$, $(\alpha + \beta)\mathbf{u} = \alpha\mathbf{u} + \beta\mathbf{u}$.
(S_3) For all $\alpha, \beta \in \mathscr{R}$ and $\mathbf{u} \in \mathscr{R}_n$, $(\alpha\beta)\mathbf{u} = \alpha(\beta\mathbf{u})$.
(S_4) For all $\mathbf{u} \in \mathscr{R}_n$, $1\mathbf{u} = \mathbf{u}$.

● **EXERCISES**

2. Verify properties (S_1)–(S_4).
3. Show that $0\mathbf{u} = \mathbf{0}$, $-1\mathbf{u} = -\mathbf{u}$, and $-(\mathbf{u} + \mathbf{v}) = (-\mathbf{u}) + (-\mathbf{v})$ for $\mathbf{u}, \mathbf{v} \in \mathscr{R}_n$.

We will call the elements of $\mathscr{R}_n$ either vectors or points. When $n = 3$, we have the familiar three-dimensional Euclidean space if we make the following identification: we identify the point P whose coordinates are (u_1, u_2, u_3), with the vector $\mathbf{u} \in \mathscr{R}_3$ given by

$$\mathbf{u} = \begin{bmatrix} u_1 \\ u_2 \\ u_3 \end{bmatrix}.$$

We can further identify this vector $\mathbf{u} \in \mathscr{R}_3$ with the directed line segment from the origin to the point P with coordinates (u_1, u_2, u_3) (Figure 5.1). The reader will recall that the vectors

$$\mathbf{e}_1 = \begin{bmatrix} 1 \\ 0 \\ 0 \end{bmatrix}, \quad \mathbf{e}_2 = \begin{bmatrix} 0 \\ 1 \\ 0 \end{bmatrix}, \quad \mathbf{e}_3 = \begin{bmatrix} 0 \\ 0 \\ 1 \end{bmatrix},$$

5.1 Real Euclidean n-Dimensional Space

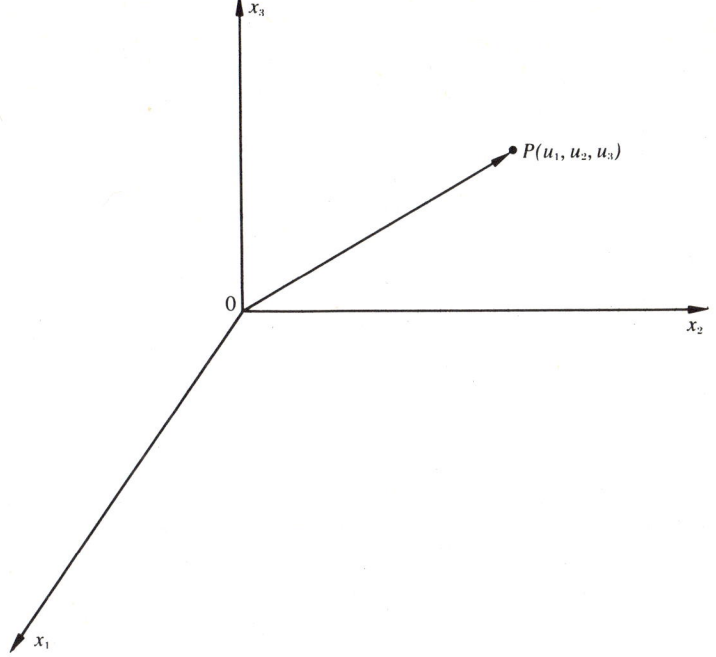

Figure 5.1

called **unit vectors in the direction of each coordinate axis** (usually denoted by **i, j**, and **k**, respectively) play an important role in three-dimensional geometry.

If $n = 2$, $\mathscr{R}_2$ is the familiar Euclidean plane when a similar identification is made. From calculus or physics courses the reader is perhaps familiar with vectors defined as directed line segments and with vector addition defined by the parallelogram law. In $\mathscr{R}_2$ we can picture this law as shown in Figure 5.2.

● **EXERCISE**

4. Using our definition of addition in $\mathscr{R}_2$, prove the parallelogram law. [*Hint:* Prove that the quadrilateral in Figure 5.2 is really a parallelogram.]

By analogy with $\mathscr{R}_2$ and $\mathscr{R}_3$, we can use the language of analytic geometry in $\mathscr{R}_n$; for example, we define a point P in n-dimensional Euclidean space with coordinates $(u_1, \ldots, u_n)$ to be the vector **u** in $\mathscr{R}_n$ given by

$$\mathbf{u} = \begin{bmatrix} u_1 \\ \vdots \\ u_n \end{bmatrix}$$

106   *Vector Spaces*

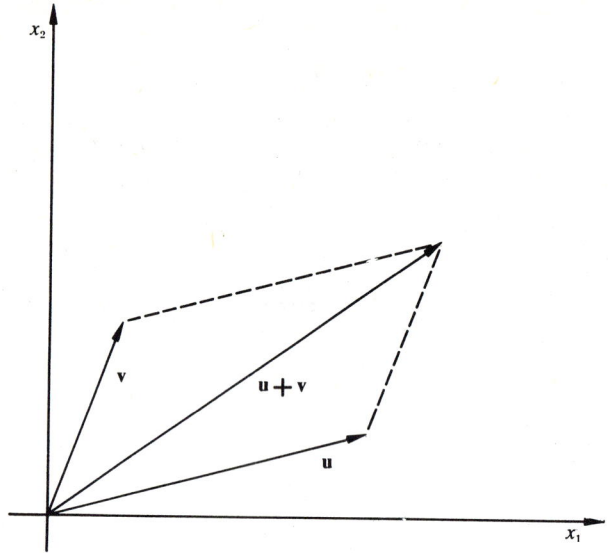

Figure 5.2

Similarly, we single out the unit vectors in $\mathscr{R}_n$:

$$\mathbf{e}_1 = \begin{bmatrix} 1 \\ 0 \\ \vdots \\ 0 \end{bmatrix}, \mathbf{e}_2 = \begin{bmatrix} 0 \\ 1 \\ 0 \\ \vdots \\ 0 \end{bmatrix}, \ldots, \mathbf{e}_n = \begin{bmatrix} 0 \\ \vdots \\ 0 \\ 1 \end{bmatrix},$$

which we will have occasion to use frequently. Also, writing column vectors rather than row vectors is a matter of convenience; that is, we could have used $\mathscr{R}_{1n}$ instead of $\mathscr{R}_{n1}$.

Finally, we define **complex Euclidean n space** to be the set $\mathscr{C}_{n1}$ of all column vectors with n complex components; instead of writing $\mathscr{C}_{n1}$ we shall write $\mathscr{C}_n$. As we have done previously, we shall use $\mathscr{F}_n$ to designate $\mathscr{R}_n$ or $\mathscr{C}_n$.

● **EXERCISE**

5. Show that the properties (A_1)–(A_4) for addition and (S_1)–(S_4) for scalar multiplication hold in $\mathscr{C}_n$.

5.2 Abstract Vector Spaces and Subspaces

In Section 5.1 we have listed the properties satisfied by elements of Euclidean space $\mathscr{R}_n$. Many other objects satisfy the same laws, and, as a matter of fact, very often in mathematics the laws are much more important than the objects. We shall call any set whose elements satisfy properties (A_1)–(A_4) and (S_1)–(S_4) **a vector space**, and we shall now give a precise definition.

Definition 1. *A vector space V over $\mathscr{F}$ is a collection of elements called vectors for which two operations, called addition and scalar multiplication, are defined:*

For every pair of vectors $\mathbf{u}, \mathbf{v} \in V$ there is a unique vector $\mathbf{u} + \mathbf{v} \in V$ called the sum of $\mathbf{u}$ and $\mathbf{v}$.
For every vector $\mathbf{u} \in V$ and for every scalar $\alpha \in \mathscr{F}$, there is a unique vector $\alpha\mathbf{u} \in V$ called the scalar product of α and $\mathbf{u}$.
Addition and scalar multiplication satisfy the following properites:

- (A_1) $(\mathbf{u} + \mathbf{v}) + \mathbf{w} = \mathbf{u} + (\mathbf{v} + \mathbf{w})$ *for all vectors* $\mathbf{u}, \mathbf{v}, \mathbf{w} \in V$
- (A_2) *There exists in V a unique vector denoted by $\mathbf{0}$ such that $\mathbf{u} + \mathbf{0} = \mathbf{u}$ for each vector $\mathbf{u} \in V$.*
- (A_3) *For every $\mathbf{u} \in V$ there is a unique vector $\mathbf{v}$ such that $\mathbf{u} + \mathbf{v} = \mathbf{0}$; $\mathbf{v}$ is denoted by $-\mathbf{u}$ and is called the additive inverse.*
- (A_4) $\mathbf{u} + \mathbf{v} = \mathbf{v} + \mathbf{u}$ *for all vectors* $\mathbf{u}, \mathbf{v} \in V$.
- (S_1) *For all $\alpha \in \mathscr{F}$ and $\mathbf{u}, \mathbf{v} \in V$, $\alpha(\mathbf{u} + \mathbf{v}) = \alpha\mathbf{u} + \alpha\mathbf{v}$.*
- (S_2) *For all $\alpha, \beta \in \mathscr{F}$ and $\mathbf{u} \in V$, $(\alpha + \beta)\mathbf{u} = \alpha\mathbf{u} + \beta\mathbf{u}$.*
- (S_3) *For all $\alpha, \beta \in \mathscr{F}$ and $\mathbf{u} \in V$, $(\alpha\beta)\mathbf{u} = \alpha(\beta\mathbf{u})$.*
- (S_4) *For all $\mathbf{u} \in V$, $1\mathbf{u} = \mathbf{u}$.*

By using the rules of vector addition and scalar multiplication we can show that $0\mathbf{u} = \mathbf{0}$ and $(-1)\mathbf{u} = -\mathbf{u}$. For example, we easily have the following sequence of steps:

$$0 = 0 + 0,$$
$$0\mathbf{u} = (0 + 0)\mathbf{u} = 0\mathbf{u} + 0\mathbf{u},$$
$$0\mathbf{u} + (-0\mathbf{u}) = (0\mathbf{u} + 0\mathbf{u}) + (-0\mathbf{u}),$$
$$\mathbf{0} = 0\mathbf{u},$$

which proves the first of these.

108   *Vector Spaces*

● **EXERCISES**

    1. Justify each step in the above argument.
    2. Prove that $(-1)\mathbf{u} = -\mathbf{u}$ for every $\mathbf{u} \in V$. [*Hint:* $(1 + (-1))\mathbf{u} = 0\mathbf{u}$.]

In Section 5.1, we considered $\mathscr{R}_n$ and $\mathscr{C}_n$ as examples of a vector space. To become thoroughly familiar with the notion of a vector space, we now consider several additional examples.

Example 1. Let m and n be fixed positive integers. Consider the set $\mathscr{F}_{mn}$ of all $m \times n$ matrices over $\mathscr{F}$ with addition and scalar multiplication defined in Section 2.2. Then the content of Theorems 1 and 2 (Section 2.2) is that $\mathscr{F}_{mn}$ is a vector space over $\mathscr{F}$.

Example 2. Let $\mathscr{I}$ be an arbitrary interval on the real line. The interval $\mathscr{I}$ may be open, closed, half-open, and either bounded or unbounded. Consider the collection $\Phi(\mathscr{R})$ of all real-valued functions defined on the interval $\mathscr{I}$. If $f, g \in \Phi(\mathscr{R})$, we define their sum $(f + g)$ to be the function in $\Phi(\mathscr{R})$ whose values are given by

$$(f + g)(t) = f(t) + g(t) \tag{5.1}$$

for every $t \in \mathscr{I}$. If α is any real number and $f \in \Phi(\mathscr{R})$, we define the product of α and f to be the function $\alpha f \in \Phi(\mathscr{R})$ whose values are given by

$$(\alpha f)(t) = \alpha f(t) \tag{5.2}$$

for every $t \in \mathscr{I}$. It is now an easy matter to verify that these operations satisfy properties (A_1)–(A_4) and (S_1)–(S_4), where the zero element of $\Phi(\mathscr{R})$ is the function that is identically zero on $\mathscr{I}$. Therefore, $\Phi(\mathscr{R})$ is a vector space over $\mathscr{R}$. For example, to verify the property (A_4), if $f, g \in \Phi(\mathscr{R})$, then for every $t \in \mathscr{I}$ we have, by the definition of addition and the commutative law of addition of real numbers, $(f + g)(t) = f(t) + g(t) = g(t) + f(t) = (g + f)(t)$; therefore, $f + g = g + f$.

Remember that by the "function e^t" we mean the function whose value at t is the number e^t. Even though it is not strictly correct, we will continue to use this terminology. Similarly, we may often refer to the function f whose value at t is $f(t)$ as the "function $f(t)$."

Example 3. Let $\mathscr{I}$ be an arbitrary interval on the real line, and consider the collection $\Phi(\mathscr{C})$ of all complex-valued functions defined on the interval $\mathscr{I}$. (For example, $e^{2it} = \cos 2t + i \sin 2t$ is an element of $\Phi(\mathscr{C})$.) With addition defined by (5.1) and scalar multiplication defined by (5.2), but with α now a complex number, it is trivial to verify by the same steps as in Example 2 that $\Phi(\mathscr{C})$ is a vector space over $\mathscr{C}$.

In subsequent sections $\Phi(\mathscr{F})$ will mean either $\Phi(\mathscr{R})$ or $\Phi(\mathscr{C})$, depending on the choice of $\mathscr{F}$ in the particular context.

Example 4. Let $\mathscr{I}$ be an arbitrary interval on the real line. Consider the collection $C(\mathscr{R})$ of all real-valued functions continuous on the interval $\mathscr{I}$. Addition and scalar multiplication are again defined by (5.1) and (5.2), respectively. Moreover, if $f, g \in C(\mathscr{R})$, then, by the theorem of calculus that the sum of two continuous functions is a continuous function, $f + g \in C(\mathscr{R})$. Similarly, if $f \in C(\mathscr{R})$ and α is a real number, then $\alpha f \in C(\mathscr{R})$, since a constant multiple of a continuous function is continuous. The properties (A_1)–(A_4) and (S_1)–(S_4) are easily verified as before. Therefore, $C(\mathscr{R})$ is a vector space over $\mathscr{R}$.

In a similar way we can define $C(\mathscr{C})$, the set of continuous complex-valued functions on $\mathscr{I}$, and show that it is a vector space over $\mathscr{C}$. By $C(\mathscr{F})$, we will mean either $C(\mathscr{R})$ or $C(\mathscr{C})$.

Example 5. Let $\mathscr{I}$ be an arbitrary interval on the real line. Then each of the following sets is a vector space over $\mathscr{R}$, with addition and scalar multiplication defined by (5.1) and (5.2), respectively.
  (a) The set $C'(\mathscr{R})$ of real-valued functions having a derivative which is continuous at every point of $\mathscr{I}$.
  (b) The set $C^{17}(\mathscr{R})$ of real-valued functions having derivatives of order up to and including 17, each of which is continuous at every point of $\mathscr{I}$.
  (c) The set $C^k(\mathscr{R})$ of real-valued functions having derivatives of order up to and including k, each of which is continuous at every point of $\mathscr{I}$. Here, k is a fixed integer, $k \geq 1$.
  (d) The set $P(\mathscr{R})$ of all polynomials of any degree with real coefficients. (Recall that a polynomial of degree k is a function defined for every real number t by an expression of the form

  $$p(t) = \alpha_0 + \alpha_1 t + \cdots + \alpha_k t^k,$$

  where the coefficients $\alpha_0, \alpha_1, \ldots, \alpha_k$ are real numbers and $\alpha_k \neq 0$.)
  (e) The set $P^k(\mathscr{R})$ of all polynomials with real coefficients of degree up to and including k, where k is a fixed integer, $k \geq 0$. Note that $P^0(\mathscr{R}) = \mathscr{R}$.

● **EXERCISES**

  3. Show that each of the sets in Example 5 is a vector space over $\mathscr{R}$.
  4. Replace $\mathscr{R}$ by $\mathscr{C}$ in each of the sets in Example 5, and show that the resultant sets are vector spaces over $\mathscr{C}$.

110   Vector Spaces

As before, we shall use the terminology $C'(\mathscr{F})$, $C^k(\mathscr{F})$, $P(\mathscr{F})$, $P^k(\mathscr{F})$ where $\mathscr{F}$ is either $\mathscr{R}$ or $\mathscr{C}$. These spaces will be of importance in this and subsequent chapters.

Example 6. Let $\mathscr{I}$ be an arbitrary interval on the real line. Consider the set $\Phi(\mathscr{R}_n)$ of all vector functions with n real components defined on the interval $\mathscr{I}$. Thus, if $\mathbf{f} \in \Phi(\mathscr{R}_n)$,

$$\mathbf{f}(t) = \begin{bmatrix} f_1(t) \\ \vdots \\ f_n(t) \end{bmatrix},$$

for each $t \in \mathscr{I}$, and $f_1, \ldots, f_n$ are real-valued functions on $\mathscr{I}$. Thus, $\mathbf{f}$ may be regarded as a function that maps each point of the interval $\mathscr{I}$ into a point of real Euclidean n-space $\mathscr{R}_n$. This is the reason for the notation $\Phi(\mathscr{R}_n)$. (See Figure 5.3 for the picture in the space $\Phi(\mathscr{R}_3)$.) We define addition in the space $\Phi(\mathscr{R}_n)$ as follows. If $\mathbf{f}, \mathbf{g}, \in \Phi(\mathscr{R}_n)$, we define their sum $\mathbf{f} + \mathbf{g}$ to be the function in $\Phi(\mathscr{R}_n)$ whose values are given by

$$(\mathbf{f} + \mathbf{g})(t) = \mathbf{f}(t) + \mathbf{g}(t) \tag{5.3}$$

for every $t \in \mathscr{I}$. (Note that if $n = 1$, then (5.3) becomes (5.1) of Example 2.) If α is any real number and $\mathbf{f} \in \Phi(\mathscr{R}_n)$, we define the product of α and $\mathbf{f}$ to be the function $\alpha\mathbf{f} \in \Phi(\mathscr{R}_n)$ whose values are given by

$$(\alpha\mathbf{f})(t) = \alpha\mathbf{f}(t) \tag{5.4}$$

for every $t \in \mathscr{I}$. (Compare with (5.2), in the case $n = 1$.) The reader should note that the zero element of the space $\Phi(\mathscr{R}_n)$ is the identically zero function from $\mathscr{I}$ into $\mathscr{R}_n$. Moreover, two elements $\mathbf{f}, \mathbf{g}$ of $\Phi(\mathscr{R}_n)$ are equal if and only if $\mathbf{f}(t) = \mathbf{g}(t)$ for every $t \in \mathscr{I}$; that is, if and only if $f_k(t) = g_k(t)$ for $k = 1, \ldots, n$

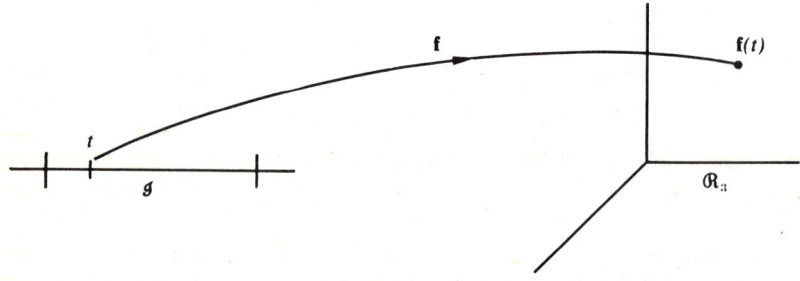

Figure 5.3

5.2 Abstract Vector Spaces and Subspaces

and every $t \in \mathscr{I}$. It is now an easy matter to verify that these operations satisfy properties (A$_1$)–(A$_4$) and (S$_1$)–(S$_4$) of a vector space. Thus, $\Phi(\mathscr{R}_n)$ is a vector space over $\mathscr{R}$.

● **EXERCISE**

5. Verify that $\Phi(\mathscr{R}_n)$ is a vector space over $\mathscr{R}$.

Example 7. Let $\mathscr{I}$ be an arbitrary interval on the real line. Then each of the following sets is a vector space over $\mathscr{R}$, with addition and scalar multiplication defined by (5.3) and (5.4), respectively.

(a) The set $C(\mathscr{R}_n)$ of all vector functions with n real components, each of which is continuous on the interval $\mathscr{I}$. Thus, if $\mathbf{f} \in C(\mathscr{R}_n)$ and

$$\mathbf{f} = \begin{bmatrix} f_1 \\ \vdots \\ f_n \end{bmatrix},$$

then each of the functions $f_1, \ldots, f_n$ is continuous on $\mathscr{I}$. (This statement is to be taken as the definition of continuity of a vector function.)

(b) The set $C'(\mathscr{R}_n)$ of all vector functions with n real components, each of which has a continuous derivative at every point of $\mathscr{I}$. Thus, if $\mathbf{f} \in C'(\mathscr{R}_n)$ we shall designate by $\mathbf{f}'$ the vector whose value at t is given by

$$\mathbf{f}'(t) = \begin{bmatrix} f'_1(t) \\ \vdots \\ f'_n(t) \end{bmatrix} \tag{5.5}$$

(c) The set $C^k(\mathscr{R}_n)$ of all vector functions with n real components, each of which has continuous derivatives of order up to and including k at every point of $\mathscr{I}$. We shall designate by $\mathbf{f}^{(j)}$, where $j = 0, 1, \ldots, k$, the vector whose value at t is given by

$$\mathbf{f}^{(j)}(t) = \begin{bmatrix} f_1^{(j)}(t) \\ \vdots \\ f_n^{(j)}(t) \end{bmatrix}. \tag{5.6}$$

● **EXERCISES**

6. Show that the sets $C(\mathscr{R}_n)$, $C'(\mathscr{R}_n)$, and $C^k(\mathscr{R}_n)$ are vector spaces over $\mathscr{R}$.

7. Replace $\mathscr{R}_n$ by $\mathscr{C}_n$ in each of the sets of Exercise 6 and show that the sets $C(\mathscr{C}_n)$, $C'(\mathscr{C}_n)$, and $C^k(\mathscr{C}_n)$ are vector spaces over $\mathscr{C}$.

As has been our practice, we shall use the notation $C(\mathscr{F}_n)$, $C^1(\mathscr{F}_n)$, and $C^k(\mathscr{F}_n)$, where $\mathscr{F}_n$ is either $\mathscr{R}_n$ or $\mathscr{C}_n$.

• EXERCISE

8. Explain why each of the following is not a vector space over $\mathscr{R}$.
(a) The set of all integers.
(b) The set of all nonsingular matrices in $\mathscr{R}_{22}$.
(c) The set of all singular matrices in $\mathscr{R}_{nn}$.
(d) The set of all functions f in $C(\mathscr{R})$ such that $\int_1^2 f(t)\,dt = 7$.
(e) The set of all functions f in $C(\mathscr{R})$ such that $f(0) = 1$.
(f) The set of all polynomials of degree four with real coefficients.
(g) The set of all functions f in $C^2(\mathscr{R})$ such that $f''(t) = \sin t$.

5.3 Subspaces

Lines and planes in Euclidean 3-dimensional space play an important role in geometry. Subsets of vector spaces that are themselves vector spaces play an equally important role. We, therefore, formulate the following definition.

Definition 1. *Let V be a vector space over $\mathscr{F}$. A subset W of V (notation $W \subseteq V$) is said to be a subspace of V if and only if W is itself a vector space over $\mathscr{F}$ (the same $\mathscr{F}$) with the same operations of addition and scalar multiplication as V.*

Before giving examples, we present a simple criterion for testing whether a given subset of a vector space is a subspace.

Theorem 1. *Let W be a subset of a vector space V over $\mathscr{F}$. Then W is a subspace of V if and only if $\alpha\mathbf{u} + \beta\mathbf{v} \in W$ for every $\alpha, \beta \in \mathscr{F}$ and every $\mathbf{u}, \mathbf{v} \in W$ (that is, if and only if every linear combination of vectors in W belongs to W).*

Proof. If W is a subspace of V, it is a vector space over $\mathscr{F}$. Thus, addition and scalar multiplication are defined in W; for every $\mathbf{u}, \mathbf{v} \in W$ and $\alpha, \beta \in \mathscr{F}$, $\alpha\mathbf{u} \in W$, $\beta\mathbf{v} \in W$, and $\alpha\mathbf{u} + \beta\mathbf{v} \in W$. Conversely, if $\alpha\mathbf{u} + \beta\mathbf{v} \in W$ for every $\mathbf{u}, \mathbf{v} \in W$ and $\alpha, \beta \in \mathscr{F}$, in particular $\mathbf{u} + \mathbf{v} \in W$ and $\alpha\mathbf{u} \in W$. Moreover, the properties (A_1)–(A_4) (Section 5.2) and (S_1)–(S_4) (Section 5.2) hold in V, and therefore they hold in the set $W \subseteq V$. ∎

Corollary. *Let W be a subspace of V, and let $\mathbf{v}_1, \ldots, \mathbf{v}_k$ belong to W. Then $\alpha_1\mathbf{v}_1 + \cdots + \alpha_k\mathbf{v}_k \in W$ for all $\alpha_1, \ldots, \alpha_k \in \mathscr{F}$.*

• EXERCISE

1. Prove this corollary. [*Hint:* Use induction.]

Example 1. In $\mathscr{R}_2$ a line L through the origin is a subspace of $\mathscr{R}_2$. For if L is a line through the origin in $\mathscr{R}_2$, then L can be described as

$$L = \left\{ \mathbf{x} = \begin{bmatrix} x_1 \\ x_2 \end{bmatrix} \,\middle|\, \lambda x_1 = \mu x_2 \right\}$$

for some real λ and μ, not both zero. If $\mathbf{u}, \mathbf{v} \in L$, and if

$$\mathbf{u} = \begin{bmatrix} u_1 \\ u_2 \end{bmatrix} \quad \text{and} \quad \mathbf{v} = \begin{bmatrix} v_1 \\ v_2 \end{bmatrix},$$

then $\lambda u_1 = \mu u_2$ and $\lambda v_1 = \mu v_2$. Now, take a linear combination of $\mathbf{u}$ and $\mathbf{v}$. If $\alpha, \beta \in \mathscr{R}$, let

$$\mathbf{w} = \alpha \mathbf{u} + \beta \mathbf{v} = \begin{bmatrix} w_1 \\ w_2 \end{bmatrix}.$$

Then $w_1 = \alpha u_1 + \beta v_1$ and $w_2 = \alpha u_2 + \beta v_2$. But $\lambda w_1 = \lambda(\alpha u_1 + \beta v_1) = \alpha(\lambda u_1) + \beta(\lambda v_1) = \alpha(\mu u_2) + \beta(\mu v_2) = \mu(\alpha u_2 + \beta v_2) = \mu w_2$. Therefore, $\mathbf{w} \in L$, and by Theorem 1, L is a subspace of $\mathscr{R}_2$.

We have previously identified a point in the plane with the vector joining the origin to that point. Thus, we can obtain all points of a line L through the origin by taking all scalar multiples of a fixed nonzero vector $\mathbf{u}$ on L. With this interpretation, Example 1 becomes obvious.

Example 2. Let $A \in \mathscr{R}_{mn}$. Let W be the set of all solutions $\mathbf{w}$ of the linear system of equations $A\mathbf{x} = \mathbf{0}$. Then W is a subspace of $\mathscr{R}_n$.

To see this, let $\mathbf{u}, \mathbf{v} \in W$. (Note that $\mathbf{u}, \mathbf{v} \in \mathscr{R}_n$, so that W is a subset of $\mathscr{R}_n$.) Let α, β be any real numbers. Then $\alpha \mathbf{u} + \beta \mathbf{v} \in W$ because $A(\alpha \mathbf{u} + \beta \mathbf{v}) = \alpha(A\mathbf{u}) + \beta(A\mathbf{v}) = \alpha \cdot \mathbf{0} + \beta \cdot \mathbf{0} = \mathbf{0}$. By Theorem 1, W is a subspace of $\mathscr{R}_n$. This example will be of considerable importance in Section 5.6.

Example 3. Let $\mathscr{I}$ be an arbitrary interval. Then $C'(\mathscr{R})$ is a subspace of $C(\mathscr{R})$.

By a theorem in calculus, every differentiable function is continuous. Therefore, $C'(\mathscr{R})$ is a subset of $C(\mathscr{R})$. Moreover, if $f, g \in C'(\mathscr{R})$ and $\alpha, \beta \in \mathscr{R}$, then by elementary theorems in calculus (which ones?), $\alpha f + \beta g \in C'(\mathscr{R})$. Therefore, by Theorem 1, $C'(\mathscr{R})$ is a subspace of $C(\mathscr{R})$.

Example 4. Let $\mathscr{I}$ be the real line $-\infty < t < \infty$. Consider the set S of functions

$$\{f \in C^2(\mathscr{R}) \mid f''(t) + 4f(t) \equiv 0\}.$$

Then S is a subspace of $C^2(\mathscr{R})$.

Obviously, S is a subset of $C^2(\mathscr{R})$. Let $f, g \in S$ and $\alpha, \beta \in \mathscr{R}$. Let $h = \alpha f + \beta g$. Then by calculus $h \in C^2(\mathscr{R})$, and

$$h''(t) + 4h(t) = [\alpha f''(t) + \beta g''(t)] + 4[\alpha f(t) + \beta g(t)]$$
$$= \alpha[f''(t) + 4f(t)] + \beta[g''(t) + 4g(t)].$$

Since $f, g \in S$, $f''(t) + 4f(t) \equiv 0$ and $g''(t) + 4g(t) \equiv 0$. Therefore, $h''(t) + 4h(t) \equiv 0$ and $\alpha f + \beta g \in S$. By Theorem 1, S is a subspace of $C^2(\mathscr{R})$.

● EXERCISES

2. Show that any plane through the origin in $\mathscr{R}_3$ is a subspace of $\mathscr{R}_3$. [*Hint:* A plane through the origin in $\mathscr{R}_3$ is a set of the form

$$\{\mathbf{x} = (x_1, x_2, x_3) \mid a_1 x_1 + a_2 x_2 + a_3 x_3 = 0\},$$

where a_1, a_2, a_3 are given real numbers. See also Example 2.]

3. A hyperplane through the origin in $\mathscr{R}_n$ is a set of points

$$S = \{\mathbf{x} \in \mathscr{R}_n \mid \sum_{j=1}^{n} a_j x_j = 0\},$$

where a_j ($j = 1, \ldots, n$) are given real numbers, not all zero. Show that a hyperplane through the origin in $\mathscr{R}_n$ is a subspace of $\mathscr{R}_n$.

4. Which of the following sets of vectors $\mathbf{x} = [x_1, \ldots, x_n]$ in $\mathscr{R}_n$ are subspaces of $\mathscr{R}_n$, ($n \geq 3$)?
(a) $\{\mathbf{x} \mid x_1 < 0\}$.
(b) $\{\mathbf{x} \mid x_1 - x_2 = x_3\}$.
(c) $\{\mathbf{x} \mid x_1 = x_2{}^2\}$.
(d) $\{\mathbf{x} \mid x_1 x_2 = 0\}$.
(e) $\{\mathbf{x} \mid x_1 \text{ is rational}\}$.

5. Which of the following sets of functions are subspaces of $\Phi(\mathscr{R})$?
(a) $\{f \mid f(t^2) = [f(t)]^2\}$.
(b) $\{f \mid f(0) = f(1)\}$.
(c) $\{f \mid f(2) = f(0) + 1\}$.
(d) $\{f \mid f(0) = 0\}$.

6. We have seen in Example 1, Section 5.2, that $\mathscr{F}_{nn}$ is a vector space over $\mathscr{F}$. Which of the following sets of matrices in $\mathscr{F}_{nn}$ are subspaces of $\mathscr{F}_{nn}$?
(a) All nonsingular matrices.
(b) All singular matrices.
(c) $\{A \in \mathscr{F}_{nn} \mid AB = BA\}$, where B is a given matrix in $\mathscr{F}_{nn}$.
(d) $\{A \in \mathscr{F}_{nn} \mid A^2 = A\}$.

(e) All symmetric matrices (that is, $a_{ij} = a_{ji}$; $i, j = 1, \ldots, n$).

7. Given the following list of vector spaces, pick out all possible pairs such that one of the pair is a subspace of the other:

$$C(\mathscr{R}), \quad C'(\mathscr{R}), \quad C^k(\mathscr{R}) \quad (k \geq 2), \quad \{f \in C^3(\mathscr{R}) \,|\, f'''(t) + \cos t\, f(t) \equiv 0\},$$
$$C(\mathscr{R}_n), \quad C'(\mathscr{R}_n), \quad C^k(\mathscr{R}_n) \quad (k \geq 2), \quad \{\mathbf{f} \in C'(\mathscr{R}_n) \,|\, \mathbf{f}'(t) \equiv A(t)\mathbf{f}(t)\},$$

where $A(t)$ is a given continuous matrix. (For a defiintion of $\mathbf{f}'(t)$, see Eq. 5.5, Section 5.2. For definitions of the above spaces, see Section 5.2.)

5.4 Span, Linear Dependence, and Linear Independence

We begin with two important new concepts. Let V be a vector space over $\mathscr{F}$, Let $\mathbf{v}_1, \ldots, \mathbf{v}_k$ be any vectors in V.

Definition 1. *Any vector of the form* $\alpha_1 \mathbf{v}_1 + \alpha_2 \mathbf{v}_2 + \cdots + \alpha_k \mathbf{v}_k$, *where* $\alpha_1, \alpha_2, \ldots, \alpha_k$ *are fixed elements of* $\mathscr{F}$, *is called a* linear combination *of the vectors* $\mathbf{v}_1, \ldots, \mathbf{v}_k$.

Definition 2. *The set of all linear combinations of the vectors* $\mathbf{v}_1, \ldots, \mathbf{v}_k$ *is called the* span *of the set of vectors* $\{\mathbf{v}_1, \mathbf{v}_2, \ldots, \mathbf{v}_k\}$. (The set of all linear combinations of $\mathbf{v}_1, \ldots, \mathbf{v}_k$ is generated by letting $\alpha_1, \ldots, \alpha_k$ in Definition 1 range over all elements in $\mathscr{F}$.) When we say that vectors $\mathbf{v}_1, \ldots, \mathbf{v}_k$ span a subspace W we mean that W is the span of the vectors $\mathbf{v}_1, \ldots, \mathbf{v}_k$.

Theorem 1. *The span S of any set* $\mathbf{v}_1, \ldots, \mathbf{v}_k$ *of vectors in V is a subspace of V. In fact, it is the smallest subspace of V that contains the vectors* $\mathbf{v}_1, \ldots, \mathbf{v}_k$, *in the sense that every subspace of V which contains* $\mathbf{v}_1, \ldots, \mathbf{v}_k$ *also contains S.*

Proof. Let $\mathbf{x}, \mathbf{y}$ be any vectors in S. Then there exist constants $\alpha_1, \ldots, \alpha_k$, $\beta_1, \ldots, \beta_k \in \mathscr{F}$ such that

$$\mathbf{x} = \alpha_1 \mathbf{v}_1 + \alpha_2 \mathbf{v}_2 + \cdots + \alpha_k \mathbf{v}_k, \quad \mathbf{y} = \beta_1 \mathbf{v}_1 + \beta_2 \mathbf{v}_2 + \cdots + \beta_k \mathbf{v}_k.$$

Therefore, if $\lambda, \mu \in \mathscr{F}$, we easily obtain

$$\lambda \mathbf{x} + \mu \mathbf{y} = (\lambda \alpha_1 + \mu \beta_1)\mathbf{v}_1 + \cdots + (\lambda \alpha_k + \mu \beta_k)\mathbf{v}_k \in S.$$

By Theorem 1, Section 5.3, S is a subspace. If T is a subspace of V containing $\mathbf{v}_1, \ldots, \mathbf{v}_k$, then T contains every linear combination of $\mathbf{v}_1, \ldots, \mathbf{v}_k$; hence, T contains S, and S is the smallest such subspace. ∎

Example 1. Let S be the subspace of $\mathscr{R}_3$ spanned by the vectors

$$\mathbf{v}_1 = \begin{bmatrix} 1 \\ 0 \\ 0 \end{bmatrix}, \quad \mathbf{v}_2 = \begin{bmatrix} 0 \\ 1 \\ 0 \end{bmatrix}, \quad \mathbf{v}_3 = \begin{bmatrix} 1 \\ 1 \\ 0 \end{bmatrix}.$$

(a) Determine whether the vector

$$\mathbf{b} = \begin{bmatrix} 2 \\ 1 \\ -1 \end{bmatrix}$$

belongs to S.

(b) Give a characterization of S.

(a) A vector in S has the form $\mathbf{b} = \alpha_1 \mathbf{v}_1 + \alpha_2 \mathbf{v}_2 + \alpha_3 \mathbf{v}_3$ for some α_1, α_2, $\alpha_3 \in \mathscr{R}$.

Let A be the matrix in $\mathscr{R}_{33}$ whose columns are the vectors $\mathbf{v}_1$, $\mathbf{v}_2$, $\mathbf{v}_3$. Thus,

$$A = [\mathbf{v}_1, \mathbf{v}_2, \mathbf{v}_3] = \begin{bmatrix} 1 & 0 & 1 \\ 0 & 1 & 1 \\ 0 & 0 & 0 \end{bmatrix}.$$

A vector $\mathbf{b} \in S$ if and only if

$$\mathbf{b} = \alpha_1 \mathbf{v}_1 + \alpha_2 \mathbf{v}_2 + \alpha_3 \mathbf{v}_3 = A\boldsymbol{\alpha}$$

for some

$$\boldsymbol{\alpha} = \begin{bmatrix} \alpha_1 \\ \alpha_2 \\ \alpha_3 \end{bmatrix} \in \mathscr{R}_3.$$

Thus, the given vector $\mathbf{b}$ lies in the subspace S if and only if the system of linear algebraic equation $A\boldsymbol{\alpha} = \mathbf{b}$ has a solution $\boldsymbol{\alpha}$. In our particular case this gives the system

$$\begin{bmatrix} 1 & 0 & 1 \\ 0 & 1 & 1 \\ 0 & 0 & 0 \end{bmatrix} \begin{bmatrix} \alpha_1 \\ \alpha_2 \\ \alpha_3 \end{bmatrix} = \begin{bmatrix} 2 \\ 1 \\ -1 \end{bmatrix}.$$

5.4 Span, Linear Dependence, and Linear Independence

Since the rank of the augmented matrix

$$[A \mid \mathbf{b}] = \begin{bmatrix} 1 & 0 & 1 & | & 2 \\ 0 & 1 & 1 & | & 1 \\ 0 & 0 & 0 & | & -1 \end{bmatrix}$$

is 3, it exceeds the rank of A (which is 2), and therefore the system $A\alpha = \mathbf{b}$ is inconsistent (see Theorem 1, Section 3.7). Hence, the given vector $\mathbf{b}$ is not in S.

(b) From the above calculation the system $A\alpha = \mathbf{b}$ is consistent if and only if $b_3 = 0$. Hence, $\mathbf{b} \in S$ if and only if $b_3 = 0$ and S consists of all vectors whose third component is zero.

• **EXERCISES**

1. Does the vector

$$\begin{bmatrix} 3 \\ -1 \\ 0 \\ -1 \end{bmatrix}$$

lie in the supspace of $\mathscr{R}_4$ spanned by the vectors

$$\begin{bmatrix} -1 \\ 1 \\ 1 \\ -3 \end{bmatrix}, \begin{bmatrix} 1 \\ 1 \\ 9 \\ 5 \end{bmatrix}, \begin{bmatrix} 1 \\ -2 \\ 3 \\ 2 \end{bmatrix}?$$

2. Let W be the set of all solutions $\mathbf{w} \in \mathscr{R}_5$ of the system

$$\begin{aligned} 2x_1 - x_2 + \tfrac{4}{3}x_3 - x_4 &= 0 \\ 9x_1 - 3x_2 + 6x_3 - 3x_4 - 3x_5 &= 0 \\ x_1 \phantom{{}-3x_2} + \tfrac{2}{3}x_3 \phantom{{}-3x_4} - x_5 &= 0. \end{aligned}$$

Show that W is a subspace of $\mathscr{R}_5$ and find a set of vectors that span W.

In general, the problem of characterizing the span of a given set of vectors in $\mathscr{R}_n$ can be stated in terms of linear equations as follows.

Example 2. Let $\mathbf{v}_1, \ldots, \mathbf{v}_k$ be given vectors in $\mathscr{R}_n$. Let S be their span. Then a vector $\mathbf{b} \in \mathscr{R}_n$ lies in S if and only if we can find scalars $\alpha_1, \ldots, \alpha_k$ such

118   Vector Spaces

that $\mathbf{b} = \alpha_1 \mathbf{v}_1 + \cdots + \alpha_k \mathbf{v}_k$. Define the matrix $A = [\mathbf{v}_1, \ldots, \mathbf{v}_k] \in \mathcal{R}_{nk}$. If

$$\alpha = \begin{bmatrix} \alpha_1 \\ \vdots \\ \alpha_k \end{bmatrix},$$

then $A\alpha = \alpha_1 \mathbf{v}_1 + \cdots + \alpha_k \mathbf{v}_k$. Hence, $\mathbf{b} \in S$ if and only if the system of equations $A\alpha = \mathbf{b}$ is consistent. As we did in Example 1 and the above exercises in a particular case, we now apply Theorem 2, Section 3.7, to test the consistency of the system $A\alpha = \mathbf{b}$.

To introduce the important concepts of linear dependence and linear independence of vectors we first illustrate the fact (already evident in several preceding examples) that a particular subspace of a vector space, given as the span of a specific set of vectors, may actually be spanned by fewer vectors. This will be the case whenever a vector of the given set is a linear combination of one or more of the others.

Example 3.   Let S be the subspace of $\mathcal{R}_3$ spanned by the vectors

$$\mathbf{v}_1 = \begin{bmatrix} 1 \\ 0 \\ 0 \end{bmatrix}, \quad \mathbf{v}_2 = \begin{bmatrix} 0 \\ 1 \\ 0 \end{bmatrix}, \quad \mathbf{v}_3 = \begin{bmatrix} 1 \\ 1 \\ 0 \end{bmatrix}.$$

As we have seen in Example 1, S consists of all vectors $\mathbf{x} \in \mathcal{R}_3$ with $x_3 = 0$. Observe that if

$$\mathbf{u} = \begin{bmatrix} \alpha_1 \\ \alpha_2 \\ \alpha_3 \end{bmatrix}$$

is any vector in S, then $\alpha_3 = 0$ and $\mathbf{u} = \alpha_1 \mathbf{v}_1 + \alpha_2 \mathbf{v}_2$. Thus, $\mathbf{v}_1$ and $\mathbf{v}_2$ also span S. Moreover, $\mathbf{v}_3 = \mathbf{v}_1 + \mathbf{v}_2$.

Definition 3.   *A set of vectors* $\mathbf{v}_1, \ldots, \mathbf{v}_k$ *in a vector space V over $\mathcal{F}$ is said to be linearly dependent if and only if there exist elements of $\mathcal{F}$* $c_1, c_2, \ldots, c_k$ *not all zero, such that*

$$c_1 \mathbf{v}_1 + c_2 \mathbf{v}_2 + \cdots + c_k \mathbf{v}_k = \mathbf{0}.$$

The most significant phrase in the above definition is "*not all zero.*"

Definition 4.   *A set of vectors* $\mathbf{v}_1, \ldots, \mathbf{v}_k$ *in a vector space V over $\mathcal{F}$ is said to be linearly independent if and only if it is not linearly dependent.*

We restate Definition 4 in its most commonly used form: *A set of vectors*

5.4 Span, Linear Dependence, and Linear Independence

$v_1, \ldots, v_k$ *in a vector space V is linearly independent if and only if the relation*

$$c_1 v_1 + c_2 v_2 + \cdots + c_k v_k = 0$$

implies that $c_1 = c_2 = \cdots = c_k = 0$. Normally, we say "the vectors $v_1, \ldots, v_k$ are linearly independent (dependent)" in place of "the set of vectors $v_1, \ldots, v_k$ is independent (dependent)." When we speak of a set of vectors $v_1, \ldots, v_k$, we allow repetitions. For example, we may have $v_2 = v_1$.

Thus, the concepts of linear dependence and linear independence are mutually exclusive.

Example 4. The vectors v_1, v_2, v_3 of Example 3 are linearly dependent. For, $v_1 + v_2 - v_3 = 0$; that is, in Definition 3 we choose $c_1 = c_2 = 1, c_3 = -1$.

Example 5. The vectors in $\mathcal{R}_5$

$$e_1 = \begin{bmatrix} 1 \\ 0 \\ 0 \\ 0 \\ 0 \end{bmatrix}, \quad e_2 = \begin{bmatrix} 0 \\ 1 \\ 0 \\ 0 \\ 0 \end{bmatrix}, \quad e_3 = \begin{bmatrix} 0 \\ 0 \\ 1 \\ 0 \\ 0 \end{bmatrix}$$

are linearly independent. Suppose that there exist real numbers $\alpha_1, \alpha_2, \alpha_3$, such that

$$\alpha_1 e_1 + \alpha_2 e_2 + \alpha_3 e_3 = 0.$$

Writing this down in terms of components we have the system of equations

$$\begin{aligned} \alpha_1 + 0\alpha_2 + 0\alpha_3 &= 0 \\ 0\alpha_1 + \alpha_2 + 0\alpha_3 &= 0 \\ 0\alpha_1 + 0\alpha_2 + \alpha_3 &= 0 \\ 0\alpha_1 + 0\alpha_2 + 0\alpha_3 &= 0 \\ 0\alpha_1 + 0\alpha_2 + 0\alpha_3 &= 0 \end{aligned}$$

whose only solution is $\alpha_1 = \alpha_2 = \alpha_3 = 0$. This establishes the linear independence of e_1, e_2, e_3. More generally, the vectors

$$e_1 = \begin{bmatrix} 1 \\ 0 \\ \vdots \\ 0 \end{bmatrix}, \quad e_2 = \begin{bmatrix} 0 \\ 1 \\ 0 \\ \vdots \\ 0 \end{bmatrix}, \ldots, e_n = \begin{bmatrix} 0 \\ \vdots \\ 0 \\ 1 \end{bmatrix}$$

120   Vector Spaces

in $\mathscr{R}_n$ are linearly independent in $\mathscr{R}_n$. This may be seen in the same way, or as a special case of the next example.

Example 6. Let $\mathbf{v}_1, \ldots, \mathbf{v}_k$ be given vectors in $\mathscr{R}_n$. Under what conditions are they linearly dependent?

Define the matrix $A = [\mathbf{v}_1, \ldots, \mathbf{v}_k] \in \mathscr{R}_{nk}$. The vectors are linearly dependent if and only if there exist real numbers $\alpha_1, \ldots, \alpha_k$, not all zero, such that

$$\alpha_1 \mathbf{v}_1 + \alpha_2 \mathbf{v}_2 + \cdots + \alpha_k \mathbf{v}_k = \mathbf{0}.$$

Thus, $\mathbf{v}_1, \ldots, \mathbf{v}_k$ are linearly dependent if and only if there exists a nonzero vector

$$\boldsymbol{\alpha} = \begin{bmatrix} \alpha_1 \\ \vdots \\ \alpha_k \end{bmatrix} \in \mathscr{R}_k$$

such that $A\boldsymbol{\alpha} = \mathbf{0}$. This is a linear homogeneous system of n equations and k unknowns (the numbers $\alpha_1, \ldots, \alpha_k$), which has a nontrivial solution if and only if the rank of A is less than k (see Corollary 1, Theorem 1, Section 3.6). In particular, if $k > n$, $\mathbf{v}_1, \ldots, \mathbf{v}_k$ are linearly dependent.

If we choose $k = n$, $\mathbf{v}_1 = \mathbf{e}_1$, $\mathbf{v}_2 = \mathbf{e}_2$, $\ldots$, $\mathbf{v}_n = \mathbf{e}_n$, then the matrix $A = [\mathbf{e}_1, \mathbf{e}_2, \ldots, \mathbf{e}_n] \in \mathscr{R}_{nn}$ is the identity matrix I_n. Since the rank of I_n is n, the vectors $\mathbf{e}_1, \ldots, \mathbf{e}_n$ in $\mathscr{R}_n$ are linearly independent.

Example 7. Consider the vector space $C'(\mathscr{R})$ of all continuously differentiable functions on the real line. The functions (vectors) e^t and e^{-t} are linearly independent.

To see this, suppose that there exist **constants** c_1 and c_2 such that $c_1 e^t + c_2 e^{-t}$ is the zero vector in $C'(\mathscr{R})$ (that is, the identically zero function). Then we must have

$$c_1 e^t + c_2 e^{-t} \equiv 0.$$

Hence, also (multiplying by e^{-t})

$$c_1 + c_2 e^{-2t} \equiv 0,$$

and differentiating with respect to t we obtain

$$-2c_2 e^{-2t} \equiv 0.$$

5.4 Span, Linear Dependence, and Linear Independence

Hence, $c_2 = 0$ and also $c_1 = 0$. Therefore, $c_1 e^t + c_2 e^{-t} \equiv 0$ implies that $c_1 = c_2 = 0$, and e^t and e^{-t} are linearly independent. It is reasonable to ask the following question.

Example 8. Does the function $\sin t$ lie in the span of the vectors (functions) e^t and e^{-t}?

If $\sin t$ lies in the span of e^t and e^{-t}, there exist **constants** a_1 and a_2 such that

$$\sin t \equiv a_1 e^t + a_2 e^{-t}.$$

Since all functions considered here can be differentiated as often as desired, we have successively

$$\cos t \equiv a_1 e^t - a_2 e^{-t}$$

and

$$-\sin t \equiv a_1 e^t + a_2 e^{-t}.$$

Subtracting the last identity from the first gives $2 \sin t \equiv 0$, which is absurd. Hence, $\sin t$ does not lie in the span of e^t and e^{-t}.

Example 9. Consider the vectors (functions) $1, t, t^2, t^3, t^4, t^5$ in vector space $C(\mathscr{R})$. Determine their span and show that this set of vectors is linearly independent.

The span of the given vectors is the set of all linear combinations of the form

$$\alpha_0 + \alpha_1 t + \cdots + \alpha_5 t^5.$$

Thus, the span consists of all polynomials of degree five or less that we have previously denoted by $P^5(\mathscr{R})$. To show that $1, t, \ldots, t^5$ are linearly independent, observe that by a standard theorem in elementary algebra there are at most five distinct roots of the equation

$$\alpha_0 + \alpha_1 t + \cdots + \alpha_5 t^5 = 0$$

for a fixed set of $\alpha_0, \alpha_1, \ldots, \alpha_5$. Hence, the identity

$$\alpha_0 + \alpha_1 t + \cdots + \alpha_5 t^5 \equiv 0$$

holds only if $\alpha_0 = \alpha_1 = \cdots = \alpha_5 = 0$. This means that the vectors $1, t, t^2, \ldots, t^5$ are linearly independent.

122 Vector Spaces

Example 10. Show that the span of the vector functions

$$\mathbf{g}(t) = \begin{bmatrix} e^t \\ e^{-t} \end{bmatrix}, \quad \mathbf{h}(t) = \begin{bmatrix} -e^t \\ 2e^{-t} \end{bmatrix}$$

in the vector space $C(\mathcal{R}_2)$ is the set of all vectors of the form

$$\begin{bmatrix} a_1 e^t \\ a_2 e^{-t} \end{bmatrix},$$

for real numbers a_1, a_2, and show that the given vectors are linearly independent.

To see this, we note that

$$\mathbf{f}(t) = \begin{bmatrix} f_1(t) \\ f_2(t) \end{bmatrix}$$

is in the span of $\mathbf{g}(t)$ and $\mathbf{h}(t)$ if and only if there exist constants c_1, c_2 such that

$$\mathbf{f}(t) \equiv c_1 \mathbf{g}(t) + c_2 \mathbf{h}(t),$$

for all t. Thus, in this case,

$$f_1(t) = c_1 e^t - c_2 e^t = (c_1 - c_2)e^t,$$
$$f_2(t) = c_1 e^{-t} + 2c_2 e^{-t} = (c_1 + 2c_2)e^{-t}.$$

Hence, the span is contained in the set of vectors of form

$$\begin{bmatrix} a_1 e^t \\ a_2 e^{-t} \end{bmatrix}.$$

Conversely, by solving simultaneous equations, given any real numbers a_1, a_2, we have

$$\begin{bmatrix} a_1 e^t \\ a_2 e^{-t} \end{bmatrix} = c_1 \begin{bmatrix} e^t \\ e^{-t} \end{bmatrix} + c_2 \begin{bmatrix} -e^t \\ +2e^{-t} \end{bmatrix};$$

thus, we choose $c_1 = (2a_1 + a_2)/3$, $c_2 = (a_2 - a_1)/3$, and the span is precisely as asserted.

To show the linear independence of $\mathbf{g}(t)$ and $\mathbf{h}(t)$, suppose that

$$0 = c_1 \mathbf{g}(t) + c_2 \mathbf{h}(t) = \begin{bmatrix} (c_1 - c_2)e^t \\ (c_1 + 2c_2)e^{-t} \end{bmatrix}.$$

5.4 Span, Linear Dependence, and Linear Independence

Then

$$c_1 - c_2 = 0$$
$$c_1 + 2c_2 = 0,$$

whence, $c_1 = c_2 = 0$; so vectors $\mathbf{g}(t)$ and $\mathbf{h}(t)$ are linearly independent.

When we defined the space $C(\mathscr{R})$ of functions continuous on an interval $\mathscr{I}$ (Example 4, Section 5.2) we did not emphasize the role of the interval $\mathscr{I}$. However, whether a given set of functions in the vector space $C(\mathscr{R})$ is linearly dependent or independent may sometimes depend on the interval $\mathscr{I}$ in question.

Example 11. Determine the linear dependence or independence of the functions $\phi_1(t) = t^2$, $\phi_2(t) = t|t|$ in the space $C(\mathscr{R})$ when the interval $\mathscr{I}$ is each of the following: (a) $0 \le t < \infty$, (b) $-2 < t < -1$, (c) $-\pi < t \le e$.
  (a) Here, $\phi_1(t) = t^2$, $\phi_2(t) = t^2$; thus, $\phi_1(t) - \phi_2(t) \equiv 0$ for $0 \le t < \infty$, and ϕ_1 and ϕ_2 are linearly dependent.
  (b) Here, $\phi_1(t) = t^2$, $\phi_2(t) = t(-t) = -t^2$; thus, $\phi_1(t) + \phi_2(t) \equiv 0$ for $-2 < t < -1$, and again ϕ_1 and ϕ_2 are linearly dependent.
  (c) Suppose there exist constants c_1 and c_2 such that

$$c_1 \phi_1(t) + c_2 \phi_2(t) \equiv 0 \qquad \text{for } -\pi < t \le e.$$

Choosing $t = 1$ (any value of t, $0 < t \le e$ will do as well), we obtain

$$c_1 + c_2 = 0,$$

Choosing $t = -1$, we obtain

$$c_1 - c_2 = 0.$$

The only solution of this system of equations is $c_1 = c_2 = 0$; thus ϕ_1 and ϕ_2 are linearly independent.

We conclude this section with the following general result on linear dependence, which holds in any vector space.

Theorem 2. *Let $\mathbf{v}_1, \ldots, \mathbf{v}_k$ be a given set of vectors with $\mathbf{v}_1 \ne \mathbf{0}$ in a vector space V over $\mathscr{F}$. This set is linearly dependent if and only if one of the vectors $\mathbf{v}_2, \ldots, \mathbf{v}_k$ is a linear combination of the preceding ones.*

Proof. Suppose that one of the vectors $\mathbf{v}_2, \ldots, \mathbf{v}_k$ is a linear combination of the preceding ones, for example,

$$\mathbf{v}_l = \alpha_1 \mathbf{v}_1 + \alpha_2 \mathbf{v}_2 + \cdots + \alpha_{l-1} \mathbf{v}_{l-1} \qquad (2 \le l \le k).$$

Then

$$\alpha_1 v_1 + \alpha_2 v_2 + \cdots + \alpha_{l-1} v_{l-1} + (-1)v_l + 0 \cdot v_{l+1} + \cdots + 0 \cdot v_k = 0,$$

and the set of vectors $v_1, \ldots, v_k$ is linearly dependent.

Conversely, let $v_1, \ldots, v_k$ be linearly dependent. Then, for some constants $c_1, \ldots, c_k$, not all zero, we have

$$c_1 v_1 + \cdots + c_k v_k = 0.$$

Let c_l be the last nonzero constant in this relation. If $l = 1$, the relation is $c_1 v_1 = 0$, whence $v_1 = 0$, contrary to hypothesis. Thus, $2 \leq l \leq k$. Dividing the above equation by c_l, we obtain

$$v_l = -\frac{c_1}{c_l} v_1 - \frac{c_2}{c_l} v_2 - \cdots - \frac{c_{l-1}}{c_l} v_{l-1}. \blacksquare$$

● **EXERCISES**

3. Determine the linear dependence or independence of the following sets of vectors in $\mathcal{R}_3$:

(a) $v_1 = \begin{bmatrix} 1 \\ 0 \\ 0 \end{bmatrix}$, $v_2 = \begin{bmatrix} 0 \\ 1 \\ 0 \end{bmatrix}$, $v_3 = \begin{bmatrix} 0 \\ 1 \\ 1 \end{bmatrix}$.

(b) $v_1 = \begin{bmatrix} 1 \\ 2 \\ 3 \end{bmatrix}$, $v_2 = \begin{bmatrix} 4 \\ 5 \\ 6 \end{bmatrix}$, $v_3 = \begin{bmatrix} 7 \\ 8 \\ 9 \end{bmatrix}$.

(c) $v_1 = \begin{bmatrix} 1 \\ 0 \\ 1 \end{bmatrix}$, $v_2 = \begin{bmatrix} 0 \\ 1 \\ 1 \end{bmatrix}$, $v_3 = \begin{bmatrix} -1 \\ 1 \\ 1 \end{bmatrix}$.

(d) $v_1 = \begin{bmatrix} 1 \\ 2 \\ 4 \end{bmatrix}$, $v_2 = \begin{bmatrix} 1 \\ 3 \\ 9 \end{bmatrix}$, $v_3 = \begin{bmatrix} 1 \\ 4 \\ 16 \end{bmatrix}$.

(e) $v_1 = \begin{bmatrix} 1 \\ -1 \\ 2 \end{bmatrix}$, $v_2 = \begin{bmatrix} 1 \\ 2 \\ 1 \end{bmatrix}$, $v_3 = \begin{bmatrix} 2 \\ 1 \\ 3 \end{bmatrix}$.

(f) $v_1 = \begin{bmatrix} 1 \\ 2 \\ 3 \end{bmatrix}$, $v_2 = \begin{bmatrix} 1 \\ 3 \\ 5 \end{bmatrix}$, $v_3 = \begin{bmatrix} 1 \\ 10 \\ -5 \end{bmatrix}$, $v_4 = \begin{bmatrix} 0 \\ -1 \\ 17 \end{bmatrix}$.

5.4 Span, Linear Dependence, and Linear Independence

4. Determine the linear dependence or independence of each of the following sets of functions in the space $C'(\mathscr{R})$ over the given interval $\mathscr{I}$.
   (a) $f_1(t) = e^t, f_2(t) = e^{3t}$, for $-\infty < t < \infty$.
   (b) $f_1(t) = e^{2t}, f_2(t) = te^{2t}$, for $-\infty < t < \infty$.
   (c) $f_1(t) = e^{\lambda t}, f_2(t) = e^{\mu t}$, for $-\infty < t < \infty$ where λ, μ are given constants.
   (d) $f_1(t) = 1, f_2(t) = 1 - t^2, f_3(t) = 1 + t^2$, for $-1 < t < 1$.
   (e) $f_1(t) = \sin 3t, f_2(t) = t \sin 3t$, for $-n < t < 2n$.
   (f) $f_1(t) = 1, f_2(t) = e^t, f_3(t) = e^{2t}$, for $0 \le t < \infty$.
   (g) $f_1(t) = t, f_2(t) = t^2 + 2t - 14, f_3(t) = t^2 - t^3, f_4(t) = t^3 - t^4$, for $-1 \le t < 1$.
   (h) $f_1(t) = e^{t^2} \sin t, f_2(t) = \int_0^t e^{s^2} \sin s \, ds$, for $0 < t < 1$.

5. Show that the set $\mathscr{C}$ of all complex numbers forms a vector space over the real numbers $\mathscr{R}$. Then show that the complex numbers $1 + i$ and $1 - i$ are linearly independent in this space.

6. Determine the linear dependence or independence of each of the following sets of vector functions in the space $C(\mathscr{R}_2)$ or $C(\mathscr{R}_3)$ over the given interval $\mathscr{I}$.

   (a) $\mathbf{f}_1(t) = \begin{bmatrix} e^{2t} \\ 2e^{2t} \end{bmatrix}$, $\mathbf{f}_2(t) = \begin{bmatrix} e^{-2t} \\ -2e^{-2t} \end{bmatrix}$, for $-\infty < t < \infty$.

   (b) $\mathbf{f}_1(t) = \begin{bmatrix} e^{r_1 t} \\ r_1 e^{r_1 t} \end{bmatrix}$, $\mathbf{f}_2(t) = \begin{bmatrix} e^{r_2 t} \\ r_2 e^{r_2 t} \end{bmatrix}$, for $-\infty < t < \infty$,

   where r_1 and r_2 are given real numbers.

   (c) $\mathbf{f}_1(t) = \begin{bmatrix} \cos t \\ -\sin t \end{bmatrix}$, $\mathbf{f}_2(t) = \begin{bmatrix} \sin t \\ \cos t \end{bmatrix}$, for $0 < t < \infty$.

   (d) $\mathbf{f}_1(t) = \begin{bmatrix} t \\ 1 \\ 0 \end{bmatrix}$, $\mathbf{f}_2(t) = \begin{bmatrix} t^3 \\ t^2 \\ -5t \end{bmatrix}$, $\mathbf{f}_3(t) = \begin{bmatrix} t^2 \\ 0 \\ 1 \end{bmatrix}$, for $-5 < t \le 5$.

   (e) $\mathbf{f}_1(t) = \begin{bmatrix} e^{r_1 t} \\ r_1 e^{r_1 t} \\ r_1^2 e^{r_1 t} \end{bmatrix}$, $\mathbf{f}_2(t) = \begin{bmatrix} e^{r_2 t} \\ r_2 e^{r_2 t} \\ r_2^2 e^{r_2 t} \end{bmatrix}$, $\mathbf{f}_3(t) = \begin{bmatrix} e^{r_3 t} \\ r_3 e^{r_3 t} \\ r_3^2 e^{r_3 t} \end{bmatrix}$,

   for $-\infty < t < \infty$ where r_1, r_2, r_3 are given real numbers.

7. Determine a number m so that each of the following sets of vectors in $\mathscr{R}_3$ is linearly dependent.

   (a) $\mathbf{v}_1 = \begin{bmatrix} 1 \\ 2 \\ 3 \end{bmatrix}$, $\mathbf{v}_2 = \begin{bmatrix} 2 \\ 3 \\ 2 \end{bmatrix}$, $\mathbf{v}_3 = \begin{bmatrix} m \\ 1 \\ 1 \end{bmatrix}$.

   (b) $\mathbf{v}_1 = \begin{bmatrix} 1 - m \\ 2 \\ 7 \end{bmatrix}$, $\mathbf{v}_2 = \begin{bmatrix} 0 \\ -2 - m \\ 12 \end{bmatrix}$, $\mathbf{v}_3 = \begin{bmatrix} 0 \\ 0 \\ -m \end{bmatrix}$.

8. For each of the sets of vectors in Exercise 3 preceding, determine whether the vector

$$\mathbf{v} = \begin{bmatrix} 1 \\ 1 \\ 1 \end{bmatrix}$$

lies in the span of the given set.

9. For each of the sets of functions in Exercise 4, parts (a), (b), (c), determine whether the function $f(t) = e^{7t}$, $-\infty < t < \infty$ lies in the span of the given set.

10. Which of the following sets of vectors in $\mathscr{R}_3$ span all of $\mathscr{R}_3$?

(a) $\mathbf{v}_1 = \begin{bmatrix} 1 \\ 0 \\ 1 \end{bmatrix}$, $\mathbf{v}_2 = \begin{bmatrix} 0 \\ 1 \\ 1 \end{bmatrix}$, $\mathbf{v}_3 = \begin{bmatrix} 0 \\ 0 \\ 1 \end{bmatrix}$.

(b) $\mathbf{v}_1 = \begin{bmatrix} 1 \\ 2 \\ 2 \end{bmatrix}$, $\mathbf{v}_2 = \begin{bmatrix} 2 \\ 1 \\ 2 \end{bmatrix}$, $\mathbf{v}_3 = \begin{bmatrix} 2 \\ 2 \\ 1 \end{bmatrix}$.

(c) $\mathbf{v}_1 = \begin{bmatrix} 0 \\ 1 \\ 1 \end{bmatrix}$, $\mathbf{v}_2 = \begin{bmatrix} 1 \\ 1 \\ 0 \end{bmatrix}$, $\mathbf{v}_3 = \begin{bmatrix} 1 \\ 2 \\ 1 \end{bmatrix}$, $\mathbf{v}_4 = \begin{bmatrix} 1 \\ 0 \\ -1 \end{bmatrix}$.

(d) $\mathbf{v}_1 = \begin{bmatrix} 2 \\ 2 \\ 0 \end{bmatrix}$, $\mathbf{v}_2 = \begin{bmatrix} \pi \\ e \\ 0 \end{bmatrix}$, $\mathbf{v}_3 = \begin{bmatrix} -\pi \\ 2+e \\ 1 \end{bmatrix}$.

11. Show that in any vector space a set of vectors containing the zero vector is linearly dependent.

12. Show that in the space $C(\mathscr{R})$ over an interval $\mathscr{I}$, two functions are linearly dependent if and only if one function is a constant multiple of the other.

13. Let $k > 1$ and let $\mathbf{v}_1, \ldots, \mathbf{v}_k$ be a set of vectors in a vector space V over $\mathscr{F}$. Then the set $\mathbf{v}_1, \ldots, \mathbf{v}_k$ is linearly dependent if and only if one of the vectors is a linear combination of the remaining ones. [*Hint:* Examine Theorem 2.]

14. Show that in any vector space V over $\mathscr{F}$ a subset of a linearly independent set of vectors is itself linearly independent.

15. Show that for every integer k the functions $1, t, t^2, \ldots, t^k$ in $C(\mathscr{R})$ are linearly independent for $-\infty < t < \infty$.

5.5 Basis and Dimension of a Vector Space

We now ask the following question: Given a vector space V and a set of vectors $\mathbf{v}_1, \mathbf{v}_2, \ldots, \mathbf{v}_k$ spanning V, can we omit some of these vectors and still have the remaining vectors span V? In other words, are some of the vectors $\mathbf{v}_1, \ldots, \mathbf{v}_k$ superfluous in determining the span of the set? As we shall see,

5.5 Basis and Dimension of a Vector Space

the answer depends on whether the given set of vectors is linearly dependent or linearly independent. We therefore give the following definition.

Definition 1. *Let V be a vector space over $\mathscr{F}$. A set $v_1, \ldots, v_k$ of vectors in V is a basis of V if and only if*
 (i) *the set $v_1, \ldots, v_k$ is linearly independent, and*
 (ii) *the set $v_1, \ldots, v_k$ spans V.*

Example 1. The vectors

$$e_1 = \begin{bmatrix} 1 \\ 0 \\ 0 \end{bmatrix}, \quad e_2 = \begin{bmatrix} 0 \\ 1 \\ 0 \end{bmatrix}, \quad e_3 = \begin{bmatrix} 0 \\ 0 \\ 1 \end{bmatrix}.$$

form a basis for $\mathscr{R}_3$. More generally, the vectors

$$e_1 = \begin{bmatrix} 1 \\ 0 \\ \vdots \\ 0 \end{bmatrix}, \quad e_2 = \begin{bmatrix} 0 \\ 1 \\ 0 \\ \vdots \\ 0 \end{bmatrix}, \ldots, e_n = \begin{bmatrix} 0 \\ \vdots \\ 0 \\ 1 \end{bmatrix}$$

form a basis for $\mathscr{R}_n$.

As we have seen (Example 6, Section 5.4), the vectors $e_1, \ldots, e_n$ are linearly independent in $\mathscr{R}_n$, and for any given vector

$$a = \begin{bmatrix} \alpha_1 \\ \alpha_2 \\ \vdots \\ \alpha_n \end{bmatrix},$$

we have

$$a = \alpha_1 e_1 + \alpha_2 e_2 + \cdots + \alpha_n e_n.$$

Thus, $e_1, e_2, \ldots, e_n$ also span $\mathscr{R}_n$ and, therefore, form a basis for $\mathscr{R}_n$.

Example 2. The vectors

$$e_1 = \begin{bmatrix} 1 \\ 0 \\ 0 \end{bmatrix}, \quad e_3 = \begin{bmatrix} 0 \\ 0 \\ 1 \end{bmatrix}$$

do not form a basis for $\mathscr{R}_3$.

Vector Spaces

To see this we note that e_1, e_3 are obviously linearly independent in $\mathscr{R}_3$, but they do not span $\mathscr{R}_3$ since, for example, the vector e_2 in $\mathscr{R}_3$ is not in the span of e_1 and e_3.

The general situation for n vectors in $\mathscr{R}_n$ is as follows.

Theorem 1. *Let* $v_1, \ldots, v_n$ *be n vectors in* $\mathscr{R}_n$. *Let A be the matrix:*

$$A = [v_1, v_2, \ldots, v_n] \in \mathscr{R}_{nn}.$$

Then $v_1, v_2, \ldots, v_n$ *form a basis for* $\mathscr{R}_n$ *if and only if A is nonsingular.* (The same result holds for $\mathscr{C}_n$.)

Proof. Suppose $v_1, \ldots, v_n$ form a basis for $\mathscr{R}_n$. Then certainly $v_1, \ldots, v_n$ are linearly independent. As we have shown in Example 6, Section 5.4, this implies that the rank of A is n. Hence, by Theorem 1, Section 3.8, A is nonsingular.

Conversely, suppose A is nonsingular; again by Theorem 1, Section 3.8, A has rank n. Using Example 6, Section 5.4, once more we see that the vectors $v_1, v_2, \ldots, v_n$ are linearly independent. By Theorem 1, Section 3.7, since A has rank n, the system $A\alpha = b$ is consistent for every b in $\mathscr{R}_n$. By Example 2, Section 5.4, every vector b in $\mathscr{R}_n$ is in the span of $v_1, v_2, \ldots, v_n$. Therefore, $v_1, \ldots, v_n$ span $\mathscr{R}_n$ and, hence, form a basis for $\mathscr{R}_n$. ∎

● **EXERCISE**

1. Determine whether the following sets of vectors form a basis for the space indicated.

(a) $v_1 = \begin{bmatrix} 1 \\ 2 \\ 3 \end{bmatrix}$, $v_2 = \begin{bmatrix} 0 \\ 1 \\ 5 \end{bmatrix}$, $v_3 = \begin{bmatrix} 0 \\ 0 \\ 4 \end{bmatrix}$ in $\mathscr{R}_3$.

(b) $v_1 = \begin{bmatrix} 2 \\ 1 \\ 6 \end{bmatrix}$, $v_2 = \begin{bmatrix} -1 \\ 0 \\ 1 \end{bmatrix}$, $v_3 = \begin{bmatrix} 3 \\ 2 \\ 13 \end{bmatrix}$ in $\mathscr{R}_3$.

(c) $v_1 = \begin{bmatrix} 1 \\ 2 \\ 2 \end{bmatrix}$, $v_2 = \begin{bmatrix} 2 \\ 1 \\ 1 \end{bmatrix}$, $v_3 = \begin{bmatrix} 3 \\ 2 \\ 2 \end{bmatrix}$, $v_4 = \begin{bmatrix} 0 \\ 1 \\ 1 \end{bmatrix}$ in $\mathscr{R}_3$.

(d) $v_1 = \begin{bmatrix} 0 \\ 2 \\ 3 \\ 0 \end{bmatrix}$, $v_2 = \begin{bmatrix} 0 \\ 0 \\ 1 \\ 0 \end{bmatrix}$, $v_3 = \begin{bmatrix} 2 \\ 1 \\ 0 \\ 0 \end{bmatrix}$, $v_4 = \begin{bmatrix} 0 \\ 1 \\ 1 \\ 0 \end{bmatrix}$ in $\mathscr{R}_4$.

(e) $\mathbf{v}_1 = \begin{bmatrix} 1 \\ 1 \\ 0 \end{bmatrix}$, $\mathbf{v}_2 = \begin{bmatrix} 0 \\ 0 \\ 7 \end{bmatrix}$ in $\mathscr{R}_3$.

(f) $\mathbf{e}_1, \mathbf{e}_2, \mathbf{e}_3, \mathbf{0}$ in $\mathscr{R}_3$.

We now turn our attention to vector spaces other than $\mathscr{R}_n$.

Example 3. The functions $f_0(t) = 1$, $f_1(t) = t$, $f_2(t) = t^2$, $f_3(t) = t^3$ form a basis for the vector space $P^3(R)$ (that is, the space of all polynomials of degree ≤ 3 with real coefficients), but not for the vector space $P^k(R)$ for $k \geq 4$, nor for the vector space $P(R)$ of all polynomials with real coefficients, nor for the vector space $C(R)$ of all real continuous functions or $-\infty < t < \infty$.

To understand this fact we note that by the same argument used in Example 9, Section 5.4, the functions $1, t, t^2, t^3$ are linearly independent in each of the above-mentioned vector spaces. To show that they form a basis for $P^3(R)$ we must demonstrate that they span $P^3(R)$. Letting $a_0 + a_1 t + a_2 t^2 + a_3 t^3$ be an arbitrary polynomial of degree three with real coefficients, we have

$$a_0 + a_1 t + a_2 t^2 + a_3 t^3 \equiv a_0 f_0(t) + a_1 f_1(t) + a_2 f_2(t) + a_3 f_3(t),$$

and therefore f_0, f_1, f_2, f_3 form a basis for $P^3(R)$. But the function $f(t) = t^4$ lies in each of the spaces $P^k(R), k \geq 4$ and $P(R)$ and $C(R)$. Yet it is impossible to express $f(t)$ as a linear combination of $f_0(t), f_1(t), f_2(t), f_3(t)$ valid for all t. Thus, f is not in the span of f_0, f_1, f_2, f_3, and the conclusion follows.

● **EXERCISES**

2. Show that the functions

$$p_0(t) = 1, \quad p_1(t) = t, \quad p_2(t) = \tfrac{1}{2}(3t^2 - 1), \quad p_3(t) = \tfrac{3}{2}(\tfrac{5}{3}t^3 - t)$$

also form a basis for $P^3(R)$ (the polynomials p_0, p_1, p_2, p_3 are the first four Legendre polynomials).

3. Show that the functions

$$q_0(t) = 1, \quad q_1(t) = t, \quad q_2(t) = 2t^2 - 1, \quad q_3(t) = 4t^3 - 3t$$

also form a basis for $P^3(R)$ (these are the first four Tchebychev polynomials).

Theorem 1, Example 3, and Exercises 2 and 3 show that a given vector space has many different bases. The Legendre and Tchebychev polynomials of Exercises 2 and 3 play important roles in various applications.

130 *Vector Spaces*

To explore whether a given set of vectors form a basis for a vector space V, we have the following analog of Theorem 1 (Theorem 1 applies only to $\mathscr{R}_n$ and $\mathscr{C}_n$).

Theorem 2. *Let V be a vector space over $\mathscr{F}$ and let $\mathbf{v}_1, \mathbf{v}_2, \ldots, \mathbf{v}_k$ be k vectors in V. Then $\mathbf{v}_1, \mathbf{v}_2, \ldots, \mathbf{v}_k$ form a basis for V if and only if every vector $\mathbf{v} \in V$ can be written uniquely as $\mathbf{v} = \alpha_1 \mathbf{v}_1 + \alpha_2 \mathbf{v}_2 + \cdots + \alpha_k \mathbf{v}_k$* (uniqueness means that if $\mathbf{v} = \alpha_1 \mathbf{v}_1 + \cdots + \alpha_k \mathbf{v}_k = \beta_1 \mathbf{v}_1 + \cdots + \beta_k \mathbf{v}_k$, then $\alpha_1 = \beta_1, \ldots, \alpha_k = \beta_k$).

Proof. Let $\mathbf{v}_1, \mathbf{v}_2, \ldots, \mathbf{v}_k$ be a basis for V. Let $\mathbf{v}$ be an arbitrary vector in V. Since $\mathbf{v}_1, \ldots, \mathbf{v}_k$ spans V, there exist scalars $\alpha_1, \alpha_2, \ldots, \alpha_k \in \mathscr{F}$ such that

$$\mathbf{v} = \alpha_1 \mathbf{v}_1 + \alpha_2 \mathbf{v}_2 + \cdots + \alpha_k \mathbf{v}_k.$$

Suppose we also have scalars $\beta_1, \beta_2, \ldots, \beta_k \in \mathscr{F}$, such that

$$\mathbf{v} = \beta_1 \mathbf{v}_1 + \beta_2 \mathbf{v}_2 + \cdots + \beta_k \mathbf{v}_k.$$

Then, subtracting, we obtain

$$(\alpha_1 - \beta_1)\mathbf{v}_1 + (\alpha_2 - \beta_2)\mathbf{v}_2 + \cdots + (\alpha_k - \beta_k)\mathbf{v}_k = \mathbf{0}.$$

But $\mathbf{v}_1, \mathbf{v}_2, \ldots, \mathbf{v}_k$ are linearly independent. Therefore,

$$\alpha_1 - \beta_1 = 0, \alpha_2 - \beta_2 = 0, \ldots, \alpha_k - \beta_k = 0,$$

and so $\alpha_1 = \beta_1, \alpha_2 = \beta_2, \ldots, \alpha_k = \beta_k$, which proves uniqueness.

Conversely, suppose every vector $\mathbf{v} \in V$ can be represented uniquely as

$$\mathbf{v} = \alpha_1 \mathbf{v}_1 + \alpha_2 \mathbf{v}_2 + \cdots + \alpha_k \mathbf{v}_k.$$

Thus, clearly $\mathbf{v}_1, \mathbf{v}_2, \ldots, \mathbf{v}_k$ span V. To prove that they are linearly independent, let

$$\mathbf{0} = \alpha_1 \mathbf{v}_1 + \alpha_2 \mathbf{v}_2 + \cdots + \alpha_k \mathbf{v}_k.$$

Since we obviously also have

$$\mathbf{0} = 0\mathbf{v}_1 + 0\mathbf{v}_2 + \cdots + 0\mathbf{v}_k,$$

our uniqueness assumption implies that $\alpha_1 = 0, \alpha_2 = 0, \ldots, \alpha_k = 0$, which proves linear independence of the vectors $\mathbf{v}_1, \mathbf{v}_2, \ldots, \mathbf{v}_k$. Hence, these vectors form a basis for V. ∎

If $v_1, v_2, \ldots, v_k$ is a basis for V and if $v \in V$ is expressed as

$$v = \alpha_1 v_1 + \cdots + \alpha_k v_k,$$

then the scalars $\alpha_1, \alpha_2, \ldots, \alpha_k \in \mathscr{F}$ are called **the coordinates of v with respect to the basis** $v_1, v_2, \ldots, v_k$. By Theorem 2 the coordinates of a given vector with respect to a particular basis are unique. This is a familiar concept in $\mathscr{R}_3$ from geometry.

• EXERCISES

4. (a) Show that the vectors

$$v_1 = \begin{bmatrix} 2 \\ -1 \\ 5 \end{bmatrix}, \quad v_2 = \begin{bmatrix} 0 \\ 7 \\ -2 \end{bmatrix}, \quad v_3 = \begin{bmatrix} 0 \\ 0 \\ -1 \end{bmatrix}$$

form a basis for $\mathscr{R}_3$.
(b) Find the coordinates of v_1, v_2, v_3 with respect to the basis e_1, e_2, e_3 of $\mathscr{R}_3$.
(c) Find the coordinates of e_1, e_2, e_3 with respect to the basis v_1, v_2, v_3 of $\mathscr{R}_3$.
5. Let $v_1, v_2, \ldots, v_n$ be a basis for $\mathscr{R}_n$. Let A be the matrix

$$A = [v_1, v_2, \ldots, v_n] \in \mathscr{R}_{nn}.$$

Let $b \in \mathscr{R}_n$. Show that the coordinates of b with respect to the basis $v_1, v_2, \ldots, v_n$ are the components of the vector $A^{-1}b$.
6. Find the coordinates of each of the functions $q_i(t)$, where $i = 0, 1, 2, 3$ in Exercise 3 with respect to the basis p_0, p_1, p_2, p_3 of $P^3(R)$ in Exercise 2.
7. Find the coordinates of each of the functions $p_i(t)$, where $i = 0, 1, 2, 3$ in Exercise 2 with respect to the basis q_0, q_1, q_2, q_3 of $P^3(R)$ in Exercise 3.
8. Consider the vector space $\mathscr{R}_{mn}$ of all $m \times n$ matrices with real entries. Let E_{ij} be the matrix in $\mathscr{R}_{mn}$, all of whose entries are zero except for the entry in the ith row and jth column, which is one. Show that the mn matrices E_{ij} ($i = 1 \ldots, m$; $j = 1, \ldots, n$) $\in \mathscr{R}_{mn}$ form a basis of $\mathscr{R}_{mn}$.

We have shown in Example 6, Section 5.4, that a set of k vectors in $\mathscr{R}_n$, where $k > n$, is linearly dependent. We have also shown in Example 1 that $\mathscr{R}_n$ has a basis, namely, $e_1, \ldots, e_n$ consisting of exactly n vectors. A similar result holds in every vector space. In order to prove this result, we first establish the following lemma.

Lemma 1. *Let $v_1, \ldots, v_n$ be a basis for a vector space V over $\mathscr{F}$. Let* u

132 *Vector Spaces*

be a nonzero vector in V. Then for some i, $1 \leq i \leq n$, the vectors $\mathbf{v}_1, \mathbf{v}_2, \ldots, \mathbf{v}_{i-1}, \mathbf{v}_{i+1}, \ldots, \mathbf{v}_n, \mathbf{u}$ form a basis of V. (In other words, some one of the original basis vectors, for instance $\mathbf{v}_i$, can be replaced by the vector $\mathbf{u}$, and the resulting set of vectors is still a basis.)

Proof. Since $\mathbf{v}_1, \ldots, \mathbf{v}_n$ span V, there exist scalars $\alpha_1, \ldots, \alpha_n \in \mathscr{F}$, such that

$$\mathbf{u} = \alpha_1 \mathbf{v}_1 + \cdots + \alpha_n \mathbf{v}_n.$$

Since $\mathbf{u} \neq \mathbf{0}$, not all of the α_j are zero. Let $i \geq 1$ be the smallest integer such that $\alpha_i \neq 0$ so

$$\mathbf{u} = \sum_{j=i}^{n} \alpha_j \mathbf{v}_j.$$

Then from this relation, we have

$$\mathbf{v}_i = \frac{1}{\alpha_i}\left(\mathbf{u} - \sum_{j=i+1}^{n} \alpha_j \mathbf{v}_j\right).$$

We now show that the set of vectors $\mathbf{v}_1, \ldots, \mathbf{v}_{i-1}, \mathbf{v}_{i+1}, \ldots, \mathbf{v}_n, \mathbf{u}$ spans V. For, let $\mathbf{w} \in V$. Then for suitable scalars $\beta_1, \ldots, \beta_n \in \mathscr{F}$, we have

$$\mathbf{w} = \sum_{j=1}^{n} \beta_j \mathbf{v}_j = \sum_{j=1}^{i-1} \beta_j \mathbf{v}_j + \beta_i \mathbf{v}_i + \sum_{j=i+1}^{n} \beta_j \mathbf{v}_j$$

$$= \sum_{j=1}^{i-1} \beta_j \mathbf{v}_j + \frac{\beta_i}{\alpha_i}\left(\mathbf{u} - \sum_{j=i+1}^{n} \alpha_j \mathbf{v}_j\right) + \sum_{j=i+1}^{n} \beta_j \mathbf{v}_j$$

$$= \sum_{j=1}^{i-1} \beta_j \mathbf{v}_j + \sum_{j=i+1}^{n} \left(\beta_j - \frac{\beta_i}{\alpha_i}\alpha_j\right)\mathbf{v}_j + \frac{\beta_i}{\alpha_i}\mathbf{u},$$

which shows that this set spans V. To prove that the set $\mathbf{v}_1, \ldots, \mathbf{v}_{i-1}, \mathbf{v}_{i+1}, \ldots, \mathbf{v}_n, \mathbf{u}$ is linearly independent, suppose there exist scalars $\gamma_1, \ldots, \gamma_{i-1}, \gamma_{i+1}, \ldots, \gamma_n, \gamma_0 \in \mathscr{F}$ such that

$$\gamma_1 \mathbf{v}_1 + \cdots + \gamma_{i-1}\mathbf{v}_{i-1} + \gamma_{i+1}\mathbf{v}_{i+1} + \cdots + \gamma_n \mathbf{v}_n + \gamma_0 \mathbf{u} = \mathbf{0}.$$

Using $\mathbf{u} = \sum_{j=i}^{n} \alpha_j \mathbf{v}_j$, we have

$$\mathbf{0} = \sum_{j=1}^{i-1} \gamma_j \mathbf{v}_j + \sum_{j=i+1}^{n} \gamma_j \mathbf{v}_j + \gamma_0\left(\sum_{j=i}^{n} \alpha_j \mathbf{v}_j\right)$$

$$= \sum_{j=1}^{i-1} \gamma_j \mathbf{v}_j + \gamma_0 \alpha_i \mathbf{v}_i + \sum_{j=i+1}^{n} (\gamma_0 \alpha_j + \gamma_j)\mathbf{v}_j.$$

Since $v_1, \ldots, v_n$ are linearly independent, each coefficient in this equation is zero. Thus, $\gamma_j = 0$, where $j = 1, \ldots, i - 1$; $\gamma_0 \alpha_i = 0$, $\gamma_0 \alpha_j + \gamma_j = 0$, where $j = i + 1, \ldots, n$. Since $\alpha_i \neq 0$, we have $\gamma_0 = 0$, which in turn implies $\gamma_j = 0$, where $j = i + 1, \ldots, n$. We now have

$$\gamma_1 = \cdots = \gamma_{i-1} = \gamma_{i+1} = \cdots = \gamma_n = \gamma_0 = 0.$$

Thus, the set $v_1, \ldots, v_{i-1}, \ldots, v_n, u$ is linearly independent and forms a basis. ∎

The reader should note that not only have we proved the lemma, but we have also given an explicit way of choosing which basis vector v_i is to be replaced by u. It might be possible to choose the vector to be replaced in other ways, but in the theorem which follows, it will be essential that v_i is chosen as in the lemma; namely, if we write $u = \sum_{j=i}^{n} \alpha_j v_j$, then i is the smallest integer such that $\alpha_i \neq 0$, and v_i is the basis vector replaced by u. It will also be essential that the new basis be given as $v_1, \ldots, v_{i-1}, v_{i+1}, \ldots, v_n, u$, with u listed last.

Theorem 3. *Let $v_1, \ldots, v_n$ be a basis for a vector space V over $\mathscr{F}$. Let $u_1, \ldots, u_k$ be a linearly independent set of vectors in V. Then $k \leq n$.*

Proof. Suppose $k > n$. By Lemma 1 there is an i, $1 \leq i \leq n$, such that the vectors $v_1, \ldots, v_{i-1}, v_{i+1}, \ldots, v_n, u_1$ form a basis of V. We apply Lemma 1 to this new basis in order to replace one of these vectors by u_2. As in the proof of the lemma, we write

$$u_2 = \beta_1 v_1 + \cdots + \beta_{i-1} v_{i-1} + \cdots + \beta_n v_n + \beta_0 u_1.$$

Since u_1, u_2 are linearly independent, u_2 is not a multiple of u_1; therefore, $\beta_1, \ldots, \beta_{i-1}, \beta_{i+1}, \ldots, \beta_n$ are not all zero. By the construction given in the proof of Lemma 1, u_2 replaces one of $v_1, \ldots, v_{i-1}, v_{i+1}, \ldots, v_n$, and **not** u_1. We thus obtain a basis consisting of $(n - 2)$ v's and u_1, u_2. We now apply Lemma 1 to this new basis. Since u_3 is not a linear combination of u_1, u_2 (Theorem 2, Section 5.4), the vector replaced is a v and not u_1 or u_2, just as in the previous step, and we obtain a basis consisting of $(n - 3)$ v's and u_1, u_2, u_3. After n steps, it is clear that we have replaced each of the vectors $v_1, \ldots, v_n$, and we have a new basis $u_1, \ldots, u_n$. But this implies that u_{n+1} is a linear combination of $u_1, \ldots, u_n$; hence, again by Theorem 2, Section 5.4, the set $u_1, \ldots, u_k$ must be linearly independent. This contradicts the hypothesis, however, and thus $k \leq n$. ∎

We remark that an equivalent form of Theorem 3 is the following.

Corollary 1. *Let* $v_1, \ldots, v_n$ *be a basis for a vector space V over $\mathscr{F}$. Then every set of k vectors, where $k > n$, is linearly dependent.*

The methods used in the proofs of Lemma 1 and Theorem 3 give the following useful result.

Theorem 4 (Theorem of Exchange). *Let V be a vector space over $\mathscr{F}$. Let $u_1, \ldots, u_k$ be a linearly independent set of vectors in V and let $v_1, \ldots, v_m$ be a set of vectors that spans V. Then there is a basis of V consisting of $u_1, \ldots, u_k$ together with some* (possibly none or possibly all) *of the v's.*

● **EXERCISE**

9. Prove Theorem 4.

A very important consequence of Theorem 3 is the fact that a vector space cannot have two bases containing a different number of vectors.

Theorem 5. *Let V be a vector space over $\mathscr{F}$ having a basis $v_1, \ldots, v_n$. Then every basis of V consists of n vectors.*

Proof. Let $u_1, \ldots, u_k$ be a basis of V. Since $v_1, \ldots, v_n$ is a basis and $u_1, \ldots, u_k$ are linearly independent, it follows by Theorem 3 that $k \leq n$. On the other hand, since $u_1, \ldots, u_k$ is a basis and $v_1, \ldots, v_n$ are linearly independent, we also have $n \leq k$. Thus, $k = n$. ∎

In view of Theorem 5, it is natural to make the following definitions.

Definition 2. *Let V be a vector space over $\mathscr{F}$. Then the dimension of V is n* (written $\dim V = n$) *if and only if V has a basis consisting of n vectors.* (By convention the vector space consisting of the zero vector alone has dimension zero.)

By Theorem 5, the dimension of V does not depend on the particular basis chosen.

Definition 3. *Let V be a vector space over $\mathscr{F}$. Then V is said to be finite dimensional if and only if there is an integer n such that $\dim V = n$. Otherwise, V is said to be infinite dimensional.*

We remark that if V is a vector space of dimension n over $\mathscr{F}$ and if W is a subspace of V, then W is finite dimensional and dim $W \le$ dim V. For, if W contains a linearly independent set of more than n vectors, then so does V, and this contradicts dim $V = n$.

By Example 1, the vector space $\mathscr{R}_n$ has dimension n. **However, not all vector spaces are finite dimensional.**

Example 4. Consider the vector space $C(\mathscr{R})$ of all continuous functions on $-\infty < t < \infty$. We have seen in Example 9, Section 5.4, that the functions $1, t, t^2, t^3, t^4, t^5$ are linearly independent in $C(\mathscr{R})$. By exactly the same argument (see Exercise 15, Section 5.4), we can show that for every positive integer k the functions $1, t, \ldots, t^k$ are linearly independent in $C(\mathscr{R})$. Suppose that $C(\mathscr{R})$ is finite dimensional, for instance, dim $C(\mathscr{R}) = n$. Then by Theorem 3, $n \ge k + 1$ for every positive integer k. This is impossible; thus $C(\mathscr{R})$ is infinite dimensional.

• **EXERCISES**

10. Compute the dimension of the vector space $\mathscr{R}_{mn}$ of all $m \times n$ real matrices. [*Hint:* See Exercise 8.]

11. Compute the dimension of the vector space $P^3(\mathscr{R})$ of polynomials with real coefficients of degree less than or equal to three. [*Hint:* See Example 3.]

12. Find the dimension of the subspace of $\mathscr{R}_4$ consisting of all solutions of the linear system of algebraic equations $A\mathbf{x} = \mathbf{0}$, where A is the matrix

$$A = \begin{bmatrix} 0 & 0 & 3 & -1 \\ 0 & -1 & 4 & 7 \\ 0 & -1 & 7 & 6 \end{bmatrix} \in \mathscr{R}_{34}.$$

13. Consider the vector space $\mathscr{S}$ of all infinite sequences of real numbers $\mathbf{s} = (s_1, s_2, \ldots)$. Addition and scalar multiplication are defined by the relations $\mathbf{s} + \mathbf{t} = (s_1 + t_1, s_2 + t_2, \ldots)$, where $\mathbf{t} = (t_1, t_2, \ldots)$, and $\alpha \mathbf{s} = (\alpha s_1, \alpha s_2, \ldots)$. The zero vector is defined by $\mathbf{0} = (0, 0, \ldots)$.
  (a) Show that $\mathscr{S}$ is a vector space over $\mathscr{R}$.
  (b) Show that $\mathscr{S}$ is infinite dimensional.
  (c) Let $\mathscr{S}_n$ be the set of vectors in $\mathscr{S}$ consisting of all sequences $\mathbf{s} = (s_1, s_2, \ldots)$ such that $s_k = 0$ for $k > n$. Show that $\mathscr{S}_n$ is a subspace of $\mathscr{S}$ and that dim $\mathscr{S}_n = n$.
  (d) Let $\mathscr{S}^*$ be the set of vectors in $\mathscr{S}$ consisting of all sequences $\mathbf{s} = (s_1, s_2, \ldots)$ with only a finite number of nonzero terms s_i. Show that $\mathscr{S}^*$ is a subspace of $\mathscr{S}$. Is $\mathscr{S}^*$ finite dimensional?
  (e) Let $\mathscr{S}_{50}$ be the set of vectors in $\mathscr{S}$ consisting of all sequences $\mathbf{s} = (s_1, s_2, \ldots)$ such that $s_1 + s_2 + \cdots + s_{50} = 0$. Show that $\mathscr{S}_{50}$ is a subspace of $\mathscr{S}$. Is $\mathscr{S}_{50}$ finite dimensional?

14. Consider the vector space $C(\mathscr{R})$ of continuous functions on the interval $-1 \leq t \leq 1$. Let W be the subset of $C(\mathscr{R})$ consisting of all functions f such that $\int_{-1}^{1} f(t)\,dt = 0$. Show that W is a subspace of $C(\mathscr{R})$. Is W finite dimensional? [*Hint:* Consider the functions t^k, where k is an odd integer.]

15. Show that each of the following vector spaces is infinite dimensional: $C'\mathscr{R}$, $C^k\mathscr{R}$, $C(\mathscr{R}_n)$, $C'(\mathscr{R}_n)$, $P(\mathscr{R})$. [*Hint:* See Example 4.]

5.6 Linear Systems of Algebraic Equations Revisited

Let $A \in \mathscr{F}_{mn}$. Let W be the set of all solutions $\mathbf{w}$ of the linear system of algebraic equations $A\mathbf{x} = \mathbf{0}$. We have seen (Example 2, Section 5.3) that W is a subspace of $\mathscr{F}_{n1}$. We now wish to show that the dimension of this subspace W of $\mathscr{F}_{n1}$ is $n - r$, where r is the row rank of A.

The temporary definition of row rank (Definition 3, Section 3.5) is not convenient for this purpose. Further, in Chapter 3, we assumed without proof that the row echelon form of a matrix is unique, and the temporary definition is given in terms of the row echelon form. We now give a new definition of row rank which, while equivalent to the temporary definition, is independent of the row echelon form and more convenient for the computation of the dimension of W.

Definition 1. Let $A \in \mathscr{F}_{mn}$. The row space of A, denoted by $R(A)$, is the subspace of $\mathscr{F}_{1n}$ spanned by the rows of A (that is, it is the set of all linear combinations of the rows $a_{1*}, a_{2*}, \ldots, a_{m*}$ of A, considered as vectors in $\mathscr{F}_{1n}$).

The reader will note that the row space of A is a subspace of $\mathscr{F}_{1n}$ by Theorem 1, Section 5.4.

Definition 2. Let $A \in \mathscr{F}_{mn}$. The row rank of A is the dimension of the row space of A.

By the remark immediately following Definition 3, Section 5.5, the row rank of A is no greater than the dimension of $\mathscr{F}_{n1}$, which is n.

Lemma 1. Let $A \in \mathscr{F}_{mn}$ and let $P \in \mathscr{F}_{mm}$. Then
(i) $R(PA) \subseteq R(A)$.
(ii) $R(PA) = R(A)$ if P is nonsingular.

Proof. By the definition of matrix multiplication, the rows of PA are linear combinations of the rows of A. Hence, $R(PA) \subseteq R(A)$, which is (i). If P is nonsingular, it follows from (i) that $R(P^{-1}PA) \subseteq R(PA)$. Thus, $R(A) \subseteq R(PA)$, whence $R(PA) = R(A)$. ∎

5.6 Linear Systems of Algebraic Equations Revisited

Corollary 1. Let $A, B \in \mathscr{F}_{mn}$. If $B \stackrel{R}{\sim} A$, then $R(B) = R(A)$.

Proof. If $B \stackrel{R}{\sim} A$, there is a nonsingular matrix $P \in \mathscr{F}_{mm}$ such that $B = PA$ (Exercise 2, Section 3.4). The result now follows from Lemma 1. ∎

Corollary 2. Let $A, B \in \mathscr{F}_{mn}$. If $B \stackrel{R}{\sim} A$, then row rank B = row rank A. The proof follows immediately from Corollary 1 because $R(B) = R(A)$. ∎

Theorem 1. Let $A \in \mathscr{F}_{mn}$ and let A_R be the row echelon form of A, having r nonzero rows. Then row rank $A = r$.

Proof. By Corollary 2 of Lemma 1, row rank A = row rank A_R. Let

$$B = A_R = \begin{bmatrix} b_{1*} \\ b_{2*} \\ \vdots \\ b_{m*} \end{bmatrix}.$$

The nonzero rows $b_{1*}, \ldots, b_{r*}$ of B span $R(B)$. To show that they form a basis of $R(B)$, we must prove that they are linearly independent. Suppose there are constants $\alpha_1, \ldots, \alpha_r \in \mathscr{F}$ such that

$$\alpha_1 b_{1*} + \alpha_2 b_{2*} + \cdots + \alpha_r b_{r*} = \mathbf{0}.$$

Each row $b_{1*}, b_{2*}, \ldots, b_{r*}$ has a leading entry. Let $b_{i,k(i)}$ be the leading entry of b_{i*}, where $i = 1, \ldots, r$. By the definition of the row echelon form (Definition 2, Section 35.), $b_{i,k(i)} = 1$, for $i = 1, \ldots, r$, and $b_{j,k(i)} = 0$, for $j \neq i$, $i = 1, \ldots, r$ (that is, a column containing a leading entry has all other entries zero). Therefore, the $k(i)$th component of $\alpha_1 b_{1*} + \cdots + \alpha_r b_{r*}$ is

$$0 = \alpha_1 b_{1,k(i)} + \cdots + \alpha_i b_{i,k(i)} + \cdots + \alpha_r b_{r,k(i)} = \alpha_i.$$

Hence, $\alpha_i = 0$, for $i = 1, \ldots, r$, and the vectors $b_{1*}, \ldots, b_{r*}$ are linearly independent. Thus, the vectors $b_{1*}, b_{2*}, \ldots, b_{r*}$ form a basis for $R(B)$ and $\dim R(B) = r$. ∎

Theorem 1 says that the new definition of row rank is equivalent to the temporary definition given in Section 3.5.

We now return to the linear system $A\mathbf{x} = \mathbf{0}$, and we restate Theorem 1, Section 3.6, in the present context. The reader is advised to recall Section 3.6 before proceeding.

Theorem 2. Let $A \in \mathscr{F}_{mn}$ have row rank r. Then the solution set W of the system $A\mathbf{x} = \mathbf{0}$ is a subspace of $\mathscr{F}_{n1}$ of dimension $n - r$.

Proof. In Theorem 1, Section 3.6, we have shown that every solution $\mathbf{u}$ of $A\mathbf{x} = \mathbf{0}$ is of the form

$$\mathbf{u} = c_1 \mathbf{u}_1 + c_2 \mathbf{u}_2 + \cdots + c_{n-r} \mathbf{u}_{n-r},$$

where $c_1, c_2, \ldots, c_{n-r}$ are suitably chosen constants in $\mathscr{F}$, and where $\mathbf{u}_j$ is a vector in $\mathscr{F}_{n1}$ such that the jth component of Class 1 is one and the remaining components of Class 1 are zero, with $j = 1, \ldots, n - r$. Hence, the solutions $\mathbf{u}_1, \ldots, \mathbf{u}_{n-r}$ span W, and W is a subspace of $\mathscr{F}_{n1}$ by Theorem 1, Section 5.4. Let u_{kj} denote the kth component of $\mathbf{u}_j$, where $k = 1, \ldots, n; j = 1, \ldots, n - r$. Suppose the indices of the components of Class 1 are $\alpha(1), \alpha(2), \ldots, \alpha(n - r)$. Then, for $j = 1, 2, \ldots, n - r$, we have

$$u_{\alpha(j), j} = 1, \qquad u_{\alpha(i), j} = 0, \qquad \text{if } i \neq j;$$

the remaining r components of $\mathbf{u}_j$ of Class 2 do not enter in the following argument. We wish to show that the vectors $\mathbf{u}_1, \mathbf{u}_2, \ldots, \mathbf{u}_{n-r}$ are linearly independent. Suppose there exist constants $\beta_1, \beta_2, \ldots, \beta_{n-r}$ in $\mathscr{F}$ such that

$$\beta_1 \mathbf{u}_1 + \beta_2 \mathbf{u}_2 + \cdots + \beta_{n-r} \mathbf{u}_{n-r} = \mathbf{0}.$$

The $\alpha(j)$th component of this linear combination is

$$0 = \beta_1 u_{\alpha(j), 1} + \beta_2 u_{\alpha(j), 2} + \cdots + \beta_j u_{\alpha(j), j} + \cdots + \beta_{n-r} u_{\alpha(j), n-r} = \beta_j$$
$$(j = 1, 2, \ldots, n - r).$$

This establishes the linear independence and thus the solutions $\mathbf{u}_1, \mathbf{u}_2, \ldots, \mathbf{u}_{n-r}$ form a basis of W. Thus, the dimension of W is $n - r$. ∎

To conclude this section we consider the following question. Let $A \in \mathscr{F}_{mn}$; let the row rank of A be r. For what vector $\mathbf{b} \in \mathscr{F}_{m1}$ is the system $A\mathbf{x} = \mathbf{b}$ consistent? Let $\mathscr{B}$ be the set of vectors $\mathbf{b} \in \mathscr{F}_{m1}$ such that the system $A\mathbf{x} = \mathbf{b}$ is consistent. It is easy to verify that $\mathscr{B}$ is a subspace of $\mathscr{F}_{m1}$.

● **EXERCISE**

1. Show that $\mathscr{B}$ is a subspace of $\mathscr{F}_{m1}$.

We wish to determine the dimension of $\mathscr{B}$.

5.6 Linear Systems of Algebraic Equations Revisited

Theorem 3. *Let $A \in \mathscr{F}_{m,n}$; let the row rank A be r. Then the set $\mathscr{B}$ of vectors $\mathbf{b} \in \mathscr{F}_{m1}$ such that the system $A\mathbf{x} = \mathbf{b}$ is consistent is a subspace of $\mathscr{F}_{m1}$ of dimension r.*

Proof. Let A_R be the REF of A. Then there exists a nonsingular matrix P such that $A_R = PA$. Let $\mathbf{b}_R = P\mathbf{b}$. Then the system $A\mathbf{x} = \mathbf{b}$ is equivalent to the system $A_R \mathbf{x} = \mathbf{b}_R$. By Theorem 1, Section 3.7, the latter system is consistent if and only if the vector $\mathbf{b}_R$ has the last $m - r$ components equal zero. Every such vector $\mathbf{b}_R$ is expressible as the linear combination

$$\mathbf{b}_R = c_1 \mathbf{e}_1 + c_2 \mathbf{e}_2 + \cdots + c_r \mathbf{e}_r,$$

where $\mathbf{e}_i$, $i = 1, \ldots, r$ are the unit vectors in $\mathscr{F}_{m1}$. Let $\mathbf{v}_i = P^{-1}\mathbf{e}_i$, $i = 1, \ldots, r$. Clearly, $\mathbf{v}_1, \mathbf{v}_2, \ldots, \mathbf{v}_r$ span $\mathscr{B}$.

● **EXERCISE**

2. Show that $\mathbf{v}_1, \mathbf{v}_2, \ldots, \mathbf{v}_r$ span $\mathscr{B}$

We claim that $\mathbf{v}_1, \mathbf{v}_2, \ldots, \mathbf{v}_r$ are linearly independent. Suppose that for some constants $c_1, \ldots, c_r \in \mathscr{F}$,

$$c_1 \mathbf{v}_1 + c_2 \mathbf{v}_2 + \cdots + c_r \mathbf{v}_r = \mathbf{0}.$$

Then

$$\begin{aligned} \mathbf{0} &= c_1 \mathbf{v}_1 + \cdots + c_r \mathbf{v}_r \\ &= c_1 P^{-1} \mathbf{e}_1 + \cdots + c_r P^{-1} \mathbf{e}_r \\ &= P^{-1}(c_1 \mathbf{e}_1 + \cdots + c_r \mathbf{e}_r). \end{aligned}$$

Since P^{-1} is nonsingular, $c_1 \mathbf{e}_1 + \cdots + c_r \mathbf{e}_r = \mathbf{0}$. But, since $\mathbf{e}_1, \ldots, \mathbf{e}_r$ are linearly independent in $\mathscr{F}_{m1}$, we have $c_1 = c_2 = \cdots = c_r = 0$. Therefore, $\mathbf{v}_1, \ldots, \mathbf{v}_r$ are linearly independent, and form a basis of $\mathscr{B}$. Thus, $\dim \mathscr{B} = r$. ∎

By analogy with the definition of the row space $R(A)$ of a matrix $A \in \mathscr{F}_{mn}$, we may define the column space $C(A)$ of a matrix A to be the subspace of $\mathscr{F}_{m1}$ spanned by the columns of A. We may likewise define the column rank of A to be the dimension of $C(A)$. Since a vector $\mathbf{b} \in \mathscr{F}_{m1}$ is a linear combination of the columns of A if and only if $\mathbf{b} = A\boldsymbol{\alpha}$ for some vector

$\alpha \in \mathscr{F}_{n1}$, the column space $C(A)$ is precisely the subspace $\mathscr{B}$ defined in Theorem 3. Thus, Theorem 3 says that the row rank of a matrix is equal to the column rank. For this reason we may speak of the rank of a matrix without ambiguity.

• **EXERCISES**

    3. Find the dimension of the solution space, and a basis for the solution space, of each of the systems $A\mathbf{x} = \mathbf{0}$ in Exercises 2 and 3, Section 3.6.

    4. For each of the matrices A in Exercises 2 and 3, Section 3.6, find the dimension of the subspace $\mathscr{B}$ of Theorem 3. Also, find a basis for $\mathscr{B}$. [*Hint:* Examine the proof of Theorem 3.]

5.7 Linear Transformations

The concept of a linear transformation of one vector space to another vector space is of great importance in the study of both linear algebra and linear differential equations. Let V and W be sets. By a function f from V into W we mean a correspondence that assigns to each $v \in V$ a unique element $w \in W$ denoted by $w = f(v)$. We shall now specialize the sets V and W to be vector spaces, and we shall study a special class of functions.

Definition 1. *Let V and W be two vector spaces over $\mathscr{F}$. A linear transformation T from V into W is a function T from V into W which is linear; that is,*
    (a) $T(\mathbf{v}_1 + \mathbf{v}_2) = T(\mathbf{v}_1) + T(\mathbf{v}_2)$
    (b) $T(\alpha \mathbf{v}) = \alpha T(\mathbf{v})$
for every $\alpha \in \mathscr{F}$ and all $\mathbf{v}_1, \mathbf{v}_2, \mathbf{v} \in V$.

We remark that conditions (a) and (b) are equivalent to the single condition

$$T(\alpha_1 \mathbf{v}_1 + \alpha_2 \mathbf{v}_2) = \alpha_1 T(\mathbf{v}_1) + \alpha_2 T(\mathbf{v}_2)$$

for all $\alpha_1, \alpha_2 \in \mathscr{F}$ and all $\mathbf{v}_1, \mathbf{v}_2 \in V$.

Example 1. Let $V = \mathscr{R}_2$, $W = \mathscr{R}_3$, and let

$$A = \begin{bmatrix} 1 & -2 \\ 0 & 3 \\ 4 & 7 \end{bmatrix} \in \mathscr{R}_{32}.$$

Then the transformation T defined by $T(\mathbf{v}) = A\mathbf{v}$ is a linear transformation from $\mathscr{R}_2$ into $\mathscr{R}_3$. To verify this, we note that the properties of matrix multiplication give

$$T(\mathbf{u} + \mathbf{v}) = A(\mathbf{u} + \mathbf{v}) = A\mathbf{u} + A\mathbf{v} = T(\mathbf{u}) + T(\mathbf{v}),$$
$$T(\alpha \mathbf{v}) = A(\alpha \mathbf{v}) = \alpha A\mathbf{v} = \alpha T(\mathbf{v}).$$

Example 1 is a special case of the following general situation.

Example 2. Let $V = \mathscr{F}_{n1}$, $W = \mathscr{F}_{m1}$, and let A be a given matrix in $\mathscr{F}_{mn}$. (The case $m = n$ is not excluded.) Define the function T by the relation $T(\mathbf{v}) = A\mathbf{v}$ for every $\mathbf{v} \in V$. Then T is a linear transformation from $\mathscr{F}_{n1}$ into $\mathscr{F}_{m1}$. To verify this, we proceed exactly as in Example 1, using the properties of matrix multiplication, which do not make use of the particular entries of the matrix A.

● **EXERCISES**

1. Show that the transformation T of Example 2 is linear.
2. Consider the function f from $\mathscr{R}_1$ into $\mathscr{R}_1$ defined by $f(t) = t^2$ for all $t \in R_1$. (Here, $V = W = \mathscr{R}_1$.) Show that f is *not* a linear transformation.
3. Consider the function g from R_1 into $\mathscr{R}_1$ defined by $g(t) = 2 + 3t$ for all $t \in \mathscr{R}$. Is g a linear transformation from $\mathscr{R}_1$ into $\mathscr{R}_1$? Prove your answer.
4. Consider the function h from $\mathscr{R}_1$ into $\mathscr{R}_1$ defined by $h(t) = a + bt$ for all $t \in \mathscr{R}_1$. For what values of a and b, if any, is h a linear transformation from $\mathscr{R}_1$ into $\mathscr{R}_1$?

Example 3. Let $\mathscr{I}$ be a fixed interval on the real line. Let V be the vector space $C'(\mathscr{R})$ of real-valued continuously differentiable functions on $\mathscr{I}$, and let W be the vector space $C(\mathscr{R})$ of real-valued continuous functions on $\mathscr{I}$. Define the function T from V into W by $T(f) = f'$, where f' is the derivative of f, that is, the function whose value at t is $f'(t)$. Then T is a linear transformation from V into W. For, by elementary properties of derivatives,

$$T(f + g) = (f + g)' = f' + g' = T(f) + T(g),$$

and

$$T(\alpha f) = (\alpha f)' = \alpha f' = \alpha T(f)$$

for every $\alpha \in \mathscr{R}$ and all $f, g \in C'(\mathscr{R})$.

Example 4. Let $\mathscr{I}$ be a fixed interval $a \leq t \leq b$ on the real line. Let V be the vector space $C(\mathscr{R})$ of real-valued continuous functions on $\mathscr{I}$, and let W be the same vector space $C(\mathscr{R})$. Define the function T from V into W

by $T(f) = g$, where $g(t) = \int_a^t f(s)\, ds$ for $a \le t \le b$. Then T is a linear transformation from V into W. By elementary properties of definite integrals, if $T(f_1) = g_1$ and $T(f_2) = g_2$, then

$$T(f_1 + f_2) = h,$$

where

$$h(t) = \int_a^t [f_1(s) + f_2(s)]\, ds = \int_a^t f_1(s)\, ds + \int_a^t f_2(s)\, ds = g_1(t) + g_2(t),$$

so that $T(f_1 + f_2) = T(f_1) + T(f_2)$. Also, if $T(\alpha f) = g$, then $g(t) = \int_a^t \alpha f(s)\, ds = \alpha \int_a^t f(s)\, ds$, so that $T(\alpha f) = \alpha T(f)$.

• **EXERCISE**

5. Let $\mathscr{I}$ be a fixed interval $a \le t < b$ on the real line. In each of the following, is T a linear transformation from the given vector space V into the given vector space W?
   (a) $V = C^2(\mathscr{R})$, $W = C(\mathscr{R})$, $T(f) = 3f'' - 5f' + 2f$.
   (b) $V = C'(\mathscr{R})$, $W = C(\mathscr{R})$, $T(f) = f' + p(t)f$, where $p(t)$ is a given continuous function on $\mathscr{I}$.
   (c) $V = C^2(\mathscr{R})$, $W = C(\mathscr{R})$, $T(f) = f'' + f + 1$.
   (d) $V = C(\mathscr{R})$, $W = C(\mathscr{R})$, $T(f) = g$, where $g(t) = \int_a^t sf(s)\, ds + t\int_a^t f(s)\, ds$.
   (e) $V = C(\mathscr{R})$, $W = C(\mathscr{R})$, $T(f) = g$, where $g(t) = \int_a^t K(t-s)f(s)\, ds$.
   Here, $K(u)$ is a given function continuous for $0 \le u \le b - a$.
   (f) $V = C(\mathscr{R})$, $W = C(\mathscr{R})$, $T(f) = g$, where $g(t) = \int_a^t f(s)\, ds + t$.
   (g) $V = C(\mathscr{R})$, $W = \mathscr{R}$, $T(f) = \int_a^b f(s)\, ds$.

We have seen in Example 2 that a matrix $A \in \mathscr{F}_{mn}$ defines a linear transformation from $\mathscr{F}_{n1}$ into $\mathscr{F}_{m1}$. We shall now show how to associate a matrix in $\mathscr{F}_{mn}$ with any given linear transformation from $\mathscr{F}_{n1}$ into $\mathscr{F}_{m1}$. As the correspondence depends only on the choices of bases for $\mathscr{F}_{n1}$ and $\mathscr{F}_{m1}$, and not on any specific properties of these vector spaces, we shall, in fact, show how to associate a matrix $A \in \mathscr{F}_{mn}$ with a given linear transformation of an n-dimensional vector space V over $\mathscr{F}$ into an m-dimensional vector space W over $\mathscr{F}$ with respect to specific bases for V and W.

Definition 2. *Let V be an n-dimensional vector space over $\mathscr{F}$, let W be an m-dimensional vector space over $\mathscr{F}$, and let T be a linear transformation from V into W. Let $\mathbf{v}_1, \ldots, \mathbf{v}_n$ be a basis of V and let $\mathbf{w}_1, \ldots, \mathbf{w}_m$ be a basis of W. Then the matrix $A \in \mathscr{F}_{mn}$ of the linear transformation T with respect to the* bases $\{v_1, \ldots, v_n\}$ of V and $\{w_1, \ldots, w_m\}$ of W is given by

$$T(v_j) = \sum_{i=1}^{m} a_{ij} w_i \quad (j = 1, \ldots, n). \tag{5.7}$$

The reader should observe that for each j, $T(v_j)$ is an element of W and, therefore, has a unique expansion (5.7) in terms of the basis $w_1, \ldots, w_m$ of W. Thus, the elements a_{ij}, where $i = 1, \ldots, m$; $j = 1, \ldots, n$, of A are well determined by the linear transformation T and the particular bases $\{v_1, \ldots, v_n\}$ and $\{w_1, \ldots, w_m\}$ chosen. Note also that the order in which the basis elements are written is important.

Example 5. Let T be the linear transformation from $\mathscr{R}_2$ into $\mathscr{R}_2$ given by $T(x) = Bx$ for all $x \in \mathscr{R}_2$, where

$$B = \begin{bmatrix} 1 & 1 \\ -1 & -1 \end{bmatrix} \in \mathscr{R}_{22}.$$

Note that here V and W are the same vector space. Let us choose the standard basis $\{e_1, e_2\}$ for both V and W.

(a) Calculate the matrix of T with respect to the bases $\{e_1, e_2\}$ of V and $\{e_1, e_2\}$ of W. We have

$$Te_1 = \begin{bmatrix} 1 & 1 \\ -1 & -1 \end{bmatrix} \begin{bmatrix} 1 \\ 0 \end{bmatrix} = \begin{bmatrix} 1 \\ -1 \end{bmatrix} = e_1 - e_2,$$

$$Te_2 = \begin{bmatrix} 1 & 1 \\ -1 & -1 \end{bmatrix} \begin{bmatrix} 0 \\ 1 \end{bmatrix} = \begin{bmatrix} 1 \\ -1 \end{bmatrix} = e_1 - e_2.$$

By comparison with (5.7), $(m = n = 2)$, $a_{11} = 1, a_{21} = -1, a_{12} = 1, a_{22} = -1$. Hence, the matrix of T is the given matrix B.

(b) Now let the basis for V be

$$v_1 = \begin{bmatrix} 1 \\ 1 \end{bmatrix}, \quad v_2 = \begin{bmatrix} 1 \\ -1 \end{bmatrix},$$

and let the basis for W be e_1, e_2, as before. Calculate the matrix of T with respect to these bases. Since

$$T(v_1) = Bv_1 = \begin{bmatrix} 1 & 1 \\ -1 & -1 \end{bmatrix} \begin{bmatrix} 1 \\ 1 \end{bmatrix} = \begin{bmatrix} 2 \\ -2 \end{bmatrix} = 2e_1 - 2e_2$$

$$T(v_2) = Bv_2 = \begin{bmatrix} 1 & 1 \\ -1 & -1 \end{bmatrix} \begin{bmatrix} 1 \\ -1 \end{bmatrix} = \begin{bmatrix} 0 \\ 0 \end{bmatrix} = 0e_1 + 0e_2,$$

144   Vector Spaces

comparison with (5.7) again shows that the matrix of T with respect to the bases $\{v_1, v_2\}$ of V and $\{e_1, e_2\}$ of W is

$$\begin{bmatrix} 2 & 0 \\ -2 & 0 \end{bmatrix}.$$

● EXERCISE

6. Calculate the matrix C of the linear transformation T of Example 5 with respect to the bases $\{v_1, v_2\}$ of V and $\{v_1, v_2\}$ of W.

The situation described in Example 5 is quite general. By the same approach we can prove the following result:

Theorem 1. *Let $V = \mathscr{R}_n$, $W = \mathscr{R}_m$ and let T be the linear transformation from V into W defined by $T(\mathbf{x}) = B\mathbf{x}$ for a given matrix $B \in \mathscr{R}_{mn}$. Then the matrix of T with respect to the bases $\{e_1, \ldots, e_n\}$ of V and $\{e_1, \ldots, e_m\}$ of W is B.*

● EXERCISE

7. Prove Theorem 1.

Example 6. Let $V = W = P^2(\mathscr{R})$, the vector space of polynomials with real coefficients of degree less than or equal to two. Choose the basis $\{1, t, t^2\}$ for both V and W. Let T be the linear transformation from V into W defined by $T(f) = f'$. Since

$$T(1) = 0 = 0 \cdot 1 + 0 \cdot t + 0 \cdot t^2, \qquad T(t) = 1 = 1 \cdot 1 + 0 \cdot t + 0 \cdot t^2,$$
$$T(t^2) = 2t = 0 \cdot 1 + 2 \cdot t + 0 \cdot t^2,$$

the matrix of T with respect to the given bases is

$$\begin{bmatrix} 0 & 1 & 0 \\ 0 & 0 & 2 \\ 0 & 0 & 0 \end{bmatrix}.$$

● EXERCISE

8. Let $V = P^2(\mathscr{R})$, $W = P^3(\mathscr{R})$, and let T be the linear transformation from V into W defined by

$$T(\alpha_0 + \alpha_1 t + \alpha_2 t^2) = \alpha_0 t + \alpha_1 \frac{t^2}{2} + \alpha_2 \frac{t^3}{3}.$$

Find the matrix of T with respect to the bases $\{1, t, t^2\}$ of V and $\{1, t, t^2, t^3\}$ of W.

We now specialize to the case where $V = W$ and the same basis is chosen for both vector spaces.

Definition 3. *Let V be an n-dimensional vector space over $\mathscr{F}$ and let T be a linear transformation of V into V. Let $\mathbf{v}_1, \ldots, \mathbf{v}_n$ be a basis of V. Then the matrix $A \in \mathscr{F}_{nn}$ of the linear transformation T with respect to the basis $\{\mathbf{v}_1, \ldots, \mathbf{v}_n\}$ of V is given by*

$$T(\mathbf{v}_j) = \sum_{i=1}^{n} a_{ij} \mathbf{v}_i \qquad (j = 1, \ldots, n). \tag{5.8}$$

We emphasize that the matrix A given by (5.8) depends on the choice of the basis $\{\mathbf{v}_1, \ldots, \mathbf{v}_n\}$. We shall examine the effect on the matrix of a different choice of basis for the **same** transformation T. In order to do this, we must first prove the following preliminary result.

Lemma 1. *Let $\{\mathbf{v}_1, \ldots, \mathbf{v}_n\}$ and $\{\mathbf{u}_1, \ldots, \mathbf{u}_n\}$ be two bases of an n-dimensional vector space V over $\mathscr{F}$. Then there is a unique nonsingular matrix $P \in \mathscr{F}_{nn}$ with $Q = P^{-1}$ such that*

$$\mathbf{u}_j = \sum_{i=1}^{n} p_{ij} \mathbf{v}_i \qquad (j = 1, \ldots, n), \tag{5.9}$$

$$\mathbf{v}_i = \sum_{k=1}^{n} q_{ki} \mathbf{u}_k \qquad (i = 1, \ldots, n). \tag{5.10}$$

Proof. Since $\{\mathbf{v}_1, \ldots, \mathbf{v}_n\}$ is a basis of V, each $\mathbf{u}_j$, where $j = 1, \ldots, n$, can be expressed uniquely in the form (5.9), and each $\mathbf{v}_i$, where $i = 1, \ldots, n$, can be expressed uniquely in the form (5.10). Thus, there exist unique matrices P and Q satisfying (5.9) and (5.10). We need only prove $QP = I$. To do this, we substitute (5.10) into (5.9) and obtain

$$\mathbf{u}_j = \sum_{i=1}^{n} p_{ij} \mathbf{v}_i = \sum_{i=1}^{n} p_{ij} \left(\sum_{k=1}^{n} q_{ki} \mathbf{u}_k \right)$$

$$= \sum_{k=1}^{n} \left(\sum_{i=1}^{n} q_{ki} p_{ij} \right) \mathbf{u}_k = \sum_{k=1}^{n} r_{kj} \mathbf{u}_k, \qquad (j = 1, \ldots, n),$$

where $R = QP$. Since $\{\mathbf{u}_1, \ldots, \mathbf{u}_n\}$ is a basis, the expression for $\mathbf{u}_j$ is unique. Thus, $r_{kj} = 1$ if $j = k$ and $r_{kj} = 0$ if $j \neq k$, and hence $QP = R = I$. ∎

146 Vector Spaces

Theorem 2. *Let V be an n-dimensional vector space over $\mathscr{F}$, and let T be a linear transformation of V into V. If $A \in \mathscr{F}_{nn}$ and $B \in \mathscr{F}_{nn}$ are the matrices of T with respect to two bases of V, then there exists a nonsingular $P \in \mathscr{F}_{nn}$ such that $B = P^{-1}AP$.*

Proof. Let $\{\mathbf{v}_1, \ldots, \mathbf{v}_n\}$ be a basis of V and let A be the matrix of T with respect to $\{\mathbf{v}_1, \ldots, \mathbf{v}_n\}$. Let $\{\mathbf{u}_1, \ldots, \mathbf{u}_n\}$ be a second basis of V and let B be the matrix of T with respect to $\{\mathbf{u}_1, \ldots, \mathbf{u}_n\}$. By (5.8), we have

$$T(\mathbf{v}_j) = \sum_{i=1}^{n} a_{ij} \mathbf{v}_i, \qquad (j = 1, \ldots, n) \tag{5.11}$$

$$T(\mathbf{u}_l) = \sum_{k=1}^{n} b_{kl} \mathbf{u}_k, \qquad (l = 1, \ldots, n). \tag{5.12}$$

By Lemma 1, there exists a nonsingular matrix P and $Q = P^{-1}$ such that (5.9) and (5.10) hold. Substituting (5.10) into (5.11), we have

$$T(\mathbf{v}_j) = \sum_{i=1}^{n} a_{ij} \mathbf{v}_i = \sum_{i=1}^{n} a_{ij} \left(\sum_{k=1}^{n} q_{ki} \mathbf{u}_k \right)$$

$$= \sum_{k=1}^{n} \left(\sum_{i=1}^{n} q_{ki} a_{ij} \right) \mathbf{u}_k = \sum_{k=1}^{n} r_{kj} \mathbf{u}_k \qquad (j = 1, \ldots, n),$$

where $R = QA = P^{-1}A$. On the other hand, using (5.10), the linearity of T, and (5.12), we have

$$T(\mathbf{v}_j) = T\left(\sum_{l=1}^{n} q_{lj} \mathbf{u}_l \right)$$

$$= \sum_{l=1}^{n} q_{lj} T(\mathbf{u}_l)$$

$$= \sum_{l=1}^{n} q_{lj} \left(\sum_{k=1}^{n} b_{kl} \mathbf{u}_k \right)$$

$$= \sum_{k=1}^{n} \left(\sum_{l=1}^{n} b_{kl} q_{lj} \right) \mathbf{u}_k = \sum_{k=1}^{n} s_{kj} \mathbf{u}_k \qquad (j = 1, \ldots, n),$$

where $S = BQ = BP^{-1}$. Since $\{\mathbf{u}_1, \ldots, \mathbf{u}_n\}$ is a basis of V, the expression for $T(\mathbf{v}_j)$ in terms of $\mathbf{u}_1, \ldots, \mathbf{u}_n$ is unique for $j = 1, \ldots, n$. Hence, $S = R$, and $BP^{-1} = P^{-1}A$, which gives $P^{-1}AP = B$. ∎

Theorem 2 leads to the following important definition.

Definition 4. *Let A and B be matrices in $\mathscr{F}_{nn}$. Then A is similar to B, written $A \sim B$, if there exists a nonsingular matrix $P \in \mathscr{F}_{nn}$ such that $B = P^{-1}AP$.*

● **EXERCISE**

9. Prove that similarity is an equivalence relation; that is, show that
   (a) $A \sim A$.
   (b) If $A \sim B$, then $B \sim A$.
   (c) If $A \sim B$ and $B \sim C$, then $A \sim C$.

In view of Exercise 9, instead of saying that A is similar to B, we may now say that A and B are similar. In the language of similarity, the content of Theorem 2 is that the matrices of a linear transformation of a finite-dimensional vector space into itself with respect to two bases are similar. Incidentally, the converse of Theorem 2 is also true. That is, similar matrices represent the same linear transformation with respect to different bases. This notion will reappear in Chapter 7.

● **EXERCISES**

10. Show by direct computation that the matrix B of Example 4 is similar to the matrix C of Exercise 6.

11. Are the matrices

$$\begin{bmatrix} 1 & 1 \\ 0 & 2 \end{bmatrix} \quad \text{and} \quad \begin{bmatrix} 1 & 0 \\ 0 & 2 \end{bmatrix}$$

similar?

● **MISCELLANEOUS EXERCISES**

1. (a) Is $\mathscr{C}$ a vector space over $\mathscr{R}$?
   (b) Is $\mathscr{R}$ a vector space over $\mathscr{C}$?
   (c) Show that $1 + i$ and $1 - i$ are linearly dependent in the vector space $\mathscr{C}$ over $\mathscr{C}$. (Contrast this with Exercise 5 of Section 5.4.)

2. Which of the following are subspaces of $C^5(\mathscr{R})$, for the interval $0 \le t \le 1$?
   (a) $\{f \mid f'' = 2f - 3\}$.
   (b) $\{f \mid f''' = f''\}$.
   (c) $\{f \mid f(t) = 0 \text{ for some } t \text{ between 0 and 1}\}$.
   (d) $\{f \mid f(\tfrac{1}{2}) = 0\}$.

3. If V is a vector space over $\mathscr{F}$ and $v_1, v_2, v_3 \in V$ such that the span of $\{v_1, v_2\}$ equals the span of $\{v_1, v_2, v_3\}$, prove that v_1, v_2, v_3 are linearly dependent.

4. In $P^2(\mathscr{R})$; describe the subspace spanned by $2t - 3t^2$ and $9t^2 - 6t$.

5. If V is a vector space of dimension n over $\mathscr{F}$, show that
   (a) any linearly independent set of n vectors is a basis;
   (b) any set of n vectors which span V is a basis.

6. Decide which of the following sets of functions are linearly dependent and which are independent in $C(\mathscr{R})$ for the interval $-\infty < t < \infty$.
   (a) $f_1(t) = e^t, f_2(t) = e^{t+1}$.
   (b) $f_1(t) = |t|^{1/2}\, f_2(t) = t$.
   (c) $f_1(t) = 1, f_2(t) = e^t, f_3(t) = e^{-t}$.
   (d) $f_1(t) = t^2, f_2(t) = t^2 \sin t$.
   (e) $f_1(t) = 1, f_2(t) = \begin{cases} 0 & \text{if } t \leq 0, \\ t & \text{if } t > 0 \end{cases} \quad f_3(t) = \begin{cases} 0 & \text{if } t \leq 0 \\ t^2 & \text{if } t > 0. \end{cases}$

7. (a) Show that

$$\begin{bmatrix} 1 \\ 2 \\ 1 \end{bmatrix}, \begin{bmatrix} 1 \\ 1 \\ 2 \end{bmatrix}, \begin{bmatrix} 2 \\ 1 \\ 1 \end{bmatrix}$$

form a basis for $\mathscr{R}_3$
   (b) Find a basis for $\mathscr{R}_3$ consisting of two vectors from part (a) and

$$\begin{bmatrix} -1 \\ 0 \\ -3 \end{bmatrix}.$$

8. Find the dimension of the subspace of $\mathscr{R}_{22}$ spanned by each of the following sets:

(a) $\begin{bmatrix} 1 & 0 \\ 0 & 1 \end{bmatrix}, \begin{bmatrix} 1 & 1 \\ 0 & 1 \end{bmatrix}, \begin{bmatrix} 2 & 1 \\ 0 & 2 \end{bmatrix}.$

(b) $\begin{bmatrix} 2 & 0 \\ 0 & 2 \end{bmatrix}, \begin{bmatrix} 2 & 0 \\ 0 & 0 \end{bmatrix}, \begin{bmatrix} 0 & 1 \\ 1 & 0 \end{bmatrix}, \begin{bmatrix} 1 & 1 \\ 1 & 1 \end{bmatrix}.$

(c) $\begin{bmatrix} -1 & 0 \\ 2 & 0 \end{bmatrix}, \begin{bmatrix} 0 & 0 \\ 0 & 0 \end{bmatrix}, \begin{bmatrix} 0 & 0 \\ 0 & 3 \end{bmatrix}, \begin{bmatrix} 5 & 4 \\ 0 & 0 \end{bmatrix}.$

9. Compute the row rank of each of the following matrices and describe the corresponding row space.

(a) $\begin{bmatrix} 1 & 2 & 2 \\ 2 & 0 & 0 \\ 3 & 5 & 5 \end{bmatrix}.$ (b) $\begin{bmatrix} 1 & 2 & 3 \\ 1 & 5 & 7 \\ 0 & -9 & -12 \end{bmatrix}.$ (c) $\begin{bmatrix} 1 & 1/2 & 1/4 \\ -6 & -3 & -3/2 \\ 20 & 10 & 5 \end{bmatrix}.$

Miscellaneous Exercises 149

10. Find the coordinates of

$$\begin{bmatrix} 1 \\ 3 \\ 2 \end{bmatrix}$$

with respect to each of the following bases for $\mathscr{R}_3$:

(a) $\begin{bmatrix} 1 \\ 0 \\ 0 \end{bmatrix}, \begin{bmatrix} 0 \\ 1 \\ 0 \end{bmatrix}, \begin{bmatrix} 0 \\ 0 \\ 1 \end{bmatrix}$.

(b) $\begin{bmatrix} 1 \\ 1 \\ 1 \end{bmatrix}, \begin{bmatrix} 1 \\ 1 \\ 0 \end{bmatrix}, \begin{bmatrix} 1 \\ 0 \\ 0 \end{bmatrix}$.

(c) $\begin{bmatrix} 2 \\ 0 \\ 1 \end{bmatrix}, \begin{bmatrix} 3 \\ 0 \\ 2 \end{bmatrix}, \begin{bmatrix} 1 \\ 1 \\ 0 \end{bmatrix}$.

11. Find the coordinates of $3t^2 + 2t - 10$ with respect to each of the following bases for $P^2(\mathscr{R})$:
(a) $3t^2$, $2t$, 5.
(b) $t^2 + t$, $t^2 - t$, $2t - 3$.
(c) $5t^2 + t + 1$, $5t^2 + 1$, $t + 2$.

12. Find a basis for the space of vectors **b** such that $A\mathbf{x} = \mathbf{b}$ is consistent, in each of the following cases:

(a) $A = \begin{bmatrix} 2 & 3 & 2 \\ 2 & 1 & 4 \\ 1 & -2 & 2 \end{bmatrix}$. (b) $A = \begin{bmatrix} 2 & 4 & -1 \\ -1 & -2 & \frac{1}{2} \\ 1 & -2 & 2 \end{bmatrix}$.

(c) $A = \begin{bmatrix} 2 & 5 & 1 \\ 1 & 3 & 4 \\ 2 & 0 & 1 \end{bmatrix}$.

13. Which of the following represent linear transformations?
(a) $T(a_0 + a_1 t + a_2 t^2) = a_0 + a_1 + a_2$
(from $P^2(\mathscr{R})$ to $\mathscr{R}$).
(b) $T(a_0 + a_1 t + a_2 t^2) = a_2 + a_1 t + a_0 t^2$
(from $P^2(\mathscr{R})$ to $P^2(\mathscr{R})$).
(c) $T(a_0 + a_1 t + a_2 t^2) = a_0 \cdot a_1 \cdot a_2$
(from $P^2(\mathscr{R})$ to $\mathscr{R}$).

14. (a) Show that $T(p(t)) = \int_0^t p(s)\,ds$ gives a linear transformation from $P^2(\mathscr{R})$ to $P^3(\mathscr{R})$.

(b) Give the matrix representation of T with respect to the bases $1, t, t^2$ and $1, t, t^2, t^3$ for $P^2(\mathscr{R})$ and $P^3(\mathscr{R})$ respectively.

15. Determine which of the following pairs of matrices are similar and for those pairs find a matrix P such that $B = P^{-1}AP$.

(a) $A = \begin{bmatrix} 1 & 1 \\ 0 & 1 \end{bmatrix}, \quad B = \begin{bmatrix} 1 & 0 \\ 1 & 1 \end{bmatrix}.$

(b) $A = \begin{bmatrix} 1 & 0 & 2 \\ 0 & 1 & 0 \\ 0 & 0 & 1 \end{bmatrix}, \quad B = \begin{bmatrix} 1 & 0 & 0 \\ 0 & 1 & 2 \\ 0 & 0 & 1 \end{bmatrix}.$

(c) $A = \begin{bmatrix} 1 & 0 & 2 \\ 0 & 1 & 0 \\ 0 & 0 & 0 \end{bmatrix}, \quad B = \begin{bmatrix} 1 & 0 & 0 \\ 0 & 1 & 2 \\ 0 & 0 & 1 \end{bmatrix}.$

Chapter 6 | LINEAR SYSTEMS OF DIFFERENTIAL EQUATIONS

In this chapter we will study the theory of systems of linear differential equations. In this study, we will apply the material on vector spaces developed in Chapter 5. As special cases of these results, we shall obtain all of the theory for linear scalar differential equations of second (or higher) order. In Chapter 1 we have already seen that problems of this type arise in a number of physical situations (see Examples 5, 6, 7, and 8).

6.1 Introduction

We shall consider *systems of first-order linear differential equations* of the form

$$\begin{aligned} y_1' &= a_{11}(t)y_1 + a_{12}(t)y_2 + \cdots + a_{1n}(t)y_n + g_1(t) \\ y_2' &= a_{21}(t)y_1 + a_{22}(t)y_2 + \cdots + a_{2n}(t)y_n + g_2(t) \\ &\vdots \qquad \vdots \qquad \vdots \qquad \qquad \vdots \qquad \vdots \\ y_n' &= a_{n1}(t)y_1 + a_{n2}(t)y_2 + \cdots + a_{nn}(t)y_n + g_n(t), \end{aligned} \qquad (6.1)$$

where the given functions $a_{ij}(t)$, where $i, j = 1, \ldots, n$, and $g_i(t)$, where $i = 1, \ldots, n$, are continuous on some fixed interval $\mathscr{I}$. If $n = 1$, we have the important special case of a scalar first-order equation, which we write in the form

$$y' = p(t)y + q(t), \tag{6.2}$$

where $p(t)$ and $q(t)$ are given functions.

System (6.1) is linear in $y_1, y_2, \ldots, y_n$, and $y'_1, y'_2, \ldots, y'_n$. The scalar equation $y' = 2y^2$ is an example of a nonlinear differential equation. These much more complicated equations will not be considered in this book.

Example 1. Consider the system

$$\begin{aligned} y'_1 &= y_1 - ty_2 + e^t \\ y'_2 &= t^2 y_1 - y_3 \\ y'_3 &= y_1 + y_2 - y_3 + 2e^{-t}, \end{aligned} \tag{6.3}$$

where $\mathscr{I}$ is the real line, $\{t \mid -\infty < t < \infty\}$. Here, $n = 3$, and in the notation of (6.1)

$$\begin{array}{llll} a_{11}(t) = 1 & a_{12}(t) = -t & a_{13}(t) = 0 & g_1(t) = e^t \\ a_{21}(t) = t^2 & a_{22}(t) = 0 & a_{23}(t) = -1 & g_2(t) = 0 \\ a_{31}(t) = 1 & a_{32}(t) = 1 & a_{33}(t) = -1 & g_3(t) = 2e^{-t}. \end{array}$$

Consider now the array

$$A(t) = \begin{bmatrix} 1 & -t & 0 \\ t^2 & 0 & -1 \\ 1 & 1 & -1 \end{bmatrix}. \tag{6.4}$$

$A(t)$ is a matrix whose entries are functions. The properties of matrix addition, multiplication by scalars, and matrix multiplication derived in Chapter 2 for matrices with constant entries (in $\mathscr{F}$) also hold for matrices whose entries are functions defined on a common interval $\mathscr{I}$. Let **y** and **y**' be the column vectors

$$\mathbf{y} = \begin{bmatrix} y_1 \\ y_2 \\ y_3 \end{bmatrix}, \quad \mathbf{y}' = \begin{bmatrix} y'_1 \\ y'_2 \\ y'_3 \end{bmatrix},$$

and let $\mathbf{g}(t)$ be the vector

$$\mathbf{g}(t) = \begin{bmatrix} e^t \\ 0 \\ 2e^{-t} \end{bmatrix}. \tag{6.5}$$

Then, observing that

$$A(t)\mathbf{y} = \begin{bmatrix} y_1 - ty_2 \\ t^2 y_1 - y_3 \\ y_1 + y_2 - y_3 \end{bmatrix}.$$

we see that system (6.3) may be represented conveniently in the matrix vector form

$$\mathbf{y}' = A(t)\mathbf{y} + \mathbf{g}(t),$$

where $A(t)$ and $\mathbf{g}(t)$ are given respectively by (6.4) and (6.5).

Returning to the general case of system (6.1), we define the $n \times n$ matrix

$$A(t) = \begin{bmatrix} a_{11}(t) & a_{12}(t) & \cdots & a_{1n}(t) \\ a_{21}(t) & a_{22}(t) & \cdots & a_{2n}(t) \\ \vdots & \vdots & & \vdots \\ a_{n1}(t) & a_{n2}(t) & \cdots & a_{nn}(t) \end{bmatrix}, \qquad (6.6)$$

whose entries are the n^2 functions $a_{ij}(t)$, where $i, j = 1, \ldots, n$. Next, define the vectors $\mathbf{g}(t)$, $\mathbf{y}$, $\mathbf{y}'$ by the relations

$$\mathbf{g}(t) = \begin{bmatrix} g_1(t) \\ g_2(t) \\ \vdots \\ g_n(t) \end{bmatrix}, \qquad \mathbf{y} = \begin{bmatrix} y_1 \\ y_2 \\ \vdots \\ y_n \end{bmatrix}, \qquad \mathbf{y}' = \begin{bmatrix} y_1' \\ y_2' \\ \vdots \\ y_n' \end{bmatrix}, \qquad (6.7)$$

Then the system (6.1) can be written in the form

$$\mathbf{y}' = A(t)\mathbf{y} + \mathbf{g}(t). \qquad (6.8)$$

● **EXERCISE**

1. Define the matrix $A(t)$ and the vectors $\mathbf{y}$, $\mathbf{y}'$, $\mathbf{g}(t)$ for the system (1.23) in Example 8, Chapter 1, and write this system in the form (6.8).

Before proceeding with the definition of a solution and a discussion of the system (6.8), we need the following definitions.

Definition 1. *A matrix* (such as $A(t)$) *or a vector* (such as $\mathbf{g}(t)$) *is continuous on an interval $\mathscr{I}$ if and only if each of its entries is a continuous function at each point of $\mathscr{I}$.*

Definition 2. An $n \times n$ matrix $B(t)$ or a vector $\mathbf{u}(t)$ with n components, defined on an interval $\mathscr{I}$ and given respectively by

$$B(t) = \begin{bmatrix} b_{11}(t) & b_{12}(t) & \cdots & b_{1n}(t) \\ b_{21}(t) & b_{22}(t) & \cdots & b_{2n}(t) \\ \vdots & \vdots & & \vdots \\ b_{n1}(t) & b_{n2}(t) & \cdots & b_{nn}(t) \end{bmatrix}, \quad \mathbf{u}(t) = \begin{bmatrix} u_1(t) \\ u_2(t) \\ \vdots \\ u_n(t) \end{bmatrix},$$

is differentiable on $\mathscr{I}$ if and only if each of its entries is differentiable at every point of $\mathscr{I}$. Their derivatives are given by

$$B'(t) = \begin{bmatrix} b'_{11}(t) & b'_{12}(t) & \cdots & b'_{1n}(t) \\ b'_{21}(t) & b'_{22}(t) & \cdots & b'_{2n}(t) \\ \vdots & \vdots & & \vdots \\ b'_{n1}(t) & b'_{n2}(t) & \cdots & b'_{nn}(t) \end{bmatrix}, \quad \mathbf{u}'(t) = \begin{bmatrix} u'_1(t) \\ u'_2(t) \\ \vdots \\ u'_n(t) \end{bmatrix},$$

respectively. Similarly, the matrix $B(t)$ or the vector $\mathbf{u}(t)$ is integrable on an interval (c, d) if and only if each of its entries is integrable on the interval (c, d). Their integrals are given by

$$\int_c^d B(t)\, dt = \begin{bmatrix} \int_c^d b_{11}(t)\, dt & \int_c^d b_{12}(t)\, dt & \cdots & \int_c^d b_{1n}(t)\, dt \\ \int_c^d b_{21}(t)\, dt & \int_c^d b_{22}(t)\, dt & \cdots & \int_c^d b_{2n}(t)\, dt \\ \vdots & \vdots & & \vdots \\ \int_c^d b_{n1}(t)\, dt & \int_c^d b_{n2}(t)\, dt & \cdots & \int_c^d b_{nn}(t)\, dt \end{bmatrix}$$

$$\int_c^d \mathbf{u}(t)\, dt = \begin{bmatrix} \int_c^d u_1(t)\, dt \\ \int_c^d u_2(t)\, dt \\ \vdots \\ \int_c^d u_n(t)\, dt \end{bmatrix}.$$

6.1 Introduction

● **EXERCISES**

2. Evaluate the derivatives of each of the following vectors or matrices:

(a) $B(t) = \begin{bmatrix} t & e^{-t} & 7 \\ \sin t & 0 & \cos t \\ t^2 & t & 1 \end{bmatrix}$ for $-\infty < t < \infty$.

(b) $B(t) = \begin{bmatrix} \cos t & \sin t \\ -\sin t & \cos t \end{bmatrix}$ for $-\infty < t < \infty$.

(c) $B(t) = \begin{bmatrix} e^{2t} & te^{2t} \\ 2e^{2t} & (2t+1)e^{2t} \end{bmatrix}$ for $-\infty < t < \infty$.

(d) $\mathbf{u}(t) = \begin{bmatrix} \log t \\ t \log t \\ t^2 \log t \end{bmatrix}$ for $0 < t < \infty$ (and where $\log t$ is the natural logarithm of t).

(e) $\mathbf{u}(t) = \begin{bmatrix} 1 \\ 2 \\ 3 \\ 4 \end{bmatrix}$ for $-1 < t < 2$.

3. Evaluate $\int_1^2 B(t)\, dt$ or $\int_1^2 \mathbf{u}(t)\, dt$ for each of the matrices $B(t)$ or vectors $\mathbf{u}(t)$ in Exercise 1. [*Hint:* In parts (c) and (d), integrate by parts.]

4. Is the vector

$$\mathbf{u}(t) = \begin{bmatrix} 1/t \\ t^2 \end{bmatrix}$$

continuous on the interval $1 \leq t \leq 2$?
Is it continuous on the interval $-1 < t < 1$? *Explain.*

We are now ready to say what is meant by a solution of system (6.8).

Definition 3. *Let $A(t)$ be a continuous $n \times n$ matrix on an interval $\mathscr{I}$. Let $\mathbf{g}(t)$ be a continuous vector with n components on the same interval $\mathscr{I}$. A solution of the system*

$$\mathbf{y}' = A(t)\mathbf{y} + \mathbf{g}(t) \tag{6.8}$$

on some interval $\mathscr{J}$ (where $\mathscr{J}$ is contained in $\mathscr{I}$) is a vector $\mathbf{u}(t)$ whose derivative $\mathbf{u}'(t)$ is continuous on the interval $\mathscr{J}$ and such that

$$\mathbf{u}'(t) = A(t)\mathbf{u}(t) + \mathbf{g}(t)$$

for every t on $\mathcal{J}$. (In the notation of Example 7, Section 5.2, a solution of system (6.8) is an element of the vector space $C'(\mathcal{R}_n)$ on the interval $\mathcal{J}$. Note that the interval $\mathcal{J}$ is not necessarily the same as $\mathcal{I}$.)

Example 2. Consider the scalar ($n = 1$) differential equation $y' = -y + 1$. Then $u(t) = e^{-t} + 1$ is a solution on the interval $-\infty < t < \infty$. For, $u'(t)$ is continuous on $-\infty < t < \infty$ and $u'(t) = -e^{-t}$. Thus, $u'(t) = e^{-t} = -u(t) + 1$.

In Chapter 1, we saw that with a differential equation, one usually associates a particular initial condition. For example, the solution $u(t) = e^{-t} + 1$ of Example 2 satisfies at $t = 0$ the initial condition $u(0) = 2$. More generally, suppose we consider the system (6.8) together with the initial condition $\mathbf{y}(t_0) = \mathbf{y}_0$, where t_0 is a given number in the the interval $\mathcal{J}$ and where $\mathbf{y}_0$ is a given vector in $\mathcal{R}_n$.

Definition 4. *By a solution of the initial value problem,*

$$\mathbf{y}' = A(t)\mathbf{y} + \mathbf{g}(t), \qquad \mathbf{y}(t_0) = \mathbf{y}_0, \tag{6.9}$$

we mean a solution $\mathbf{u}(t)$ *of the system* $\mathbf{y}' = A(t)\mathbf{y} + \mathbf{g}(t)$ *on an interval* $\mathcal{J}$ *containing the point* t_0, *such that* $\mathbf{u}(t_0) = \mathbf{y}_0$.

Our object will be to learn as much as possible about such initial value problems. As a matter of fact, when $n = 1$ the initial value problem (6.9) can always be solved, and we now obtain a formula for the solution.

Example 3. Solve the scalar equation

$$y' = p(t)y + q(t), \tag{6.10}$$

where $p(t), q(t)$ are continuous on an interval $\mathcal{I}$, subject to the initial condition $y(t_0) = y_0$ for some t_0 in the interval $\mathcal{I}$ with y_0 a given number. Suppose such a solution exists; call it $u(t)$. Then

$$u'(t) - p(t)u(t) = q(t).$$

Observe that by calculus, since $d/dt \int_{t_0}^{t} p(s)\, ds = p(t)$,

$$\frac{d}{dt}\left[u(t) \exp\left(-\int_{t_0}^{t} p(s)\, ds\right)\right] = [u'(t) - p(t)u(t)] \exp\left(-\int_{t_0}^{t} p(s)\, ds\right).$$

Since $\exp\left(-\int_{t_0}^{t} p(s)\, ds\right)$ is not zero in the interval $\mathcal{I}$, we have

$$\exp\left(-\int_{t_0}^{t} p(s)\, ds\right)[u'(t) - p(t)u(t)] = \exp\left(-\int_{t_0}^{t} p(s)\, ds\right) q(t).$$

Thus,

$$\frac{d}{dt}\left[u(t)\exp\left(-\int_{t_0}^t p(s)\,ds\right)\right] = \exp\left(-\int_{t_0}^t p(s)\,ds\right)q(t).$$

Integrating both sides with respect to t, we obtain

$$u(t)\exp\left(-\int_{t_0}^t p(s)\,ds\right) - u(t_0) = \int_{t_0}^t \exp\left(-\int_{t_0}^\sigma p(s)\,ds\right)q(\sigma)\,d\sigma,$$

and finally using $u(t_0) = y_0$,

$$u(t) = y_0 \exp\left(\int_{t_0}^t p(s)\,ds\right) + \exp\left(\int_{t_0}^t p(s)\,ds\right)$$
$$\times \int_{t_0}^t \exp\left(-\int_{t_0}^\sigma p(s)\,ds\right)q(\sigma)\,d\sigma, \qquad (6.11)$$

as a candidate for the solution. That is, we have shown that if (6.10) has a solution satisfying the initial condition $y(t_0) = y_0$, then it must have the form (6.11). On the other hand, the function $u(t)$ defined by (6.11) is a solution of (6.10) satisfying the initial condition $u(t_0) = y_0$ on the interval $\mathscr{I}$. For if p and q are continuous on $\mathscr{I}$, we have by the fundamental theorem of calculus,

$$\frac{d}{dt}\left[\int_{t_0}^t p(s)\,ds\right] = p(t)$$

and, therefore, differentiating (6.11)

$$u'(t) = y_0\, p(t) \exp\left(\int_{t_0}^t p(s)\,ds\right)$$
$$+ \exp\left(\int_{t_0}^t p(s)\,ds\right)\exp\left(-\int_{t_0}^t p(s)\,ds\right)q(t)$$
$$+ p(t)\exp\left(\int_{t_0}^t p(s)\,ds\right)\int_{t_0}^t \exp\left(-\int_{t_0}^\sigma p(s)\,ds\right)q(\sigma)\,d\sigma$$
$$= p(t)\bigg[y_0 \exp\left(\int_{t_0}^t p(s)\,ds\right) + \exp\left(\int_{t_0}^t p(s)\,ds\right)$$
$$\times \int_{t_0}^t \exp\left(-\int_{t_0}^\sigma p(s)\,ds\right)q(\sigma)\,d\sigma\bigg] + q(t)$$
$$= p(t)u(t) + q(t).$$

158 Linear Systems of Differential Equations

Note that although (6.11) is the solution of the problem, we may not be able to evaluate the integrals for some particular functions p and q.

To take a specific problem, we consider the following.

Example 4. Solve the initial value problem $y' + 2ty = \sin t$, $y(0) = y_0$. If $\phi(t)$ is a solution, so that $\phi'(t) + 2t\phi(t) = \sin t$, we can make the left side of the equation the derivative of $\exp(t^2)\phi(t)$ by multiplying by $\exp(t^2)$. If we multiply the whole equation through by $\exp(t^2)$, we obtain

$$d/dt[\exp(t^2)]\phi(t)] = [\exp(t^2)] \sin t.$$

But now we can integrate to obtain

$$[\exp(t^2)]\phi(t) - \phi(0) = \int_0^t [\exp(s^2)] \sin s \, ds$$

or

$$[\exp(t^2)]\phi(t) = y_0 + \int_0^t [\exp(s^2)] \sin s \, ds.$$

This gives $\phi(t) = y_0 \exp(-t^2) + \exp(-t^2) \int_0^t [\exp(s^2)] \sin s \, ds$ for $-\infty < t < \infty$. The integral $\int_0^t [\exp(s^2)] \sin s \, ds$ cannot be evaluated in terms of elementary functions; nevertheless, it defines a differentiable function, whose value for $t = 0$ is zero and whose derivative is $[\exp(t^2)] \sin t$. We now check our result by substituting this expression for ϕ in the differential equation $y' + 2ty = \sin t$. We obtain, using the product rule and the fundamental theorem of calculus,

$$\phi'(t) = -2t[\exp(-t^2)]y_0 + \exp(-t^2)\frac{d}{dt}\int_0^t \exp(s^2) \sin s \, ds$$

$$- 2t[\exp(-t^2)]\int_0^t \exp(s^2) \sin s \, ds$$

$$= -2t[\exp(-t^2)]y_0 + \sin t - 2t[\exp(-t^2)]\int_0^t \exp(s^2) \sin s \, ds$$

$$= -2t\phi(t) + \sin t.$$

This shows that ϕ is actually a solution of the differential equation.

● **EXERCISES**

5. Find the solution of $t^2 y' + 2ty = 1$ satisfying the initial condition $y(1) = 0$. On what interval is the solution valid?

6. Find the solution of $y' + 2y = e^t$ satisfying the initial condition $y(0) = 1$.

7. Find the solution of $y' + py = q$ satisfying the initial condition $y(t_0) = y_0$, where p and q are constants.

8. Find the solution of the equation $y' + a(t)y = b(t)y^k$ (Bernoulli equation) through (t_0, y_0), where k is a constant, $k \neq 1$. [*Hint:* Make the change of dependent variable $z = y^{1-k}$.]

9. The current y in an electrical circuit consisting of resistance R, inductance L, and applied voltage $E \sin \omega t$ connected in series, where R, L, E, and ω are positive constants, is governed by the differential equation $Ly' + Ry = E \sin \omega t$. Find the solution ϕ with $\phi(0) = 0$, and show that it can be written in the form

$$\phi(t) = \frac{E\omega L}{R^2 + \omega^2 L^2} e^{-Rt/L} + \frac{E}{(R^2 + \omega^2 L^2)^{1/2}} \sin(\omega t - \alpha),$$

where α is an angle defined by

$$\cos \alpha = \frac{R}{(R^2 + \omega^2 L^2)^{1/2}}, \qquad \sin \alpha = \frac{\omega L}{(R^2 + \omega^2 L^2)^{1/2}}.$$

Note that the solution for the current is the sum of two terms; the first approaches zero as $t \to +\infty$ and is called the transient solution; the second is called the steady state solution.

10. Discuss the behavior as $t \to \infty$ of the solution of the initial value problem $y' = \lambda y$, $y(t_0) = y_0$, for each of the cases $\lambda > 0$, $\lambda = 0$, $\lambda < 0$.

11. Discuss the behavior $t \to \infty$ of the solution of each of the following initial value problems:
  (a) $y' = -2y + e^{-t}$, $y(t_0) = y_0$,
  (b) $y' = -2y + e^t$, $y(t_0) = y_0$,
  (c) $y' = -2y + 1$, $y(t_0) = y_0$,
  (d) $y' = -2y + f(t)$, $y(t_0) = y_0$,
where $f(t)$ is a continuous function for which $\lim_{t \to \infty} f(t) = 0$. [*Hint:* use l'Hospital's rule to evaluate the limit.]

When we leave the scalar equation, the situation, except in very special cases, becomes more complicated.

Example 5. Show that the vector

$$\mathbf{u}(t) = \begin{bmatrix} \cos t \\ -\sin t \end{bmatrix} = \begin{bmatrix} u_1(t) \\ u_2(t) \end{bmatrix}$$

Linear Systems of Differential Equations

is a solution of the system

$$y' = \begin{bmatrix} 0 & 1 \\ -1 & 0 \end{bmatrix} y, \quad \text{where} \quad y = \begin{bmatrix} y_1 \\ y_2 \end{bmatrix},$$

on $-\infty < t < \infty$, satisfying the initial condition

$$u(0) = \begin{bmatrix} 1 \\ 0 \end{bmatrix}.$$

Obviously,

$$u(0) = \begin{bmatrix} \cos 0 \\ -\sin 0 \end{bmatrix} = \begin{bmatrix} 1 \\ 0 \end{bmatrix}.$$

Since $\cos t$ and $\sin t$ have continuous derivatives everywhere, we have

$$u'(t) = \begin{bmatrix} -\sin t \\ -\cos t \end{bmatrix} = \begin{bmatrix} u_2(t) \\ -u_1(t) \end{bmatrix} = \begin{bmatrix} 0 & 1 \\ -1 & 0 \end{bmatrix} \begin{bmatrix} u_1(t) \\ u_2(t) \end{bmatrix} = \begin{bmatrix} 0 & 1 \\ -1 & 0 \end{bmatrix} u(t).$$

● EXERCISES

12. Show that

$$v(t) = \begin{bmatrix} \sin t \\ \cos t \end{bmatrix}$$

is a solution of the system of Example 5 on $-\infty < t < \infty$ satisfying the initial condition

$$v(0) = \begin{bmatrix} 0 \\ 1 \end{bmatrix}.$$

13. Show that

$$w(t) = c_1 u(t) + c_2 v(t),$$

where $u(t)$, $v(t)$ are given in Example 5 and Exercise 12, respectively, and where c_1, c_2 are any constants, is a solution of the initial value problem

$$y' = \begin{bmatrix} 0 & 1 \\ -1 & 0 \end{bmatrix} y, \quad y(0) = \begin{bmatrix} c_1 \\ c_2 \end{bmatrix},$$

for $-\infty < t < \infty$.

14. Show that

$$\mathbf{u}(t) = \begin{bmatrix} e^t \\ e^t \end{bmatrix}$$

is a solution of the initial value problem

$$\mathbf{y}' = \begin{bmatrix} 0 & 1 \\ 1 & 0 \end{bmatrix} \mathbf{y}, \quad \mathbf{y}(0) = \begin{bmatrix} 1 \\ 1 \end{bmatrix}$$

on the interval $-\infty < t < \infty$.

15. Find a solution $\boldsymbol{\phi}$ of the initial value problem

$$y_1' = -y_1, \quad y_2' = y_1 + y_2, \qquad \mathbf{y}(0) = \begin{bmatrix} 2 \\ 1 \end{bmatrix}.$$

[*Hint:* Solve the first equation and substitute in the second equation. What is the interval of validity?]

16. Find a solution $\boldsymbol{\phi}$ of the initial value problem

$$y_1' = -y_1, \quad y_2' = y_1 + t y_2, \qquad \mathbf{y}(0) = \begin{bmatrix} 1 \\ 1 \end{bmatrix}.$$

17. Describe a method for solving the "triangular system"

$$\begin{aligned} y_1' &= a_{11} y_1 + a_{12} y_2 + \cdots + a_{1n} y_n \\ y_2' &= \phantom{a_{11} y_1 +} a_{22} y_2 + \cdots + a_{2n} y_n \\ &\vdots \\ y_{n-1}' &= a_{n-1,n-1} y_{n-1} + a_{n-1,n} y_n \\ y_n' &= a_{nn} y_n, \end{aligned}$$

where a_{ij}, with $j \geq i$, are constants; note that a_{ij}, with $j < i$, are zero.

18. Find a solution $\boldsymbol{\phi}$ of the initial value problem

$$y_1' = y_1 + y_2 + f(t), \quad y_2' = y_1 + y_2, \qquad \mathbf{y}(0) = \begin{bmatrix} 0 \\ 0 \end{bmatrix},$$

where $f(t)$ is a continuous function. [*Hint:* Define $v(t) = y_1(t) + y_2(t)$.]

Exercises 15, 16, and 17 show that "triangular" systems of first-order differential equations can be solved by successive solution of scalar first-order equations. However, a system that is not triangular cannot, in general, be solved by putting it in triangular form because the elementary row operations that triangulate the right-hand side (that is, the coefficient matrix A) destroy the isolation of the derivatives on the left-hand side. Note that this difficulty does not arise in the case of linear algebraic equations (see Chapter 3).

Examples 5 and 6 of Chapter 1 lead us to a specific linear second-order scalar differential equation with initial conditions. These can be reduced to an initial value problem for a linear system of two first-order equations of the form (6.9) by the following method.

Example 6. Consider the scalar second-order linear initial value

$$y'' + p(t)y' + q(t)y = r(t), \qquad y(t_0) = \eta_1, \qquad y'(t_0) = \eta_2, \qquad (6.12)$$

where p, q, r are given functions continuous on an interval $\mathscr{I}$, t_0 is in $\mathscr{I}$, and η_1, η_2 are given constants. In agreement with Definition 4, by a solution of (6.12) on an interval $\mathscr{J}$ contained in $\mathscr{I}$ we mean a function $w(t)$ such that $w'(t)$, $w''(t)$ exist and are continuous at each point of $\mathscr{J}$, such that $w''(t) + p(t)w'(t) + q(t)w(t) = r(t)$ for every t in $\mathscr{J}$, and such that $w(t_0) = \eta_1$, $w'(t_0) = \eta_2$. The idea is to introduce new unknowns y_1 and y_2 by means of the definitions $y_1 = y$, $y_2 = y'$. Then

$$y_1' = y_2$$
$$y_2' = y'' = -p(t)y' - q(t)y + r(t) = -p(t)y_2 - q(t)y_1 + r(t).$$

This suggests that the given initial value problem (6.12) can be described by the initial value problem

$$\mathbf{y}' = \begin{bmatrix} 0 & 1 \\ -q(t) & -p(t) \end{bmatrix} \mathbf{y} + \begin{bmatrix} 0 \\ r(t) \end{bmatrix}, \qquad \mathbf{y}(t_0) = \begin{bmatrix} \eta_1 \\ \eta_2 \end{bmatrix} = \boldsymbol{\eta}, \qquad (6.13)$$

where

$$\mathbf{y} = \begin{bmatrix} y_1 \\ y_2 \end{bmatrix}.$$

Note that (6.13) is a special case of (6.9) with $n = 2$ and $A(t)$, $\mathbf{g}(t)$ displayed in (6.13).

We will now show **the initial value problems (6.12) and (6.13) are equivalent in the sense that given a solution of either one, we can construct a solution of the other one.** More precisely, let $\psi(t)$ be a solution of (6.12) on some interval

$\mathscr{J}$ containing t_0. Define the functions ϕ_1 and ϕ_2 on $\mathscr{J}$ by the relations

$$\phi_1(t) = \psi(t), \qquad \phi_2(t) = \psi'(t).$$

Define the vector $\boldsymbol{\phi}$ by the relations

$$\boldsymbol{\phi}(t) = \begin{bmatrix} \phi_1(t) \\ \phi_2(t) \end{bmatrix}.$$

We claim that $\boldsymbol{\phi}(t)$ is a solution of (6.13) on $\mathscr{J}$. Clearly,

$$\boldsymbol{\phi}(t_0) = \begin{bmatrix} \phi_1(t_0) \\ \phi_2(t_0) \end{bmatrix} = \begin{bmatrix} \psi(t_0) \\ \psi'(t_0) \end{bmatrix} = \begin{bmatrix} \eta_1 \\ \eta_2 \end{bmatrix} = \boldsymbol{\eta};$$

moreover,

$$\boldsymbol{\phi}'(t) = \begin{bmatrix} \phi_1'(t) \\ \phi_2'(t) \end{bmatrix} = \begin{bmatrix} \psi'(t) \\ \psi''(t) \end{bmatrix} = \begin{bmatrix} \psi'(t) \\ -p(t)\psi'(t) - q(t)\psi(t) + r(t) \end{bmatrix}$$

$$= \begin{bmatrix} \phi_2(t) \\ -p(t)\phi_2(t) - q(t)\phi_1(t) + r(t) \end{bmatrix}$$

$$= \begin{bmatrix} 0 & 1 \\ -q(t) & -p(t) \end{bmatrix} \boldsymbol{\phi}(t) + \begin{bmatrix} 0 \\ r(t) \end{bmatrix},$$

for every t on $\mathscr{J}$, which shows that $\boldsymbol{\phi}(t)$ is a solution of (6.13) on $\mathscr{J}$. Conversely, let $\mathbf{u}(t)$ be a solution of (6.13) on some interval $\mathscr{J}$ containing t_0. Let

$$\mathbf{u}(t) = \begin{bmatrix} u_1(t) \\ u_2(t) \end{bmatrix}$$

and define the function w by $w(t) = u_1(t)$. We claim that w is a solution of (6.12) on $\mathscr{J}$. From the first equation in (6.13), we have $w'(t) = u_1'(t) = u_2(t)$, and from the second equation in (6.13), and the definition of w, we have

$$w''(t) = u_2'(t) = -q(t)u_1(t) - p(t)u_2(t) + r(t)$$
$$= -q(t)w(t) - p(t)w'(t) + r(t).$$

Hence, $w''(t) + p(t)w'(t) + q(t)w(t) = r(t)$ on $\mathscr{J}$. We also have $w(t_0) = u_1(t_0) = \eta_1$, and $w'(t_0) = u_2(t_0) = \eta_2$. Hence, w is a solution of (6.12).

The reader will note that the initial value problem in Example 5 is equivalent to the scalar second-order differential equation $y'' + y = 0$ with initial conditions $y(0) = 1$, $y'(0) = 0$. (Compare with (6.12) and (6.13).)

Example 7. More generally, consider the scalar nth order-linear initial value problem

$$y^{(n)} + p_1(t)y^{(n-1)} + \cdots + p_{n-1}(t)y' + p_n(t)y = r(t),$$
$$y(t_0) = \eta_1,\ y'(t_0) = \eta_2,\ \ldots,\ y^{(n-1)}(t_0) = \eta_n, \tag{6.14}$$

where $p_1, p_2, \ldots, p_n$ are given functions continuous on an interval $\mathscr{I}$, t_0 is in $\mathscr{I}$, and $\eta_1, \ldots, \eta_n$ are given constants. By an argument that parallels exactly the second-order case (letting $y_1 = y$, $y_2 = y'$, $\ldots$, $y_n = y^{(n-1)}$), we can easily show that the initial value problem (6.14) is equivalent to

$$\mathbf{y}' = \begin{bmatrix} 0 & 1 & 0 & \cdots & 0 \\ 0 & 0 & 1 & 0 & \vdots \\ \vdots & \vdots & & \ddots & 0 \\ 0 & 0 & & 0 & 1 \\ -p_n(t) & -p_{n-1}(t) & \cdots & -p_2(t) & -p_1(t) \end{bmatrix} \mathbf{y} + \begin{bmatrix} 0 \\ \vdots \\ 0 \\ r(t) \end{bmatrix},$$

$$\mathbf{y}(t_0') = \begin{bmatrix} \eta_1 \\ \vdots \\ \eta_n \end{bmatrix} = \boldsymbol{\eta}, \tag{6.15}$$

where

$$\mathbf{y} = \begin{bmatrix} y_1 \\ \vdots \\ y_n \end{bmatrix},\quad \mathbf{y}' = \begin{bmatrix} y_1' \\ \vdots \\ y_n' \end{bmatrix}.$$

We remark that whereas every nth-order scalar equation is equivalent to a system of first-order equations (as shown in Example 7), the converse is not true. For example, the system

$$\mathbf{y}' = \begin{bmatrix} 1 & 0 \\ 0 & 1 \end{bmatrix}\mathbf{y},\quad \mathbf{y} = \begin{bmatrix} y_1 \\ y_2 \end{bmatrix},$$

cannot be made equivalent to a second-order scalar equation, because the coefficient matrix has the wrong form.

6.2 The Existence and Uniqueness Theorem

● **EXERCISES**

19. For each of the following initial value problems, write an equivalent initial value problem for a first-order system:
    (a) $y'' + 2y' + 7ty = e^{-t}$, $y(1) = 7$, $y'(1) = -2$.
    (b) $2y'' - 5t^2 y' + (\cos t)y = \log t$, $y(2) = 1$, $y'(2) = 0$.
    (c) $y''' - 6y'' + 3y' + e^{-t}y = \sin t$, $y(0) = 0$, $y'(0) = 0$, $y''(0) = 0$.
    (d) $y^{(4)} + 16y = te^t$, $y(0) = 1$, $y'(0) = -1$, $y''(0) = 2$, $y'''(0) = 0$.

20. Formulate and establish the equivalence of the initial value problems (6.14) and (6.15).

21. Reduce each of the following initial value problems to an equivalent initial value problem for a first-order system:
    (a) Example 5, Chapter 1.   (b) Example 6, Chapter 1.

22. Reduce each of the following initial value problems to an equivalent initial value problem for a first-order system:
    (a) $y'' + 5z' - 7y + 6z = e^t$
    $z'' - 2y' + 13z - 15y = \cos t$,
    where

    $y(0) = 1$, $y'(0) = 0$, $z(0) = 0$, $z'(0) = 1$.

    [*Hint:* Let $w_1 = y$, $w_2 = y'$, $w_3 = z$, $w_4 = z'$.]
    (b) $y' + 5z + 2y = t^2$
    $z'' + 6y' + 11z' - 3y - z = t$,
    where

    $y(0) = 1$, $z(0) = 2$, $z'(0) = 3$.

23. Reduce the initial value problem (1.18), (1.19) of Example 7, Chapter 1, to an equivalent initial value problem for a first-order system.

6.2 The Existence and Uniqueness Theorem

In discussing the scalar first-order differential equation (Example 3, Section 6.1), we were able to give an explicit expression for the solution of the initial value problem. However, for second- or higher-order scalar differential equations and more generally for first-order systems, it is frequently not possible to give an explicit expression for the solution. For example, the scalar differential equation

$$t^2 y'' + ty' + (t^2 - p^2)y = 0,$$

where p is a constant, is called the **Bessel equation** (of index p), and arises in many problems of mathematical physics. This equation and its solutions have been studied extensively. Except for special values of p, such as $p = \frac{1}{2}$ or $p = 3/2$, these solutions cannot be expressed in terms of a finite number of elementary functions. Nevertheless, the Bessel equation does have solutions for every initial value problem with $t_0 \neq 0$. This fact is a special case of the following general theorem.

Theorem 1. *Let $A(t)$ be a continuous $n \times n$ matrix on some interval $\mathscr{I}$. Let $g(t)$ be a vector with n components continuous on the same interval $\mathscr{I}$. Then for every t_0 in $\mathscr{I}$ and every constant vector $\boldsymbol{\eta}$, the initial value problem*

$$\mathbf{y}' = A(t)\mathbf{y} + \mathbf{g}(t), \qquad \mathbf{y}(t_0) = \boldsymbol{\eta} \tag{6.9}$$

has a unique solution existing on the same interval $\mathscr{I}$.

The interested reader is referred to Appendix 2 for the proof. Our objective here will be to learn to apply the theorem. In Theorem 1, the matrix $A(t)$ and the vector $\mathbf{g}(t)$ may have real- or complex-valued entries.

Example 1. Let $n = 3$ and consider the initial value problem (6.9) with

$$A(t) = \begin{bmatrix} 1 & -t & 0 \\ \dfrac{1}{t^2-1} & 0 & -1 \\ 2 & \dfrac{1}{t^2+1} & 3 \end{bmatrix}, \quad \mathbf{g}(t) = \begin{bmatrix} e^t \\ \cos t \\ -e^t \end{bmatrix}, \quad t_0 = 0, \quad \boldsymbol{\eta} = \begin{bmatrix} 1 \\ 0 \\ -1 \end{bmatrix}$$

Determine whether this initial value problem has a unique solution and find the largest interval $\mathscr{I}$ of existence of this solution in accordance with the theorem.

The entries of $A(t)$ and $\mathbf{g}(t)$, with the exception of $1/(t^2 - 1)$, are continuous on $-\infty < t < \infty$. However, $1/(t^2 - 1)$ fails to be continuous at $t = \pm 1$. Since $t_0 = 0$, Theorem 1, therefore, tells us that the given initial value problem has a unique solution $\boldsymbol{\phi}(\boldsymbol{\phi}(0) = \boldsymbol{\eta})$, and the solution $\boldsymbol{\phi}$ exists on the interval $-1 < t < 1$. It is worth pointing out that if we choose a different t_0, for example $t_0 = 10$, the new initial value problem will also have a unique solution $\boldsymbol{\psi}$ ($\boldsymbol{\psi}(10) = \boldsymbol{\eta}$) and the solution $\boldsymbol{\psi}$ will exist for $1 < t < \infty$.

We now apply Theorem 1 to the important special case of the initial value

6.2 The Existence and Uniqueness Theorem

problem for a linear second-order scalar equation

$$y'' + p(t)y' + q(t)y = r(t), \qquad y(t_0) = \eta_1, \qquad y'(t_0) = \eta_2, \qquad (6.12)$$

where p, q, and r are functions continuous on some interval $\mathscr{I}$ and t_0 is a point of $\mathscr{I}$. In Example 6, Section 6.1, we have seen that the initial value problem (6.12) is equivalent to the initial value problem

$$\mathbf{y}' = \begin{bmatrix} 0 & 1 \\ -q(t) & -p(t) \end{bmatrix} \mathbf{y} + \begin{bmatrix} 0 \\ r(t) \end{bmatrix}, \qquad \mathbf{y}(t_0) = \begin{bmatrix} \eta_1 \\ \eta_2 \end{bmatrix} \qquad (6.13)$$

to which Theorem 1 can be applied directly. This yields the following result.

Corollary 1 to Theorem 1. *Let p, q, r be given functions continuous on an interval $\mathscr{I}$ and let t_0 be in $\mathscr{I}$. Then the initial value problem (6.12) has a unique solution w ($w(t_0) = \eta_1$, $w'(t_0) = \eta_2$) that exists on the same interval $\mathscr{I}$.*

Example 2. Consider the initial value problem

$$(t^2 + 4)y'' + ty' + (\sin t)y = 1, \qquad y(1) = 2, \qquad y'(1) = 0.$$

Determine the existence and uniqueness of the solution as well as the interval of existence. To apply Corollary 1, we must first reduce the given differential equation to the exact form of (6.12). This is accomplished by dividing by $(t^2 + 4)$. Thus, in the notation of the corollary

$$p(t) = \frac{t}{t^2 + 4}, \qquad q(t) = \frac{\sin t}{t^2 + 4}, \qquad r(t) = \frac{1}{t^2 + 4}.$$

Since these functions are continuous for $-\infty < t < \infty$, the given initial value problem has a unique solution existing on $-\infty < t < \infty$.

Similar to Corollary 1, we may consider the more general initial value problem

$$y^{(n)} + p_1(t)y^{(n-1)} + \cdots + p_{n-1}(t)y' + p_n(t)y = r(t)$$
$$y(t_0) = \eta_1, \qquad y'(t_0) = \eta_2, \ldots, y^{(n-1)}(t_0) = \eta_n \qquad (6.14)$$

for the scalar nth-order linear equation. From Example 7, Section 6.1, and from Theorem 1 we obtain immediately the following.

Corollary 2 to Theorem 1. *Let $p_1, p_2, \ldots, p_n$ be given functions continuous on an interval $\mathscr{I}$ and let t_0 be in $\mathscr{I}$. Then the initial value problem (6.14) has* a unique solution w ($w(t_0) = \eta_1$, $w'(t_0) = \eta_2, \ldots, w^{(n-1)}(t_0) = \eta_n$) that exists on the same interval $\mathscr{I}$.

Note that, unlike the study of linear systems of algebraic equations in Chapter 3, we consider only systems in which the number of equations and the number of unknown functions is the same. In Chapter 1, we observed that physical problems should lead to mathematical formulations of differential equations with exactly one solution. For a system of first-order linear differential equations, if the number of equations is not the same as the number of unknown functions, the initial value problem does not in general have a unique solution. For example, consider the initial value problem

$$y'_1 = y_2, \quad y_1(0) = 1, \quad y_2(0) = 1,$$

that is, one equation in two unknown functions. Then

$$\phi(t) = \begin{bmatrix} e^t \\ e^t \end{bmatrix}$$

is a solution, but so is

$$\psi(t) = \begin{bmatrix} t+1 \\ 1 \end{bmatrix}.$$

In fact, there are infinitely many solutions, for we can add an arbitrary differential equation for y_2,

$$y'_2 = a(t)y_1 + b(t)y_2 + g(t),$$

where a, b, and g are **arbitrary** continuous functions. With this additional differential equation, the initial value problem will have a unique solution by Theorem 1.

From this example, the reader can see that, similar to the situation for linear algebraic systems, an initial value problem for a system of linear differential equations with fewer equations than unknowns may have infinitely many solutions. Unlike the algebraic case, however, such a problem may have no solution, as is shown by the example $y'_1 + y'_2 + y'_3 = 1$, $y'_1 + y'_2 + y'_3 = 0$, $y_1(0) = y_2(0) = y_3(0) = 0$. If there are more equations than unknowns, the system may be inconsistent. For example,

$$\begin{aligned} y'_1 &= y_1 \\ y'_1 &= 0 \end{aligned} \quad y_1(0) = 1$$

is obviously inconsistent.

6.2 The Existence and Uniqueness Theorem

• **EXERCISES**

1. What does Theorem 1 tell you about each of the following initial value problems for systems of the form (6.9)?
   (a) $n = 2$, $t_0 = 1$

$$A(t) = \begin{bmatrix} t & \log t \\ -1 & t \log t \end{bmatrix}, \quad g(t) = \begin{bmatrix} e^t \\ 0 \end{bmatrix}, \quad \eta = \begin{bmatrix} 1 \\ -1 \end{bmatrix}.$$

   (b) Same as part (a), except $t_0 = -1$.
   (c) Same as part (a), excapt $t_0 = 0$.
   (d) Same as part (a), except

$$g(t) = \begin{bmatrix} 1 \\ 1 \\ t^2 - 9 \end{bmatrix}.$$

   (e) $n = 3$, $t_0 = 5$

$$A(t) = \begin{bmatrix} 1 & \cos t & -\sin t \\ e^{-t} & t^3 & 0 \\ 0 & e^t & t \end{bmatrix}, \quad g(t) = \begin{bmatrix} 0 \\ 0 \\ 0 \end{bmatrix}, \quad \eta = \begin{bmatrix} 1 \\ -1 \\ 6 \end{bmatrix}.$$

   (f) Same as part (e), except

$$\eta = \begin{bmatrix} 0 \\ 0 \\ 0 \end{bmatrix}.$$

   Can you guess the solution?

2. What does Corollary 1 to Theorem 1 tell you about each of the following initial value problems for linear, scalar second-order equations of the form (6.12)?
   (a) $y'' + ty' - y = 0$, $y(1) = 1$, $y'(1) = 1$.
   (b) Same as part (a), except $y(1) = 1$, $y'(1)$ unspecified.
   (c) Same as part (a), except $y(1) = 0$, $y'(1) = 0$. (Can you guess the solution?)
   (d) $t^2 y'' + ty' + (t^2 - 4)y = 0$, $y(-1) = 0$, $y'(-1) = 1$.
   (e) Same as part (d), except $y(0) = 0$, $y'(0) = 1$.

3. What does Theorem 1 tell you about each of the initial value problems of Exercise 20, Section 6.1?

4. Discuss, using Corollary 2 to Theorem 1, each of the following initial value problems:
   (a) $y''' + t^2 y' + (\tan t)y = 2t$, $y(0) = 0$, $y'(0) = 0$, $y''(0) = 1$.
   (b) $y^{(4)} - e^t y = 1$, $y(0) = 0$, $y'(0) = 0$, $y''(0) = 0$, $y'''(0) = 0$.

(c) $y''' + (t^2 - 1)^{1/2} y = 0$, $y(-1) = 1$, $y'(-1) = 0$, $y''(-1) = -1$.
(d) $y''' + (t^2 - 1)^{1/2} y = 0$, $y(2) = 0$, $y'(2) = 0$, $y''(2) = 1$.
5. Prove Corollary 2 of Theorem 1.

6.3 Linear Homogeneous Systems

In our study of linear systems of algebraic equations $Ax = b$ in Chapter 3, we first studied the homogeneous system $Ax = 0$ before studying the full system. We shall proceed in a similar manner to study linear systems of differential equations

$$\mathbf{y}' = A(t)\mathbf{y} + \mathbf{g}(t). \tag{6.16}$$

If $\mathbf{g}(t) \neq \mathbf{0}$, the system (6.16) is said to be nonhomogeneous. (This corresponds to the case $\mathbf{b} \neq \mathbf{0}$ of the algebraic system $Ax = b$.) With every nonhomogeneous system (6.16), we associate the corresponding homogeneous system

$$\mathbf{y}' = A(t)\mathbf{y}. \tag{6.17}$$

This section is devoted to the study of the algebraic structure of the set of all solutions of (6.17).[1] We assume throughout that the matrix $A(t)$ is continuous on some fixed interval $\mathscr{I}$, which may be finite or infinite. The entries $a_{ij}(t)$ of the matrix $A(t)$ can be real or complex valued.

By Definition 3, Section 6.1, a solution of the system (6.17) is a vector $\mathbf{u}(t)$ whose derivative $\mathbf{u}'(t)$ is continuous on $\mathscr{I}$. In the language of Chapter 5, (Example 7, Section 5.2), this means that a solution of the system (6.17) is an element of the vector space $C'(\mathscr{F}_n)$ on $\mathscr{I}$, that is, the vector space of functions with n components, real- or complex-valued (according as $\mathscr{F} = \mathscr{R}$ or $\mathscr{F} = \mathscr{C}$), and having continuous first derivatives on the interval $\mathscr{I}$.

Let $\mathbf{u}(t)$ and $\mathbf{v}(t)$ be any two solutions of (6.17) on the interval $\mathscr{I}$. Let α and β be arbitrary real or complex constants. We claim that $\alpha \mathbf{u}(t) + \beta \mathbf{v}(t)$ is also a solution of (6.17) on $\mathscr{I}$. For, by the rules of vector calculus discussed in Section 6.1, and the fact that $\mathbf{u}(t)$ and $\mathbf{v}(t)$ are solutions of (6.17),

$$[\alpha \mathbf{u}(t) + \beta \mathbf{v}(t)]' = \alpha \mathbf{u}'(t) + \beta \mathbf{v}'(t)$$
$$= \alpha A(t)\mathbf{u}(t) + \beta A(t)\mathbf{v}(t) = A(t)[\alpha \mathbf{u}(t) + \beta \mathbf{v}(t)].$$

---

[1] We are not concerned here with a specific initial value problem for (6.17). Rather, we are interested in the algebraic structure of the set of all solutions of the system of differential equations.

Since every linear combination of solutions of (6.17) is again a solution, it follows from Theorem 1, Section 5.3, that the set V of all solutions of (6.17) is a subspace of the vector space $C'(\mathscr{F}_n)$.

The reader will recall from Example 4 and Exercise 15, Section 5.5, that the vector space $C'(\mathscr{F}_n)$ is not finite dimensional. This fact emphasizes the importance of the problem of finding the dimension of the vector space V of solutions of (6.17). We have the answer in the following basic results.

Theorem 1. *If the complex $n \times n$ matrix $A(t)$ is continuous on an interval $\mathscr{I}$, then the solutions of the system*

$$\mathbf{y}' = A(t)\mathbf{y} \tag{6.17}$$

on $\mathscr{I}$ form a vector space V of dimension n over the complex numbers.

In view of the remarks preceding the statement of the theorem, it is significant that, according to Theorem 1, to find every solution of (6.17) it suffices to find a finite number of solutions, namely, a set that forms a basis for the vector space V.

Proof of Theorem 1. We have already established that the solutions form a vector space V over the complex numbers. To establish that the dimension of V is n, we need to construct a basis for V consisting of n linearly independent vectors in V, that is, of n linearly independent solutions of (6.17) on $\mathscr{I}$. We proceed as follows. Let t_0 be any point of $\mathscr{I}$ and let $\boldsymbol{\sigma}_1, \boldsymbol{\sigma}_2, \ldots, \boldsymbol{\sigma}_n$ be any n linearly independent points (vectors) in complex Euclidean n-space C_n. [For example, $\mathbf{e}_1, \mathbf{e}_2, \ldots, \mathbf{e}_n$ (see Section 5.1 and also Example 6, Section 5.4),

$$\mathbf{e}_j = \begin{bmatrix} 0 \\ \vdots \\ 0 \\ 1 \\ 0 \\ \vdots \\ 0 \end{bmatrix} \leftarrow j\text{th row}$$

are obviously n such vectors.] By Theorem 1, Section 6.2, the system (6.17) possesses n solutions $\boldsymbol{\phi}_1, \boldsymbol{\phi}_2, \ldots, \boldsymbol{\phi}_n$, each of which exists on the entire interval $\mathscr{I}$, and each solution $\boldsymbol{\phi}_j$ satisfies the initial condition

$$\boldsymbol{\phi}_j(t_0) = \boldsymbol{\sigma}_j \qquad (j = 1, 2, \ldots, n). \tag{6.18}$$

We first show that the solutions $\phi_1, \phi_2, \ldots, \phi_n$, are linearly independent on $\mathscr{I}$. Recall (see Examples 7, 8, 9, 10, 11, Section 5.4) that this involves examination of linear combinations of vector functions, but with scalar (constant) coefficients. Suppose there exist complex **constants** $a_1, a_2, \ldots, a_n$, such that

$$a_1 \phi_1(t) + a_2 \phi_2(t) + \cdots + a_n \phi_n(t) = 0 \quad \text{for every } t \text{ on } \mathscr{I}.$$

In particular, putting $t = t_0$, and using the initial conditions (6.18), we have

$$a_1 \sigma_1 + a_2 \sigma_2 + \cdots + a_n \sigma_n = 0.$$

But this implies that $a_1, a_2, \ldots, a_n$ are all zero because of the assumed linear independence of the given vectors $\sigma_1, \sigma_2, \ldots, \sigma_n$. Thus, $\phi_1, \phi_2, \ldots, \phi_n$ are linearly independent on $\mathscr{I}$.

To complete the proof we must show that these n linearly independent solutions of (6.17) span V; that is, they have the property that every solution $\psi(t)$ of (6.17) can be expressed as a linear combination of the solutions $\phi_1, \phi_2, \ldots, \phi_n$. We proceed as follows. Compute the value of the solution ψ at t_0 and let $\psi(t_0) = \sigma$. Since the constant vectors $\sigma_1, \sigma_2, \ldots, \sigma_n$ are linearly independent in Euclidean n-space C_n, they form a basis for C_n, and there exist unique constants $c_1, c_2, \ldots, c_n$ such that the constant vector σ can be represented as

$$\sigma = c_1 \sigma_1 + c_2 \sigma_2 + \cdots + c_n \sigma_n.$$

Now, consider the vector

$$\phi(t) = c_1 \phi_1(t) + c_2 \phi_2(t) + \cdots + c_n \phi_n(t).$$

Clearly, $\phi(t)$ is a solution of (6.17) on $\mathscr{I}$. (Why? Prove this.) Moreover, the initial value of ϕ is [using (6.18)]

$$\phi(t_0) = c_1 \sigma_1 + c_2 \sigma_2 + \cdots + c_n \sigma_n = \sigma.$$

Therefore, $\phi(t)$ and $\psi(t)$ are both solutions of (2.7) on $\mathscr{I}$ with $\phi(t_0) = \psi(t_0) = \sigma$. Thus, by the uniqueness part of Theorem 1, Section 6.2, $\phi(t) = \psi(t)$ for every t on $\mathscr{I}$, and the solution $\psi(t)$ is expressed as the unique linear combination

$$\psi(t) = c_1 \phi_1(t) + c_2 \phi_2(t) + \cdots + c_n \phi_n(t) \quad \text{for every } t \text{ on } \mathscr{I}. \quad (6.19)$$

6.3 Linear Homogeneous Systems 173

● **EXERCISE**

1. Show that this expression of $\psi(t)$ as a linear combination of $\phi_1(t), \ldots, \phi_n(t)$ is unique. [*Hint:* Assume $\psi(t) = d_1\phi_1(t) + \cdots + d_n\phi_n(t)$ in addition to (6.19) and show that $d_j = c_j$, where $j = 1, \ldots, n$.]

Thus, we have shown that the solutions $\phi_1, \phi_2, \ldots, \phi_n$ of (6.17) span the vector space V. Since they are also linearly independent, they form a basis for the solution space V, and the dimension of V is n. This completes the proof of Theorem 1. ∎

We often say that the linearly independent solutions $\phi_1, \ldots, \phi_n$ form a **fundamental set of solutions**. There are clearly infinitely many different fundamental sets of solutions of (6.17), namely, one corresponding to every basis $\sigma_1, \ldots, \sigma_n$ of $\mathscr{R}_n$.

● **EXERCISE**

2. Prove the following analog of Theorem 1 for systems with real coefficients. If the real $n \times n$ matrix $A(t)$ is continuous on an interval $\mathscr{I}$, then the real solutions of (6.17) on $\mathscr{I}$ form a vector space of dimension n over the real numbers.. [*Hint:* This is not a trick question; just check that the proof of Theorem 1 applies here.]

When we wish to apply Theorem 1, it is useful to restate the result in the following manner: *The system* (6.17) *possesses n linearly independent solutions on the interval $\mathscr{I}$. Moreover, every solution $\phi(t)$ of* (6.17) *on $\mathscr{I}$ can be expressed as a unique linear combination of those n solutions.* In practice, this means that it suffices to find, in any manner, n solutions of (6.17) and show that they are linearly independent. We shall devote considerable attention to this in Section 6.4 for the special case of scalar linear homogeneous differential equations, and in Chapter 7 to the case of (6.17) with $A(t)$ a constant matrix.

We now apply Theorem 1 to the scalar linear homogeneous second-order differential equation

$$y'' + p(t)y' + q(t)y = 0, \qquad (6.20)$$

where p and q are continuous functions on the interval $\mathscr{I}$.

Corollary 1 to Theorem 1. *Let p and q be continuous on the interval $\mathscr{I}$. Then equation* (6.20) *possesses two linearly independent solutions $\psi_1(t), \psi_2(t)$ on the interval $\mathscr{I}$. Moreover, if $\psi(t)$ is any solution of* (6.20) *on $\mathscr{I}$, then there exist unique constants c_1, c_2 such that*

$$\psi(t) = c_1\psi_1(t) + c_2\psi_2(t) \qquad \text{for every } t \text{ in } \mathscr{I}.$$

174   Linear Systems of Differential Equations

Proof. By the method of Example 6, Section 6.1, the scalar equation (6.20) is equivalent to the linear system

$$\mathbf{y}' = \begin{bmatrix} 0 & 1 \\ -q(t) & -p(t) \end{bmatrix} \mathbf{y}, \quad \mathbf{y} = \begin{bmatrix} y_1 \\ y_2 \end{bmatrix}, \tag{6.21}$$

which is a special case of (6.17). By Theorem 1, there exist two linearly independent (vector) solutions $\boldsymbol{\phi}_1(t)$, $\boldsymbol{\phi}_2(t)$ of (6.21) such that every solution $\boldsymbol{\phi}(t)$ of (6.21) has the form $\boldsymbol{\phi}(t) = c_1 \boldsymbol{\phi}_1(t) + c_2 \boldsymbol{\phi}_2(t)$. By the equivalence of (6.21) and (6.20),

$$\boldsymbol{\phi}_1(t) = \begin{bmatrix} \psi_1(t) \\ \psi_1'(t) \end{bmatrix}, \quad \boldsymbol{\phi}_2(t) = \begin{bmatrix} \psi_2(t) \\ \psi_2'(t) \end{bmatrix},$$

and $\psi_1(t)$, $\psi_2(t)$ are solutions of (6.20) on $\mathscr{I}$. We know that the vector solutions $\boldsymbol{\phi}_1(t)$ and $\boldsymbol{\phi}_2(t)$ are linearly independent on $\mathscr{I}$. We wish to show that $\psi_1(t)$ and $\psi_2(t)$ are linearly independent on $\mathscr{I}$. Suppose that $c_1 \psi_1(t) + c_2 \psi_2(t) = 0$ for every t in $\mathscr{I}$, then $c_1 \psi_1'(t) + c_2 \psi_2'(t) = 0$ for every t in $\mathscr{I}$. Thus, $c_1 \boldsymbol{\phi}_1(t) + c_2 \boldsymbol{\phi}_2(t) = 0$ on $\mathscr{I}$. Since $\boldsymbol{\phi}_1(t)$, $\boldsymbol{\phi}_2(t)$ are linearly independent on $\mathscr{I}$, $c_1 = 0$, $c_2 = 0$. Therefore $\psi_1(t)$ and $\psi_2(t)$ are linearly independent on $\mathscr{I}$. Also, every solution $\psi(t)$ of (6.20) is the first component of the corresponding vector solution $\boldsymbol{\phi}(t)$ of the system (6.21). Since $\boldsymbol{\phi}(t)$ has the form $\boldsymbol{\phi}(t) = c_1 \boldsymbol{\phi}_1(t) + c_2 \boldsymbol{\phi}_2(t)$, $\psi(t)$ has the form $\psi(t) = c_1 \psi_1(t) + c_2 \psi_2(t)$. ∎

The same reasoning applied to the scalar linear homogeneous differential equation of order n gives the following result.

Corollory 2 to Theorem 1.   *Let $p_1, \ldots, p_n$ be continuous on the interval $\mathscr{I}$. Then the differential equation*

$$y^{(n)} + p_1(t)y^{(n-1)} + \cdots + p_n(t)y = 0 \tag{6.22}$$

possesses n linearly independent solutions $\psi_1(t), \ldots, \psi_n(t)$ on the interval $\mathscr{I}$. Moreover, if $\psi(t)$ is any solution of (6.22) on $\mathscr{I}$, then there exist unique constants $c_1, \ldots, c_n$ such that $\psi(t) = c_1 \psi_1(t) + c_2 \psi_2(t) + \cdots + c_n \psi_n(t)$.

● **EXERCISE**

  3. Prove Corollary 2 to Theorem 1.

We can interpret Theorem 1 in a different and useful way. A matrix of n rows whose columns are solutions of (6.17) is called a **solution matrix**. Now,

6.3 Linear Homogeneous Systems

if we form an $n \times n$ matrix using n linearly independent solutions as columns, we will have a solution matrix on $\mathscr{I}$ and also its columns will be linearly independent on $\mathscr{I}$. A solution matrix whose columns are linearly independent on $\mathscr{I}$ is called a **fundamental matrix** for (6.17) on $\mathscr{I}$... Let us denote the fundamental matrix formed from the solutions $\phi_1, \phi_2, \ldots, \phi_n$ as columns by Φ. Then the statement that every solution ψ is the linear combination (6.19) for some unique choice of the constants $c_1, \ldots, c_n$ is simply that

$$\psi(t) = \Phi(t)\mathbf{c}, \tag{6.23}$$

where Φ is the fundamental matrix constructed above and $\mathbf{c}$ is the column vector with components $c_1, \ldots, c_n$. (The vector $\Phi(t)\mathbf{c}$ is obtained by forming the linear combination of columns of $\Phi(t)$ with $c_1, \ldots, c_n$ as coefficients.) It is clear that if $\tilde{\Phi}$ is any other fundamental matrix of (6.17) in $\mathscr{I}$, then the above solution ψ can be expressed as

$$\psi(t) = \tilde{\Phi}(t)\tilde{\mathbf{c}} \quad \text{for every } t \text{ on } \mathscr{I}$$

for a suitably chosen constant vector $\tilde{\mathbf{c}}$. Clearly, every solution of (6.17) on $\mathscr{I}$ can be expressed in this form by using any fundamental matrix.

We see from the discussion above that to find any solution of (6.17) we need to find a fundamental matrix. A natural question, then, is the following. Suppose we have found a solution matrix of (6.17) on some interval $\mathscr{I}$; can we test in some simple way whether this solution matrix is a fundamental matrix? The answer is contained in the following result.

Theorem 2. *A solution matrix* $\Phi(t)$ *of*

$$\mathbf{y}' = A(t)\mathbf{y} \tag{6.17}$$

is a fundamental matrix if and only if $\det \Phi(t) \neq 0$ *for every t in $\mathscr{I}$. Further, if* $\det \Phi(t_0) \neq 0$ *for some t_0 in $\mathscr{I}$, then* $\det \Phi(t) \neq 0$ *for all t in $\mathscr{I}$.*

Proof. If $\det \Phi(t) \neq 0$ for every t in $\mathscr{I}$, then the columns of the solution matrix $\Phi(t)$ are linearly independent on $\mathscr{I}$. For suppose there exist constants $c_1, \ldots, c_n$ such that

$$c_1 \phi_1(t) + c_2 \phi_2(t) + \cdots + c_n \phi_n(t) = \mathbf{0} \quad \text{for every } t \text{ in } \mathscr{I},$$

where $\phi_1(t), \ldots, \phi_n(t)$ are the columns of $\Phi(t)$. This can be written in the form

$$\Phi(t)\mathbf{c} = \mathbf{0} \quad \text{for every } t \text{ in } \mathscr{I},$$

where

$$\mathbf{c} = \begin{bmatrix} c_1 \\ \vdots \\ c_n \end{bmatrix}.$$

Fix t at $t = t_0$ in $\mathscr{I}$. Then $\Phi(t_0)\mathbf{c} = \mathbf{0}$ is a system of n algebraic equations for the n unknowns $c_1, \ldots, c_n$. Since $\det \Phi(t_0) \neq 0$, $c_1 = 0, c_2 = 0, \ldots, c_n = 0$ by Theorems 2 and 4, Section 4.5 This proves that the columns

$$\boldsymbol{\phi}_1(t), \ldots, \boldsymbol{\phi}_n(t)$$

are linearly independent; hence $\Phi(t)$ is a fundamental matrix on $\mathscr{I}$.

Conversely, suppose $\Phi(t)$ is a fundamental matrix of (6.17) on $\mathscr{I}$. Let $\boldsymbol{\phi}(t)$ be a solution of (6.17) on $\mathscr{I}$. By Equation 6.23, there exists a unique vector $\mathbf{c}$ such that $\boldsymbol{\phi}(t) = \Phi(t)\mathbf{c}$ for every t in $\mathscr{I}$. Fix t_0 in $\mathscr{I}$; then, in fact, the constant vector $\mathbf{c}$ is uniquely determined by solving the algebraic system $\Phi(t_0)\mathbf{x} = \boldsymbol{\phi}(t_0)$. Since this algebraic system has a unique solution for each right-hand side $\boldsymbol{\phi}(t_0)$, the coefficient matrix $\Phi(t)_0$ has rank n (Theorem 4, Section 3.7). By Theorem 1, Section 3.8, $\Phi(t_0)$ is nonsingular, and by Theorem 2, Section 4.5, $\det \Phi(t_0) \neq 0$. This is true for each fixed t_0 in $\mathscr{I}$, and therefore $\det \Phi(t) \neq 0$ for each t in $\mathscr{I}$. It may appear that the vector $\mathbf{c}$ depends on the choice of t_0. However, it does not for the following reason. Since $\boldsymbol{\phi}(t) = \Phi(t)\mathbf{c}$ for every t in $\mathscr{I}$, if $t_1 \neq t_0$ in $\mathscr{I}$, then $\boldsymbol{\phi}(t_1) = \Phi(t_1)\mathbf{c}$. Thus, the unique solution of the algebraic system $\Phi(t_1)\mathbf{x} = \boldsymbol{\phi}(t_1)$ is the same vector $\mathbf{c}$ obtained as the unique solution of the algebraic system $\Phi(t_0)\mathbf{x} = \boldsymbol{\phi}(t_0)$.

Finally, if $\det \Phi(t_0) \neq 0$ for some t_0 in $\mathscr{I}$, let $\boldsymbol{\sigma}_1 = \boldsymbol{\phi}_1(t_0), \ldots, \boldsymbol{\sigma}_n = \boldsymbol{\phi}_n(t_0)$. The vectors $\boldsymbol{\sigma}_1, \ldots, \boldsymbol{\sigma}_n$ are linearly independent (Example 6, Section 5.4), and therefore form a basis for C_n. We claim that the solutions $\boldsymbol{\phi}_1(t), \ldots, \boldsymbol{\phi}_n(t)$ are linearly independent on $\mathscr{I}$; for if not, there exist scalars $c_1, c_2, \ldots, c_n$ not all zero such that

$$c_1 \boldsymbol{\phi}_1(t) + c_2 \boldsymbol{\phi}_2(t) + \cdots + c_n \boldsymbol{\phi}_n(t) \equiv 0 \text{ on } \mathscr{I}.$$

Putting $t = t_0$, we obtain

$$c_1 \boldsymbol{\sigma}_1 + c_2 \boldsymbol{\sigma}_2 + \cdots + c_n \boldsymbol{\sigma}_n = 0$$

which contradicts the linear independence of $\boldsymbol{\sigma}_1, \ldots, \boldsymbol{\sigma}_n$. Hence, $\Phi(t)$ is a fundamental matrix of (6.17). Therefore, by the second part of the proof, $\det \Phi(t) \neq 0$ for every t in $\mathscr{I}$. ∎

6.3 Linear Homogeneous Systems

The reader is warned that a matrix may have its determinant identically zero on some interval, although its columns are linearly independent. Indeed, let

$$\Phi(t) = \begin{bmatrix} 1 & t & t^2 \\ 0 & 2 & t \\ 0 & 0 & 0 \end{bmatrix}.$$

Then clearly det $\Phi(t) = 0$, $-\infty < t < \infty$, and yet the columns are linearly independent. This, according to Theorem 2, cannot happen for solutions of (6.17).

Example 1. Show that

$$\Phi(t) = \begin{bmatrix} e^t & te^t \\ 0 & e^t \end{bmatrix}$$

is a fundamental matrix for the system

$$\mathbf{y}' = \begin{bmatrix} 1 & 1 \\ 0 & 1 \end{bmatrix} \mathbf{y}, \quad \text{where} \quad \mathbf{y} = \begin{bmatrix} y_1 \\ y_2 \end{bmatrix}.$$

We first show that $\Phi(t)$ is a solution matrix. Let $\boldsymbol{\phi}_1(t)$ denote the first column of $\Phi(t)$; then

$$\boldsymbol{\phi}_1'(t) = \begin{bmatrix} e^t \\ 0 \end{bmatrix} = \begin{bmatrix} 1 & 1 \\ 0 & 1 \end{bmatrix} \begin{bmatrix} e^t \\ 0 \end{bmatrix} = \begin{bmatrix} 1 & 1 \\ 0 & 1 \end{bmatrix} \boldsymbol{\phi}_1(t)$$

for $-\infty < t < \infty$. Similarly, if $\boldsymbol{\phi}_2(t)$ denotes the second column of $\Phi(t)$, we have

$$\boldsymbol{\phi}_2'(t) = \begin{bmatrix} (t+1)e^t \\ e^t \end{bmatrix} = \begin{bmatrix} 1 & 1 \\ 0 & 1 \end{bmatrix} \begin{bmatrix} te^t \\ e^t \end{bmatrix} = \begin{bmatrix} 1 & 1 \\ 0 & 1 \end{bmatrix} \boldsymbol{\phi}_2(t)$$

for $-\infty < t < \infty$. Therefore, $\Phi(t) = [\boldsymbol{\phi}_1(t), \boldsymbol{\phi}_2(t)]$ is a solution matrix for $-\infty < t < \infty$. By Theorem 2, since det $\Phi(t) = e^{2t} \neq 0$, $\Phi(t)$ is a fundamental matrix for $-\infty < t < \infty$. By Theorem 2 also, it is enough to compute det $\Phi(t)$ at one point, for instance $t = 0$. Since $\Phi(0) = I$, this gives det $\Phi(0) = 1 \neq 0$.

Linear Systems of Differential Equations

● **EXERCISES**

4. Show, with the aid of Theorem 2, that

$$\begin{bmatrix} \cos t & \sin t \\ -\sin t & \cos t \end{bmatrix}$$

is a fundamental matrix for the system $\mathbf{y}' = A\mathbf{y}$, where

$$A = \begin{bmatrix} 0 & 1 \\ -1 & 0 \end{bmatrix}.$$

5. Show, with the aid of Theorem 2, that

$$\begin{bmatrix} \exp(r_1 t) & \exp(r_2 t) \\ r_1 \exp(r_1 t) & r_2 \exp(r_2 t) \end{bmatrix}$$

is a fundamental matrix for the system $\mathbf{y}' = A\mathbf{y}$, where

$$A = \begin{bmatrix} 0 & 1 \\ -a_2 & -a_1 \end{bmatrix},$$

and r_1, r_2 are the distinct roots of the quadratic equation $z^2 + a_1 z + a_2 = 0$. (We shall learn in Chapter 7 how to construct this fundamental matrix.)

Corollary 1 to Theorem 2. *If $\Phi(t)$ is a fundamental matrix of $\mathbf{y}' = A(t)\mathbf{y}$ on an interval $\mathscr{I}$ and if $C \in \mathscr{F}_{nn}$ is a nonsingular constant matrix, then $\Phi(t)C$ is also a fundamental matrix of $\mathbf{y}' = A(t)\mathbf{y}$ on $\mathscr{I}$.*

Proof. Let $\Phi(t) = [\boldsymbol{\phi}_1(t), \boldsymbol{\phi}_2(t), \ldots, \boldsymbol{\phi}_n(t)]$. Then the columns of $\Phi(t)C$ are linear combinations of the columns of $\Phi(t)$ (by matrix multiplication). Since the columns of $\Phi(t)$ are solutions, $\Phi(t)C$ is a solution matrix on $\mathscr{I}$. By Theorem 3, Section 4.5, $\det \Phi(t)C = \det \Phi(t) \det C$. By Theorem 2, $\det \Phi(t) \neq 0$ on $\mathscr{I}$, and since C is nonsingular, $\det C \neq 0$. Thus, $\det \Phi(t)C \neq 0$ on $\mathscr{I}$ and, again by Theorem 2, $\Phi(t)C$ is a fundamental matrix on $\mathscr{I}$. ∎

● **EXERCISE**

6. Show that $C\Phi(t)$, where C is a constant matrix and $\Phi(t)$ is a fundamental matrix, need not be a solution matrix $\mathbf{y}' = A(t)\mathbf{y}$.

The converse of Corollary 1 is also true.

Corollary 2 to Theorem 2. *If $\Phi(t)$ and $\Psi(t)$ are two fundamental matrices of $y' = A(t)y$ on $\mathscr{I}$, then there exists a nonsingular constant matrix C such that $\Psi(t) = \Phi(t)C$ on $\mathscr{I}$.*

Proof. Letting ψ_j be the *j*th column of Ψ, we see from (6.23) that $\psi_j = \Phi c_j$, $j = 1, \ldots, n$, where c_j are suitable constant vectors. Therefore, if we define C as the constant matrix whose columns are the vectors $c_j, j = 1, \ldots, n$, we have at once that $\Psi(t) = \Phi(t)C$ for every t on $\mathscr{I}$. Since

$$\det \Psi(t) = \det \Phi(t) \det C$$

and since det Φ and det Ψ are both different from zero on $\mathscr{I}$ (why?), we also have det $C \neq 0$ so that C is a nonsingular constant matrix. ∎

● **EXERCISES**

7. (a) Show that

$$\Phi(t) = \begin{bmatrix} t^2 & t \\ 2t & 1 \end{bmatrix}$$

is a fundamental matrix for the system $y' = A(t)y$, where

$$A(t) = \begin{bmatrix} 0 & 1 \\ -\frac{2}{t^2} & \frac{2}{t} \end{bmatrix}$$

on any interval $\mathscr{I}$ not including the origin.
(b) Does the fact that det $\Phi(0) = 0$ contradict Theorem 2?

8. Show that if a real homogeneous system of two first-order equations has a fundamental matrix

$$\Phi(t) = \begin{bmatrix} e^{it} & e^{-it} \\ ie^{it} & -ie^{-it} \end{bmatrix},$$

then

$$\begin{bmatrix} \cos t & \sin t \\ -\sin t & \cos t \end{bmatrix}$$

is also a fundamental matrix. Can you find another real fundamental matrix? [*Hint:* Let $\Phi(t) = [\phi_1(t), \phi_2(t)]$. Show that $\mathscr{R}\phi_1(t)$ and $\mathscr{R}\phi_2(t)$ are solutions of $y' = Ay$, A real, where $\mathscr{R}$ is the real part. By the real part of a vector we mean, of course, the real part of each component. A similar result holds for the imaginary parts of $\phi_1(t)$ and $\phi_2(t)$.]

We shall now apply Theorem 2 to the scalar linear homogeneous second-order equation

$$y'' + p(t)y' + q(t)y = 0, \qquad (6.20)$$

where p and q are continuous functions on a given interval $\mathscr{I}$. As we have seen in the proof of Corollary 1 to Theorem 1, (6.20) is equivalent to the system

$$\mathbf{y}' = \begin{bmatrix} 0 & 1 \\ -q(t) & -p(t) \end{bmatrix} \mathbf{y}, \qquad \mathbf{y} = \begin{bmatrix} y_1 \\ y_2 \end{bmatrix}. \qquad (6.21)$$

If $\Phi(t)$ is a solution matrix of (6.21) on $\mathscr{I}$, then $\Phi(t) = [\phi_1(t), \phi_2(t)]$, where

$$\phi_1(t) = \begin{bmatrix} \psi_1(t) \\ \psi_1'(t) \end{bmatrix}, \qquad \phi_2(t) = \begin{bmatrix} \psi_2(t) \\ \psi_2'(t) \end{bmatrix}$$

with $\psi_1(t)$, $\psi_2(t)$ solutions of the scalar equation (6.20). By Theorem 2, $\Phi(t)$ is a fundamental matrix of (6.21) on $\mathscr{I}$ if and only if

$$\det \Phi(t) = \det \begin{bmatrix} \psi_1(t) & \psi_2(t) \\ \psi_1'(t) & \psi_2'(t) \end{bmatrix} \neq 0 \qquad \text{for } t \text{ in } \mathscr{I}.$$

This determinant is called the **Wronskian** of $\psi_1(t)$ and $\psi_2(t)$. Thus, by the proof of Corollary 1 of Theorem 1, if $\det \Phi(t) \neq 0$, then the solutions $\psi_1(t)$, $\psi_2(t)$ of the scalar equation (6.20) are linearly independent on $\mathscr{I}$, and every solution of (6.20) can be written as a linear combination of $\psi_1(t)$ and $\psi_2(t)$. This is one-half of the following result.

Corollary 3 to Theorem 2. *Two solutions ψ_1, ψ_2 of (6.20) on $\mathscr{I}$ are linearly independent on $\mathscr{I}$ if and only if their Wronskian*

$$W[\psi_1(t), \psi_2(t)] = \det \begin{bmatrix} \psi_1(t) & \psi_2(t) \\ \psi_1'(t) & \psi_2'(t) \end{bmatrix}$$

is different from zero for all t in $\mathscr{I}$.

Proof. We must still prove that if the solutions ψ_1, ψ_2 are linearly independent on $\mathscr{I}$, then their Wronskian is different from zero for every t in $\mathscr{I}$.

6.3 Linear Homogeneous Systems

Suppose there is at least one $\hat{t}$ on $\mathscr{I}$ such that $W[\psi_1(\hat{t}), \psi_2(\hat{t})] = 0$. (If no such $\hat{t}$ exists, then there is nothing to prove.) Consider the algebraic system

$$a_1\psi_1(\hat{t}) + a_2\psi_2(\hat{t}) = 0$$
$$a_1\psi_1'(\hat{t}) + a_2\psi_2'(\hat{t}) = 0,$$

for the unknowns a_1, a_2. By Theorem 2, Section 4.5 and Corollary 2, Theorem 1, Section 3.8, this system has a nontrivial solution $\hat{a}_1, \hat{a}_2$, where $\hat{a}_1, \hat{a}_2$ are not both zero. Consider the function $\psi(t) = \hat{a}_1\psi_1(t) + \hat{a}_2\psi_2(t)$. Since (6.20) is linear, ψ is a solution of (6.20) on $\mathscr{I}$, and $\psi(\hat{t}) = 0$, $\psi'(\hat{t}) = 0$. By Corollary 1 to Theorem 1, Section 6.2, there is only one solution to the initial value problem consisting of (6.20) together with the initial conditions $y(\hat{t}) = 0$, $y'(\hat{t}) = 0$. Since the identically zero function is a solution of this initial value problem, we conclude that $\psi(t) \equiv 0$ on $\mathscr{I}$. Therefore, $\hat{a}_1\psi_1(t) + \hat{a}_2\psi_2(t) = 0$ for every t in $\mathscr{I}$. Since $\hat{a}_1, \hat{a}_2$ are not both zero, ψ_1 and ψ_2 are linearly dependent on $\mathscr{I}$. Thus, if ψ_1 and ψ_2 are linearly independent on $\mathscr{I}$, there can be no such $\hat{t}$, and $W[\psi_1(t), \psi_2(t)] \neq 0$ for every t in $\mathscr{I}$. ∎

By Corollary 1 to Theorem 1 every solution of equation (6.20) on $\mathscr{I}$ has the form $c_1\psi_1 + c_2\psi_2$ for some unique choice of the constants c_1, c_2. For this reason a pair of linearly independent solutions, such as ψ_1, ψ_2, of Eq. (6.20) are said to form a **fundamental set of solutions**.

● **EXERCISES**

9. Show that e^{2t}, e^{-2t} are linearly independent solutions of $y'' - 4y = 0$ on $-\infty < t < \infty$.

10. Show that $e^{-t/2}\cos\sqrt{3}t/2$, $e^{-t/2}\sin\sqrt{3}t/2$ are linearly independent solutions of $y'' + y' + y = 0$ on $-\infty < t < \infty$.

11. Show that e^{-t}, te^{-t} are linearly independent solutions of $y'' + 2y' + y = 0$ on $-\infty < t < \infty$.

12. Show that $\sin t^2, \cos t^2$ are linearly independent solutions of $ty'' - y' + 4t^3y = 0$ on $0 < t < \infty$ or $-\infty < t < 0$. Show that $W(\sin t^2, \cos t^2)(0) = 0$. Why does this fact not contradict Corollary 3 of Theorem 2?

13. (a) Let ϕ_1, ϕ_2 be any two solutions on some interval $\mathscr{I}$, of $L(y) = a_0(t)y'' + a_1(t)y' + a_2(t)y = 0$, where a_0, a_1, a_2 are continuous on $\mathscr{I}$ and $a_0(t) \neq 0$ on $\mathscr{I}$. Show that the Wronskian $W(\phi_1, \phi_2)(t)$ satisfies the first-order linear differential equation

$$W' = -\frac{a_1(t)}{a_0(t)}W \quad (t \text{ on } \mathscr{I}.) \tag{*}$$

182 Linear Systems of Differential Equations

[Hint:

$$W'(\phi_1, \phi_2)(t) = \begin{vmatrix} \phi_1(t) & \phi_2(t) \\ \phi_1'(t) & \phi_2'(t) \end{vmatrix}' = (\phi_1\phi_2' - \phi_1'\phi_2)' = \phi_1\phi_2'' - \phi_1''\phi_2.$$

Now, use the fact that ϕ_1, ϕ_2 are solutions of $L(y) = 0$ on $\mathscr{I}$ to replace ϕ_1'' and ϕ_2'' by terms involving $\phi_1, \phi_1', \phi_2, \phi_2'$. If you then collect terms you should get Eq. (*).]

(b) By solving (*), derive **Abel's formula**:

$$W(\phi_1, \phi_2)(t) = W(\phi_1, \phi_2)(t_0) \exp\left(-\int_{t_0}^{t} \frac{a_1(s)}{a_0(s)} ds\right),$$

for t_0, t on $\mathscr{I}$. This gives another way of seeing that if the Wronskian is different from zero at one point, then it is never zero. (Compare Theorem 2.)

14. State the analog of Corollary 3 to Theorem 2 for the linear third-order differential equation

$$L_3(y) = a_0(t)y''' + a_1(t)y'' + a_2(t)y' + a_3(t)y = 0.$$

15. Show that $e^t, \cos t, \sin t$ are linearly independent solutions on $-\infty < t < \infty$ of the differential equation $y''' - y'' + y' - y = 0$.

16. Show that

$$\phi_1(t) = 1 + \sum_{m=1}^{\infty} \frac{t^{3m}}{2 \cdot 3 \cdot 5 \cdot 6 \cdots (3m-1)(3m)},$$

$$\phi_2(t) = t + \sum_{m=1}^{\infty} \frac{t^{3m+1}}{3 \cdot 4 \cdot 6 \cdot 7 \cdots (3m)(3m+1)},$$

are linearly independent solutions of $y'' - ty = 0$ on the interval $-\infty < t < \infty$. (Here you may assume that it has already been shown that ϕ_1 and ϕ_2 are solutions of $y'' - ty = 0$, but how could you verify this?)

By a similar argument we can establish the following analog of Corollary 3 to Theorem 2 for the scalar equation of order n.

Corollary 4 to Theorem 2. *A set of n solutions $\psi_1, \psi_2, \ldots, \psi_n$ on $\mathscr{I}$ of*

$$y^{(n)} + p_1(t)y^{(n-1)} + \cdots + p_n(t)y = 0, \qquad (6.22)$$

where $p_1, p_2, \ldots, p_n$ are continuous on $\mathscr{I}$, is linearly independent on $\mathscr{I}$ if and

6.4 Solution of Scalar Linear Differential Equations

only if the Wronskian

$$W[\psi_1(t),\ldots,\psi_n(t)] = \det \begin{bmatrix} \psi_1(t) & \psi_2(t) & \cdots & \psi_n(t) \\ \psi_1'(t) & \psi_2'(t) & \cdots & \psi_n'(t) \\ \vdots & \vdots & & \vdots \\ \psi_1^{(n-1)}(t) & \psi_2^{(n-1)}(t) & \cdots & \psi_n^{(n-1)}(t) \end{bmatrix}$$

is different from zero for every t on $\mathscr{I}$.

A set of n linearly independent solutions of Eq. 6.22 is said to form a **fundamental set of solutions** (see the remarks following Corollary 3 to Theorem 2 in the second-order case).

● **EXERCISE**

17. Prove Corollary 4 to Theorem 2. [*Hint:* Imitate the proof of Corollary 3 to Theorem 2.]

6.4 Solution of Scalar Linear Differential Equations with Constant Coefficients

To provide the reader with some concrete examples of solving differential equations, we solve the general second-order linear homogeneous equation with constant coefficients p, q:

$$L(y) = y'' + py' + qy = 0. \tag{6.24}$$

We then indicate how to solve the general linear homogeneous equation of order n with constant coefficients $p_1, p_2, \ldots, p_n$:

$$L_n(y) = y^{(n)} + p_1 y^{(n-1)} + \cdots + p_n y = 0. \tag{6.25}$$

The reader already familiar with this material should omit this section, as it serves only as an illustration and does not present any new general theory.[2]

We begin with the second-order equation (6.24). In accordance with the

---

[2] Note that for any functions f, g with continuous second derivatives on $-\infty < t < \infty$ and any constants a, b we have $L(af + bg) = aL(f) + bL(g)$; L is called a linear differential operator (of order two). Similarly, L_n is a linear differential operator of order n.

theory developed in the preceding section (Corollary 1 to Theorem 1, Section 6.3), our task is to find two linearly independent solutions of (6.24). Recall that for the first-order equation $y' + ry = 0$, where r is a constant, e^{-rt} is a solution. We could also find it as follows: If for some constant z, e^{zt} is to be a solution of $y' + ry = 0$, then we must have $(e^{zt})' + re^{zt} = 0$ or $(z + r)e^{zt} = 0$. Since $e^{zt} \neq 0$ we see that e^{zt} can be a solution of $y' + ry = 0$ only if $z = -r$ which gives e^{-rt} as a candidate for a solution. Direct verification shows that it is.

Let us try to find a solution of (6.24) of the form e^{zt} on $-\infty < t < \infty$. Then we must have $L(e^{zt}) = 0$. But

$$L(e^{zt}) = (e^{zt})'' + p(e^{zt})' + qe^{zt} = (z^2 + pz + q)e^{zt}.$$

Therefore, e^{zt} can be a solution of $L(y) = 0$ on $-\infty < t < \infty$ only if

$$(z^2 + pz + q)e^{zt} = 0$$

or, since $e^{zt} \neq 0$, only if z is a root of the quadratic equation

$$z^2 + pz + q = 0. \tag{6.26}$$

Equation (6.26) is called the **characteristic equation** or **auxiliary equation** associated with (6.24), and $z^2 + pz + q$ is called the **characteristic polynomial** associated with (6.24). The quadratic equation (6.26) has the roots

$$z_1 = \frac{-p + (p^2 - 4q)^{1/2}}{2}, \quad z_2 = \frac{-p - (p^2 - 4q)^{1/2}}{2}.$$

● **EXERCISE**

1. Verify that $\exp(z_1 t)$, $\exp(z_2 t)$ are solutions of (6.24).

Now, there are two possibilities:
(i) If $p^2 \neq 4q$, the roots z_1 and z_2 are distinct.
(ii) If $p^2 = 4q$, $z_1 = z_2$.

In case (i) this means that $\exp(z_1 t)$ and $\exp(z_2 t)$, where $z_1 \neq z_2$, are two distinct solutions of the differential equation (6.25) on $-\infty < t < \infty$. The only question which remains is: Are these solutions linearly independent on $-\infty < t < \infty$? Since $\exp(z_1 t)$, $\exp(z_2 t)$ are solutions of (6.24), we can establish their linear independence by Corollary 3, Theorem 2 (Section 6.3),

6.4 Solution of Scalar Linear Differential Equations

using the Wronskian. We have

$$W[\exp(z_1 t), \exp(z_2 t)] = \begin{vmatrix} \exp(z_1 t) & \exp(z_2 t) \\ z_1 \exp(z_1 t) & z_2 \exp(z_2 t) \end{vmatrix}$$

$$= (z_2 - z_1) \exp(z_1 + z_2) t \qquad (-\infty < t < \infty).$$

Since $z_1 \neq z_2$, $W[\exp(z_1 t), \exp(z_2 t)] \neq 0$ and $\exp(z_1 t)$, $\exp(z_2 t)$ are linearly independent solutions on $-\infty < t < \infty$. Therefore, in case (i), by Corollary 1 to Theorem 1, Section 6.3, every solution ϕ of Eq. (6.24) has the form

$$\phi(t) = c_1 \exp(z_1 t) + c_2 \exp(z_2 t) \qquad (-\infty < t < \infty), \qquad (6.27)$$

for some unique choice of the constants c_1, c_2. Because every solution of (6.24) has the form (6.27), (6.27) is called the **general solution** of (6.24).

Example 1. Find the general solution of $y'' - 9y = 0$.

By the method just given, the characteristic equation (6.26) is $z^2 - 9 = 0$, which has the roots $z_1 = 3$, $z_2 = -3$. Since $z_1 \neq z_2$, the solutions e^{3t}, e^{-3t} are linearly independent on $-\infty < t < \infty$ and, therefore, the general solution of $y'' - 9y = 0$ is $\phi(t) = c_1 e^{-3t} + c_2 e^{-3t}$, where c_1, c_2 are arbitrary constants.

● **EXERCISES**

2. Find the general solution of each of the following equations, and then the solution ϕ satisfying the given initial conditions:
  (a) $y'' - y = 0$, $\phi(0) = 0$, $\phi'(0) = 1$.
  (b) $y'' - 5y' + 6y = 0$, $\phi(0) = 0$, $\phi'(0) = 1$.
  (c) $y''' - 6y'' + 11y' - 6y = 0$, $\phi(0) = \phi'(0) = 0$, $\phi''(0) = 1$.

3. In Eq. (6.24) for damped oscillations (that is, $p, q > 0$), find values of p and q such that the roots of the characteristic equation are real and distinct. For such values, discuss the asymptotic behavior (behavior as $t \to +\infty$) of the solutions by computing $\lim_{t \to +\infty} \phi(t)$, where $\phi(t)$ is any solution of (6.24).

The careful reader may have noticed that in case (i), the roots z_1, z_2 will be real and distinct if $p^2 > 4q$, and will be complex conjugate (hence, distinct) if $p^2 < 4q$. For example, the differential equation

$$y'' + y' + y = 0$$

has the characteristic equation $z^2 + z + 1 = 0$, and its roots are

$$z_1 = \frac{-1 + \sqrt{3}\,i}{2}, \qquad z_2 = \frac{-1 - \sqrt{3}\,i}{2}.$$

Therefore,

$$\exp\left[\frac{-1+\sqrt{3}\,i}{2}\right]t \quad \text{and} \quad \exp\left[\frac{-1-\sqrt{3}\,i}{2}\right]t$$

should be, and in fact are, solutions. However, these functions are **complex-valued functions of the real variable** t, and since the differential equation has real coefficients, we would expect to be able to express the solutions in terms of real functions.

● EXERCISE

4. Show that the functions

$$\exp\left[\frac{-1-\sqrt{3i}}{2}\right]t \quad \text{and} \quad \exp\left[\frac{-1/\sqrt{3}\,i}{2}\right]t$$

satisfy the differential equation $y'' + y' + y = 0$ for all real t.

We now present a result on complex-valued solutions of **real** linear differential equations which enables us to obtain real solutions. Note that this result is not restricted to equations with constant coefficients.

Theorem 1. *Let ϕ be a complex-valued solution of the differential equation*

$$L(y) = a_0(t)y'' + a_1(t)y' + a_2(t)y = 0$$

on some interval $\mathcal{I}$, where a_0, a_1, a_2 are given real functions on $\mathcal{I}$. Then the real functions $u = \mathcal{R}\phi$, $v = \mathcal{I}\phi$ are themselves (real) solutions of $L(y) = 0$ on $\mathcal{I}$.

Proof. Since ϕ is a solution of $L(y) = 0$ on $\mathcal{I}$, we have

$$a_0(t)\phi''(t) + a_1(t)\phi'(t) + a_2(t)\phi(t) = 0,$$

for every t on $\mathcal{I}$. (The fact that ϕ may be complex valued does not change anything.) Since $\phi = u + iv$, we have, from the definition of derivative $\phi'(t) = u'(t) + iv'(t)$, $\phi''(t) = u''(t) + iv''(t)$. Therefore,

$$a_0(t)[u''(t) + iv''(t)] + a_1(t)[u'(t) + iv'(t)] + a_2(t)[u(t) + iv(t)] = 0.$$

Separating the left-hand side into real and imaginary parts, we obtain

6.4 Solution of Scalar Linear Differential Equations

(remember that a_0, a_1, a_2 are real) for all t on $\mathcal{I}$:

$$a_0(t)u''(t) + a_1(t)u'(t) + a_2(t)u(t) \\ + i[a_0(t)v''(t) + a_1(t)v'(t) + a_2(t)v(t)] = 0.$$

(*Note*: This also shows that $L(\phi) = L(u) + iL(v)$: this is true in general if L is a linear differential operator with real coefficients.) Since the last relation holds for every t on $\mathcal{I}$ and since a complex number is zero if and only if both its real and imaginary parts are zero, we have, for all t in $\mathcal{I}$:

$$L(u) = a_0(t)u''(t) + a_1(t)u'(t) + a_2(t)u(t) = 0$$

and

$$L(v) = a_0(t)v''(t) + a_1(t)v'(t) + a_2(t)v(t) = 0,$$

which shows that $u = \mathcal{R}\phi$ and $v = \mathcal{I}\phi$ are both solutions of $L(y) = 0$ on $\mathcal{I}$ and completes the proof. ∎

● **EXERCISE**

5. Let ϕ be a solution on some interval $\mathcal{I}$ of the differential equation

$$L(y) = a_0(t)y'' + a_1(t)y' + a_2(t)y = b(t),$$

where a_0, a_1, a_2 are real and b is complex. Show that $u = \mathcal{R}\phi$ satisfies the equation $L(y) = \mathcal{R}b$ and prove an analogous result for $v = \mathcal{I}\phi$.

Example 2. We have seen that

$$\phi_1(t) = \exp\left[\frac{-1 + \sqrt{3}\,i}{2}t\right] \quad \text{and} \quad \phi_2(t) = \exp\left[\frac{-1 - \sqrt{3}\,i}{2}t\right]$$

are both (complex-valued) solutions of $y'' + y' + y = 0$ on $-\infty < t < \infty$. They are linearly independent on $-\infty < t < \infty$; for, by Corollary 3 to Theorem 2, Section 6.3,

$$W(\phi_1, \phi_2)(t) = -\sqrt{3}\,ie^{-t} \neq 0 \quad (-\infty < t < \infty).$$

Therefore, by Corollary 1 to Theorem 1, Section 6.3, every solution ϕ (possibly complex valued) of $y'' + y' + y = 0$ on $-\infty < t < \infty$ has the form $\phi(t) = c_1\phi_1(t) + c_2\phi_2(t)$ for some unique choice of the (possibly complex) constants c_1, c_2. By Theorem 1 (applicable because the coefficients are real) the real functions $u_1(t) = \mathscr{R}\phi_1(t) = \exp[-t/2] \cos(\sqrt{3}/2)t$ and $v_1(t) = \mathscr{I}\phi_1(t) = \exp[-t/2] \sin(\sqrt{3}/2)t$ are also solutions of $y'' + y' + y = 0$ for $-\infty < t < \infty$. The same statement applies to

$$u_2(t) = \mathscr{R}\phi_2(t) = \exp[-t/2] \cos(\sqrt{3}/2)$$

and

$$v_2(t) = \mathscr{I}\phi_2(t) = -\exp[-t/2] \sin(\sqrt{3}/2)t.$$

(Note that the solutions u_2, v_2 yield the same general solution as the solutions u_1, v_1.) The reader can easily check that $W(u_1, v_1)(t) \neq 0$ on $-\infty < t < \infty$. Therefore by Corollary 3 to Theorem 2, Section 6.3, again, every solution ϕ of $y'' + y' + y = 0$ on $-\infty < t < \infty$ has the form

$$\phi(t) = a_1 \exp\left[-\frac{t}{2}\right] \cos\frac{\sqrt{3}}{2}t + a_2 \exp\left[-\frac{t}{2}\right] \sin\frac{\sqrt{3}}{2}t$$

for some unique choice of the (possibly complex) constants a_1, a_2. Starting with the complex form of the general solution ϕ we may also arrive at the "real form" as follows. Using Euler's formula, $e^{(\alpha + i\beta)t} = e^{\alpha t} e^{i\beta t} = e^{\alpha t}(\cos \beta t + i \sin \beta t)$, and collecting terms, we see that the general solution ϕ is given by

$$\phi(t) = c_1 \phi_1(t) + c_2 \phi_2(t)$$

$$= c_1 \exp\left[-\frac{t}{2}\right]\left(\cos\frac{\sqrt{3}}{2}t + i\sin\frac{\sqrt{3}}{2}t\right)$$

$$+ c_2 \exp\left[-\frac{t}{2}\right]\left(\cos\frac{\sqrt{3}}{2}t - i\sin\frac{\sqrt{3}}{2}t\right)$$

$$= (c_1 + c_2) \exp\left[-\frac{t}{2}\right] \cos\frac{\sqrt{3}}{2}t + i(c_1 - c_2) \exp\left[-\frac{t}{2}\right] \sin\frac{\sqrt{3}}{2}t.$$

If we now define $a_1 = c_1 + c_2$, $a_2 = i(c_1 - c_2)$, we obtain the desired form. It is clear from this that the solution $\phi(t)$ of the equation $y'' + y' + y = 0$ will be real if and only if $c_2 = \bar{c}_1$ (the complex conjugate of c_1). In this case, of course, a_1 and a_2 are both real.

We now return to the general equation (6.24), where p and q are real constants and summarize what we have learned up to this point.

6.4 Solution of Scalar Linear Differential Equations

Theorem 2. *Every solution ϕ of the differential equation*

$$y'' + py' + qy = 0, \tag{6.24}$$

where p, q are real constants with $p^2 \neq 4q$ is defined on $-\infty < t < \infty$ and has the form

$$\phi(t) = c_1 \exp(z_1 t) + c_2 \exp(z_2 t) \qquad (-\infty < t < \infty), \tag{6.27}$$

for some unique choice of the constants c_1 and c_2. The numbers z_1, z_2 are the distinct roots of the characteristic equation

$$z^2 + pz + q = 0. \tag{6.26}$$

If $p^2 > 4q$, z_1 and z_2 are real and distinct. If $p^2 < 4q$, the roots z_1, z_2 are complex conjugates. In this case, if $z_1 = \alpha + i\beta$ (α, β real) the solution ϕ may be expressed in the form

$$\phi(t) = e^{\alpha t}(a_1 \cos \beta t + a_2 \sin \beta t), \tag{6.28}$$

for some uniquely chosen constants a_1 and a_2. If ϕ is real, a_1 and a_2 are real.

We have already proved all of Theorem 2 except for Eq. 6.28. To prove (6.28) directly, we proceed exactly as in Example 2 above; namely, we know from Theorem 1 that $e^{\alpha t} \cos \beta t$, $e^{\alpha t} \sin \beta t$ are solutions of $y'' + py' + qy = 0$, where $\alpha + i\beta$ is a root of $z^2 + pz + q = 0$. Since these solutions are linearly independent on $-\infty < t < \infty$, formula (6.28) is a direct consequence of Corollary 1 to Theorem 1, Section 6.3. ∎

● **EXERCISES**

    6. Show that $e^{\alpha t} \cos \beta t$, $e^{\alpha t} \sin \beta t$ are linearly independent solutions on $-\infty < t < \infty$ of (6.24) when $p^2 < 4q$.

    7. Proceeding as in Example 2, show that a_1, a_2 in (6.28) are given in terms of c_1 and c_2 by the formulas $a_1 = c_1 + c_2$, $a_2 = i(c_1 - c_2)$.

    8. Find the solution ϕ satisfying the initial conditions $\phi(0) = \phi'(0) = 1$ of each of the following differential equations.
    (a) $y'' + y = 0$.   (b) $y'' - 4y' + 13y = 0$.
    (c) $y'' + 4y = 0$.   (d) $y'' + 2y' + 2y = 0$.

    9. In Eq. 7.24 with p, q nonnegative, find conditions on the constants which lead to complex roots of the characteristic equation and investigate the behavior of the solutions for various choices of these constants as $t \to +\infty$.

190 Linear Systems of Differential Equations

We now turn to the case of equal roots of the characteristic equation. In case (ii), $p^2 = 4q$, and the characteristic equation

$$z^2 + pz + q = 0$$

has the double root $z = -p/2$ and, therefore, $\exp[(-p/2)t]$ is a solution of

$$y'' + py' + qy = 0 \tag{6.24}$$

on $-\infty < t < \infty$ if $p^2 = 4q$. The theory tells us that in all cases (6.24) should have two linearly independent solutions. **We now employ a useful trick to find (guess) a second linearly independent solution.**

We know that $\exp[(-p/2)t]$ is a solution of $L(y) = y'' + py' + (p^2/4)y = 0$. This means that

$$L(e^{zt})\bigg|_{z=-p/2} = e^{zt}\left(z^2 + pz + \frac{p^2}{4}\right)\bigg|_{z=-p/2} = 0.$$

Since

$$\frac{\partial}{\partial z}(L(e^{zt})) = \frac{\partial}{\partial z}\left(e^{zt}\left(z^2 + pz + \frac{p^2}{4}\right)\right)$$

$$= te^{zt}\left(z^2 + pz + \frac{p^2}{4}\right) + e^{zt}(2z + p),$$

we see by substituting $z = -p/2$ that also

$$\frac{\partial}{\partial z}(L(e^{zt}))\bigg|_{z=-p/2} = 0.$$

Notice that $2z + p$ is the derivative of $z^2 + pz + p^2/2$ and both of these vanish at the double root $z = -p/2$. (This is a general result about multiple roots.) As the reader may verify,

$$\frac{\partial}{\partial z}\left(\frac{\partial e^{zt}}{\partial t}\right) = \frac{\partial}{\partial t}\left(\frac{\partial e^{zt}}{\partial z}\right), \quad \frac{\partial}{\partial z}\left(\frac{\partial^2 e^{zt}}{\partial t^2}\right) = \frac{\partial^2}{\partial t^2}\left(\frac{\partial e^{zt}}{\partial z}\right),$$

so that $(\partial/\partial z(L(e^{zt}))) = L(\partial/\partial z(e^{zt}))$. Therefore,

$$L\left(\frac{\partial}{\partial z}e^{zt}\right)\bigg|_{z=-p/2} = L(te^{zt})\bigg|_{z=-p/2} = 0.$$

This shows that $te^{(-p/2(t))}$ is also a solution of $y'' + py' + qy = 0$ if $q = p^2/4$.

6.4 Solution of Scalar Linear Differential Equations

Since the solutions $e^{(-p/2(t))}$, $te^{(-p/2(t))}$ are linearly independent on $-\infty < t < \infty$, as may be verified by computing the Wronskian, we have established the following result.

Theorem 3. *Let p and q be constant such that $p^2 = 4q$. Then every solution ϕ on $-\infty < t < \infty$ of*

$$y'' + py' + qy = 0 \tag{6.24}$$

has the form

$$\phi(t) = (c_1 + c_2 t) \exp\left[-\frac{p}{2}t\right] \qquad (-\infty < t < \infty),$$

where c_1 and c_2 are constants.

● **EXERCISES**

10. Find the general solution of each of the following equations. If the equation is real, express the solution in real form. Note that Theorem 3 is true if p and q are complex, and thus equations with complex coefficients can be solved.
    (a) $y'' + 9y = 0$.
    (b) $y'' - 5y' + 6y = 0$.
    (c) $y'' + 10y' + 25y = 0$.
    (d) $y'' + 2iy' + y = 0$.
    (e) $4y'' - y = 0$.
    (f) $y'' + 5y' + 10y = 0$.
    (g) $\varepsilon y'' + 2y' + y = 0 \qquad (0 < \varepsilon < 1)$.
    (h) $4y'' + 4y' + y = 0$.

11. In Eq. 6.24 with $p^2 = 4q$, investigate the behavior of the solutions as $t \to +\infty$ for various values of the constants.

12. Recall the *Definition: A function f is said to be bounded on some interval I if and only if there exists a constant $M > 0$ such that $|f(t)| \leq M$ for all t on I.* For example, $\sin t$, $\cos t$ are bounded on any interval, $1/t$ is bounded on $[1, 2]$ but not on $(0, \infty)$, e^{-t} is bounded on $[-5, \infty)$ but not on $(-\infty, -5]$.
    (a) *Determine* which differential equations in Exercise 10 have *all* their solutions bounded on $[0, \infty)$.
    (b) Repeat part (a) for the interval $(-\infty, \infty)$.

We can easily generalize the results for second-order linear differential equations with constant coefficients to equations of arbitrary order. Consider the linear homogeneous equation of order n with constant coefficients

$$L_n(y) = y^{(n)} + p_1 y^{(n-1)} + \cdots + p_{n-1} y' + p_n y = 0, \tag{6.25}$$

and look for a solution of the form e^{zt} as before. Note that Eq. 6.25 reduces to (6.24) when $n = 2$, with $p_1 = p$, $p_2 = q$. Since $L_n(e^{zt}) = p_n(z)e^{zt}$, where

$$p_n(z) = z^n + p_1 z^{n-1} + \cdots + p_{n-1} z + p_n$$

is a polynomial of degree n, called the **characteristic polynomial**, we see that the analog of Theorems 2 and 3 (although rather more involved) may be stated as follows:

Theorem 4. Let $z_1, z_2, \ldots, z_s$ where $s \leq n$ be the distinct roots of the characteristic equation (of degree n)

$$p_n(z) = z^n + p_1 z^{n-1} + \cdots + p_{n-1} z + p_n = 0,$$

and suppose the root z_i has multiplicity m_i, where $i = 1, \ldots, s$ (and $m_1 + m_2 + \cdots + m_s = n$). Then the n functions

$$e^{z_1 t}, te^{z_1 t}, \ldots, t^{m_1 - 1} e^{z_1 t},$$
$$e^{z_2 t}, te^{z_2 t}, \ldots, t^{m_2 - 1} e^{z_2 t},$$
$$\vdots$$
$$e^{z_s t}, te^{z_s t}, \ldots, t^{m_s - 1} e^{z_s t},$$

are (i) solutions of $L_n(y) = 0$ on $-\infty < t < \infty$, and are (ii) linearly independent on $-\infty < t < \infty$.

We do not prove Theorem 4 except to remark that if z_i, $1 \leq z_i \leq s$, is a root of multiplicity m_i of the polynomial equation $f_n(z) = 0$, then

$$f_n(z_i) = 0, \ f'_n(z_i) = 0, \ \ldots, \ f_n^{(m_i - 1)}(z_i) = 0,$$

but $p_n^{(m_i)}(z_i) \neq 0$. This observation enables us to prove the result much as was done in the second-order case. The linear independence of these solutions is shown by the following lemma.

Lemma. The n functions

$$e^{r_1 t}, te^{r_1 t}, \ldots, t^{k_1 - 1} e^{r_1 t},$$
$$e^{r_2 t}, te^{r_2 t}, \ldots, t^{k_2 - 1} e^{r_2 t},$$
$$\vdots \quad \vdots \qquad \vdots \quad \vdots$$
$$e^{r_i t}, te^{r_i t}, \ldots, t^{k_i - 1} e^{r_i t},$$
$$\vdots \quad \vdots \qquad \vdots \quad \vdots$$
$$e^{r_s t}, te^{r_s t}, \ldots, t^{k_s - 1} e^{r_s t},$$

6.4 Solution of Scalar Linear Differential Equations

where $k_1 + k_2 + \cdots + k_s = n$ and where $r_1, r_2, \ldots, r_s$ are distinct numbers, are linearly independent on every interval $\mathscr{I}$.

Proof. Suppose the n functions are linearly dependent on some interval $\mathscr{I}$. Then there exist n constants $a_{ij}, i = 1, 2, \ldots, s; j = 0, 1, \ldots, k_i - 1$, not all zero, such that

$$a_{10}e^{r_1 t} + a_{11}te^{r_1 t} + \cdots + a_{1,k_1-1}t^{k_1-1}e^{r_1 t} + a_{20}e^{r_2 t} + a_{21}te^{r_2 t} + \cdots$$
$$+ a_{2,k_2-1}t^{k_2-1}e^{r_2 t} + \cdots + a_{s0}e^{r_s t} + a_{s1}te^{r_s t} + \cdots$$
$$+ a_{s,k_s-1}t^{k_s-1}e^{r_s t} = 0$$

or, more compactly,

$$\sum_{i=1}^{s}(a_{i0}e^{r_i t} + a_{i1}te^{r_i t} + \cdots + a_{i,k_i-1}t^{k_i-1}e^{r_i t}) = 0$$

for all t in $\mathscr{I}$. We may define the polynomials

$$P_i(t) = a_{i0} + a_{i1}t + \cdots + a_{i,k_i-1}t^{k_i-1} \qquad (i = 1, \ldots, s),$$

to write this condition in the form

$$P_1(t)e^{r_1 t} + P_2(t)e^{r_2 t} + \cdots + P_s(t)e^{r_s t} = 0, \tag{6.29}$$

for all t in $\mathscr{I}$. Since, by assumption, the constants a_{ij} are not all zero, at least one of the polynomials $P_i(t)$ is not identically zero. It is convenient to assume that $P_s(t) \neq 0$; we can always arrange this by a suitable labeling of the numbers $r_1, r_2, \ldots, r_s$. Now, we divide Eq. 6.29 by $e^{r_1 t}$ and differentiate at most k_1 times until the first term drops out. Note that all terms in (6.29) can be differentiated as often as we wish. Then we have an equation of the form

$$Q_2(t)e^{(r_2 - r_1)t} + Q_3(t)e^{(r_3 - r_1)t} + \cdots + Q_s(t)e^{(r_s - r_1)t} = 0, \tag{6.30}$$

for every t in $\mathscr{I}$. The term $Q_i(t)e^{(r_i - r_1)t}$ in (6.30) is obtained by differentiating $P_i(t)e^{(r_i - r_1)t}$, where $(i = 2, \ldots, s)$, as often as necessary to remove the first term $P_1(t)$. Note that differentiation of a polynomial multiplied by an exponential gives a polynomial **of the same degree** multiplied by the same exponential (think of the rule for differentiation of products). Thus, the polynomial Q_s in (6.30) has the same degree as P_s, and does not vanish identically. We continue this procedure, dividing by the exponential in the first term and then differentiating often enough to remove the first term, until we are left with only one term. Then we have an equation of the form

$$R_s(t)e^{(r_s - r_{s-1})t} \equiv 0,$$

in which the polynomial R has the same degree as P_s, and does not vanish identically. However, the exponential term in this equation does not vanish, and we have a contradiction. This shows that all the constants a_{ij} must be zero, and therefore that the n given functions are linearly independent on $\mathscr{I}$. ∎

Example 3. Find the general solution of the equation $y^{(4)} + 16y = 0$. Since this equation has order four, is homogeneous, and has constant coefficients, Theorem 4 is applicable. The characteristic equation is $z^4 + 16 = 0$. To solve this equation, we write

$$z^4 = -16 = 16e^{i(\pi + 2n\pi)} \qquad (n = 0, \pm 1, \pm 2, \ldots)$$

or, letting $z = re^{i\theta}$,

$$r^4 e^{4i\theta} = 16e^{i(\pi + 2n\pi)} \qquad (n = 0, \pm 1, \pm 2, \ldots).$$

Hence, $r^4 = 16$ and $\theta = \pi/4 + (n/2)\pi$, where $n = 0, \pm 1, \pm 2, \ldots$, and the only distinct roots are $z_1 = 2 \exp [i(\pi/4)] = \sqrt{2}(1 + i)$, $z_2 = 2 \exp [3i(\pi/4)] = \sqrt{2}(-1 + i)$, $z_3 = 2 \exp [-i(\pi/4)] = \sqrt{2}(1 - i)$, $z_4 = 2 \exp [-3i(\pi/4)] = \sqrt{2}(-1 - i)$, corresponding to $n = 0$, $n = 1$, $n = -1$, $n = -2$, respectively. It is clear that the choices $n = +2, \pm 3, \ldots$, lead us back to one of the roots z_1, z_2, z_3, z_4 already listed. Thus, every solution ϕ of the equation $y^{(4)} + 16y = 0$ has, by Theorem 4 (here $n = 4$, $m_1 = m_2 = m_3 = m_4 = 1$), the form

$$\phi(t) = c_1 \exp [\sqrt{2}(1 + i)t] + c_2 \exp [\sqrt{2}(1 - i)t]$$
$$+ c_3 \exp [\sqrt{2}(-1 + i)t] + c_4 \exp [\sqrt{2}(-1 - i)t],$$

for some unique choice of the constants c_1, c_2, c_3, c_4. This may be written in the real form

$$\phi(t) = \exp [\sqrt{2} t](a_1 \cos \sqrt{2} t + a_2 \sin \sqrt{2} t)$$
$$+ \exp [-\sqrt{2} t](a_3 \cos \sqrt{2} t + a_4 \sin \sqrt{2} t),$$

for some unique choice of constants a_1, a_2, a_3, a_4.

Example 4. Find the general solution of the equation $y''' + 3y'' + 3y' + y = 0$. Again, Theorem 4 is applicable and the characteristic equation is $z^3 + 3z^2 + 3z + 1 = (z+1)^3 = 0$. Thus, $z = -1$ is a triple root and e^{-t}, te^{-t}, $t^2 e^{-t}$ are by Theorem 4 linearly independent solutions on $-\infty < t < \infty$. Hence, every solution ϕ has the form

$$\phi(t) = e^{-t}(c_1 + c_2 t + c_3 t^2),$$

for some unique choice of the constants c_1, c_2, c_3.

● **EXERCISES**

13. Find the general solution of the following differential equations.
    (a) $y''' - 27y = 0$.   (b) $y^{(4)} - 16y = 0$.
    (c) $y^{(4)} + 2y'' + y = 0$.   (d) $y^{(4)} + 5y'' + 4y = 0$.
    (e) $y^{(6)} + y = 0$.
14. Find that solution ϕ of $y^{(4)} + 16y = 0$ for which $\phi(0) = 1$, $\phi'(0) = 0$, $\phi''(0) = 0$, $\phi'''(0) = 0$. (See Example 3 above.)
15. Given the equation $y^{(4)} + \lambda y = 0$, where λ is a constant, find the general solution (in real form) in each case: (a) $\lambda = 0$, (b) $\lambda > 0$, (c) $\lambda < 0$.
16. Which of the equations in Exercise 13 have the property that (a) all their solutions tend to zero at $t \to +\infty$, (b) all their solutions are bounded on $0 \leq t < \infty$, (c) all their solutions are bounded on $-\infty < t < \infty$?

6.5 Linear Nonhomogeneous Systems

We now use the theory developed in Sections 6.2 and 6.3 to discuss the form of solutions of the nonhomogeneous system

$$\mathbf{y}' = A(t)\mathbf{y} + \mathbf{g}(t), \tag{6.31}$$

where $A(t)$ is a given continuous matrix and $\mathbf{g}(t)$ is a given continuous vector on an interval $\mathscr{I}$. The entire development rests on the assumption that we can find a fundamental matrix of the corresponding homogeneous system $\mathbf{y}' = A(t)\mathbf{y}$. The vector $\mathbf{g}(t)$ is usually referred to as a forcing term because if (6.31) describes a physical system, $\mathbf{g}(t)$ represents an external force. By Theorem 1, Section 6.2, we know that given any point $(t_0, \mathbf{\eta})$, t_0 in $\mathscr{I}$, there is a unique solution $\mathbf{\phi}$ of (6.31) existing in all of $\mathscr{I}$ such that $\mathbf{\phi}(t_0) = \mathbf{\eta}$.

To construct solutions of (6.31), we let $\Phi(t)$ be a fundamental matrix of the homogeneous system $\mathbf{y}' = A(t)\mathbf{y}$ on $\mathscr{I}$; Φ exists as a consequence of Theorem 1, Section 6.3 (see also remarks immediately following its proof). Suppose $\mathbf{\phi}_1$

and ϕ_2 are any solutions of (6.31) on $\mathscr{I}$. Then $\phi_1 - \phi_2$ is a solution of the homogeneous system on $\mathscr{I}$.

● **EXERCISE**

1. Verify this fact.

By Theorem 1, Section 6.3, and the remarks immediately following its proof (in particular, see Eq. 6.23), there exists a contant vector **c** such that

$$\phi_1 - \phi_2 = \Phi\mathbf{c}. \tag{6.32}$$

Formula (6.32) tells us that to find any solution of (6.31), we need only know one solution of (6.31). (Every other solution differs from the known one by some solution of the homogeneous system.) There is a simple method, known as variation of constants, to determine a solution of (6.31) **provided we know a fundamental matrix for the homogeneous system** $\mathbf{y}' = (At)\mathbf{y}$. Let Φ be such a fundamental matrix on $\mathscr{I}$. We attempt to find a solution ψ of (6.31) of the form

$$\psi(t) = \Phi(t)\mathbf{v}(t), \tag{6.33}$$

where **v** is a vector to be determined. (Note that if **v** is a constant vector, then ψ satisfies the homogeneous system and thus for the present purpose $\mathbf{v}(t) \equiv \mathbf{c}$ is ruled out.) Suppose such a solution exists. Then substituting (6.33) into (6.31), we find for all t on $\mathscr{I}$

$$\psi'(t) = \Phi'(t)\mathbf{v}(t) + \Phi(t)\mathbf{v}'(t) = A(t)\Phi(t)\mathbf{v}(t) + \mathbf{g}(t).$$

Since Φ is a fundamental matrix of the homogeneous system, $\Phi'(t) = A(t)\Phi(t)$, and the terms involving $A(t)\Phi(t)\mathbf{v}(t)$ cancel. Therefore, if $\psi(t) = \Phi(t)\mathbf{v}(t)$ is a solution of (6.31), we must determine $\mathbf{v}(t)$ from the relation

$$\Phi(t)\mathbf{v}'(t) = \mathbf{g}(t).$$

Since $\Phi(t)$ is nonsingular on $\mathscr{I}$ we can premultiply by $\Phi^{-1}(t)$ and we have, on integrating,

$$\mathbf{v}(t) = \int_{t_0}^{t} \Phi^{-1}(s)\mathbf{g}(s)\, ds \qquad (t_0, t \text{ on } \mathscr{I})$$

6.5 Linear Nonhomogeneous Systems

and, therefore, (6.33) becomes

$$\psi(t) = \Phi(t) \int_{t_0}^{t} \Phi^{-1}(s)g(s)\, ds \qquad (t_0, t \text{ on } \mathscr{I}). \tag{6.34}$$

Thus, if (6.31) has a solution ψ of the form (6.33), then ψ is given by (6.34). Conversely, define ψ by (6.34), where Φ is a fundamental matrix of the homogeneous system on $\mathscr{I}$. Then, differentiating (6.34) and using the fundamental theorem of calculus, we have

$$\psi'(t) = \Phi'(t) \int_{t_0}^{t} \Phi^{-1}(s)g(s)\, ds + \Phi(t)\Phi^{-1}(t)g(t)$$

$$= A(t)\Phi(t) \int_{t_0}^{t} \Phi^{-1}(s)g(s)\, ds + g(t),$$

and using (6.34) again,

$$\psi'(t) = A(t)\psi(t) + g(t)$$

of every t on $\mathscr{I}$. Obviously, $\psi(t_0) = 0$. Thus, we have proved the **variation of constants formula**:

Theorem 1. *If Φ is a fundamental matrix of* $\mathbf{y}' = A(t)\mathbf{y}$ *on $\mathscr{I}$, then the function*

$$\psi(t) = \Phi(t) \int_{t_0}^{t} \Phi^{-1}(s)g(s)\, ds$$

is the (unique) *solution of (6.31) satisfying the initial condition*

$$\psi(t_0) = \mathbf{0}$$

and valid on $\mathscr{I}$.

Combining Theorem 1 with the remarks made at the beginning of this section, we see that every solution ϕ of (6.31) on $\mathscr{I}$ has the form

$$\phi(t) = \phi_h(t) + \psi(t) \tag{6.35}$$

where ψ is the solution of Eq. 6.31 satisfying the initial condition $\psi(t_0) = \mathbf{0}$, and ϕ_h is the solution of the homogeneous system satisfying the same initial condition at t_0 as ϕ, for example $\phi_h(t_0) = \mathbf{\eta}$.

Example 1. Find the solution of the initial value problem

$$\mathbf{y}' = \begin{bmatrix} 1 & 1 \\ 0 & 1 \end{bmatrix} \mathbf{y} + \begin{bmatrix} e^{-t} \\ 0 \end{bmatrix}, \quad \mathbf{y} = \begin{bmatrix} y_1 \\ y_2 \end{bmatrix}, \quad \mathbf{y}(0) = \begin{bmatrix} -1 \\ 1 \end{bmatrix}.$$

We have seen in Example 1 Section 6.3 that

$$\Phi(t) = \begin{bmatrix} e^t & te^t \\ 0 & e^t \end{bmatrix}$$

is a fundamental matrix of the associated homogeneous system on $-\infty < t < \infty$. We have (see Section 4.5, particularly Theorem 1 and Exercise 3)

$$\Phi^{-1}(s) = \frac{\begin{bmatrix} e^s & -se^s \\ 0 & e^s \end{bmatrix}}{e^{2s}} = \begin{bmatrix} 1 & -s \\ 0 & 1 \end{bmatrix} e^{-s}.$$

Thus, by Theorem 1 the solution ψ satisfying the initial condition

$$\psi(0) = \begin{bmatrix} 0 \\ 0 \end{bmatrix}$$

is

$$\begin{aligned} \psi(t) &= \begin{bmatrix} e^t & te^t \\ 0 & e^t \end{bmatrix} \int_0^t e^{-s} \begin{bmatrix} 1 & -s \\ 0 & 1 \end{bmatrix} \begin{bmatrix} e^{-s} \\ 0 \end{bmatrix} ds \\ &= \begin{bmatrix} e^t & te^t \\ 0 & e^t \end{bmatrix} \int_0^t \begin{bmatrix} e^{-2s} \\ 0 \end{bmatrix} ds \\ &= \begin{bmatrix} e^t & te^t \\ 0 & e^t \end{bmatrix} \begin{bmatrix} \frac{1}{2}(1 - e^{-2t}) \\ 0 \end{bmatrix} = \begin{bmatrix} \frac{1}{2}(e^t - e^{-t}) \\ 0 \end{bmatrix}. \end{aligned}$$

Since $\Phi(0) = I$, the solution of the corresponding homogeneous system satisfying the initial condition

$$\mathbf{y}(0) = \begin{bmatrix} -1 \\ 1 \end{bmatrix}$$

is

$$\phi_h(t) = \Phi(t) \begin{bmatrix} -1 \\ 1 \end{bmatrix} = \begin{bmatrix} (t-1)e^t \\ e^t \end{bmatrix}.$$

6.5 Linear Nonhomogeneous Systems

By (6.35), the desired solution is

$$\phi(t) = \phi_h(t) + \psi(t) = \begin{bmatrix} (t-1)e^t \\ e^t \end{bmatrix} + \begin{bmatrix} \frac{1}{2}(e^t - e^{-t}) \\ 0 \end{bmatrix}$$

$$= \begin{bmatrix} te^t - \frac{1}{2}(e^t + e^{-t}) \\ e^t \end{bmatrix}.$$

● **EXERCISES**

2. Consider the system $y' = Ay + g(t)$, where

$$A = \begin{bmatrix} 2 & 1 \\ 0 & 2 \end{bmatrix}, \quad y = \begin{bmatrix} y_1 \\ y_2 \end{bmatrix}, \quad g(t) = \begin{bmatrix} \sin t \\ \cos t \end{bmatrix}.$$

Verify that

$$\Phi(t) = \begin{bmatrix} e^{2t} & te^{2t} \\ 0 & e^{2t} \end{bmatrix}$$

is a fundamental matrix of $y' = Ay$. Find that solution ϕ of the nonhomogeneous system for which

$$\phi(0) = \begin{bmatrix} 1 \\ -1 \end{bmatrix}.$$

3. Find the solution ϕ of the system $y' = Ay + g(t)$ with A the same as in Exercise 2 and with

$$g(t) = \begin{bmatrix} 0 \\ e^{2t} \end{bmatrix},$$

satisfying the initial condition

$$\phi(0) = \begin{bmatrix} 1 \\ -1 \end{bmatrix}.$$

4. Consider the system $y' = A(t)y + g(t)$, where

$$A(t) = \begin{bmatrix} 0 & 1 \\ -2 & 2 \\ t^2 & t \end{bmatrix} \quad g(t) = \begin{bmatrix} t^4 \\ t^3 \end{bmatrix}.$$

Find the solution ϕ satisfying the initial condition

$$\phi(2) = \begin{bmatrix} 1 \\ 4 \end{bmatrix}$$

and determine the interval of validity of this solution. [*Hint:* Use the fundamental matrix given in Exercise 7, Section 6.3.]

We now consider the form of the variation of constants formula for the scalar second-order linear nonhomogeneous differential equation

$$y'' + p(t)y' + q(t)y = r(t), \qquad (6.36)$$

where p, q, and r are continuous on an interval $\mathscr{I}$. We have seen in Corollary 1 to Theorem 1, Section 6.3, that the corresponding homogeneous equation

$$y'' + p(t)y' + q(t)y = 0 \qquad (6.20)$$

has two linearly independent solutions ψ_1, ψ_2 on $\mathscr{I}$. These solutions are the first components of the (vector) solutions,

$$\phi_1(t) = \begin{bmatrix} \psi_1(t) \\ \psi_1'(t) \end{bmatrix}, \qquad \phi_2(t) = \begin{bmatrix} \psi_2(t) \\ \psi_2'(t) \end{bmatrix}$$

of the equivalent system

$$\mathbf{y}' = \begin{bmatrix} 0 & 1 \\ -q(t) & -p(t) \end{bmatrix} \mathbf{y}, \qquad \mathbf{y} = \begin{bmatrix} y_1 \\ y_2 \end{bmatrix}.$$

We apply Theorem 1 to the system

$$\mathbf{y}' = \begin{bmatrix} 0 & 1 \\ -q(t) & -p(t) \end{bmatrix} \mathbf{y} + \begin{bmatrix} 0 \\ r(t) \end{bmatrix}, \qquad \mathbf{y} = \begin{bmatrix} y_1 \\ y_2 \end{bmatrix} \qquad (6.21)$$

which is equivalent to (6.36). Since $\psi_1(t)$, $\psi_2(t)$ are linearly independent solutions of (6.20), by the equivalence of the equations (6.20) and (6.21), the matrix

$$\Phi(t) = \begin{bmatrix} \psi_1(t) & \psi_2(t) \\ \psi_1'(t) & \psi_2'(t) \end{bmatrix}$$

6.5 Linear Nonhomogeneous Systems

is a fundamental matrix of the system (6.21) on $\mathscr{I}$. Let

$$\mathbf{u}(t) = \begin{bmatrix} u_1(t) \\ u_2(t) \end{bmatrix}$$

be the solution of (6.37) satisfying the initial condition $\mathbf{u}(t_0) = \mathbf{0}$. By Theorem 1,

$$\mathbf{u}(t) = \Phi(t) \int_{t_0}^{t} \Phi^{-1}(s) \begin{bmatrix} 0 \\ r(s) \end{bmatrix} ds.$$

From

$$\Phi^{-1}(s) = \frac{1}{\det \Phi(s)} \begin{bmatrix} \psi_2'(s) & -\psi_2(s) \\ -\psi_1'(s) & \psi_1(s) \end{bmatrix}$$

$$= \frac{1}{W[\psi_1(s), \psi_2(s)]} \begin{bmatrix} \psi_2'(s) & -\psi_2(s) \\ -\psi_1'(s) & \psi_1(s) \end{bmatrix},$$

we obtain

$$\Phi^{-1}(s) \begin{bmatrix} 0 \\ r(s) \end{bmatrix} = \frac{1}{W[\psi_1(s), \psi_2(s)]} \begin{bmatrix} -\psi_2(s)r(s) \\ \psi_1(s)r(s) \end{bmatrix},$$

and therefore

$$\mathbf{u}(t) = \Phi(t) \int_{t_0}^{t} \frac{1}{W[\psi_1(s), \psi_2(s)]} \begin{bmatrix} -\psi_2(s)r(s) \\ \psi_1(s)r(s) \end{bmatrix} ds$$

$$= \int_{t_0}^{t} \frac{1}{W[\psi_1(s), \psi_2(s)]} \begin{bmatrix} \psi_1(t) & \psi_2(t) \\ \psi_1'(t) & \psi_2'(t) \end{bmatrix} \begin{bmatrix} -\psi_2(s)r(s) \\ \psi_1(s)r(s) \end{bmatrix} ds$$

$$= \int_{t_0}^{t} \frac{1}{W[\psi_1(s), \psi_2(s)]} \begin{bmatrix} \psi_2(t)\psi_1(s) - \psi_1(t)\psi_2(s) \\ \psi_2'(t)\psi_1(s) - \psi_1'(t)\psi_2(s) \end{bmatrix} r(s) \, ds.$$

The solution of (6.36) satisfying the initial conditions $y(t_0) = 0$, $y'(t_0) = 0$ is, by the equivalence of (6.36) and (6.37), the first component $u_1(t)$ of $\mathbf{u}(t)$. Therefore, this solution is

$$u_1(t) = \int_{t_0}^{t} \frac{[\psi_2(t)\psi_1(s) - \psi_1(t)\psi_2(s)]r(s)}{W[\psi_1(s), \psi_2(s)]} ds.$$

Linear Systems of Differential Equations

Thus, we have proved the variation of constants formula for the scalar second-order linear equation:

Corollary to Theorem 1. *Let $\psi_1(t), \psi_2(t)$ be linearly independent solutions of*

$$y'' + p(t)y' + q(t)y = 0 \tag{6.20}$$

on an interval $\mathscr{I}$. Then the function

$$u_1(t) = \int_{t_0}^{t} \frac{[\psi_2(t)\psi_1(s) - \psi_1(t)\psi_2(s)]r(s)}{W[\psi_1(s), \psi_2(s)]} \, ds \tag{6.38}$$

is the (unique) *solution of*

$$y'' + p(t)y' + q(t)y = r(t) \tag{6.36}$$

on $\mathscr{I}$ satisfying the initial conditions $y(t_0) = 0$, $y'(t_0) = 0$. Moreover, every solution of (6.36) *on $\mathscr{I}$ has the form*

$$w(t) = c_1\psi_1(t) + c_2\psi_2(t) + u_1(t) \tag{6.39}$$

for some unique choice of the constants c_1, c_2.

This last expression (6.39) is called the **general solution** of (6.36).

Example 2. Find a particular solution of the differential equation

$$y'' + y = \tan t \qquad -\frac{\pi}{2} < t < \frac{\pi}{2}.$$

We apply the corollary to Theorem 1 directly using the linearly independent solutions $\psi_1(t) = \cos t$, $\psi_2(t) = \sin t$ of the homogeneous equation $y'' + y = 0$. We have

$$W(\psi_1(t), \psi_2(t)) = \begin{vmatrix} \cos t & \sin t \\ -\sin t & \cos t \end{vmatrix} \equiv 1.$$

Hence, formula (6.38) yields (using $t_0 = 0$) the particular solution

$$u_1(t) = \int_0^t (\sin t \cos s - \cos t \sin s) \tan s \, ds$$

$$= \sin t \int_0^t \sin s \, ds - \cos t \int_0^t \sin s \tan s \, ds$$

$$= \sin t(1 - \cos t) + \cos t \int_0^t (\cos s - \sec s) \, ds$$

$$= \sin t(1 - \cos t) + \cos t(\sin t - \log|\sec t + \tan t|).$$

$$= \sin t - \cos t \log|\sec t + \tan t|.$$

We note that since $\sin t$ is a solution of the homogeneous equation, the function

$$u(t) = -\cos t \log|\sec t + \tan t|$$

is also a particular solution. We also remark that we could apply Theorem 1 directly by first converting the given differential equation to an equivalent system of first-order equations as was done for Eq. 6.36; however, for second-order scalar equations it is more efficient to employ the Corollary.

● **EXERCISES**

5. Verify by direct substitution that $u_1(t)$ is a solution of (6.36) on $\mathscr{I}$ satisfying the initial conditions $u_1(t_0) = 0$, $u_1'(t_0) = 0$.

6. Find the general solution of each of the following differential equations.
   (a) $y'' + y = \sec t$ $(-\pi/2 < t < \pi/2)$.
   (b) $y'' + 4y' + 4y = \cos 2t$.
   (c) $y'' + 4y = f(t)$, where f is any continuous function on some interval $\mathscr{I}$.
   (d) $y'' - 4y' + 4y = 3e^{-t} + 2t^2 + \sin t$.

7. If ϕ is a solution of the equation $y'' + k^2 y = f(t)$, where k is a real constant different from zero and f is continuous for $0 \leq t < \infty$, show that c_1 and c_2 can be chosen so that

$$\phi(t) = c_1 \cos kt + \frac{c_2}{k} \sin kt + \frac{1}{k} \int_0^t \sin k(t-s) f(s) \, ds$$

for $0 \leq t < \infty$. (Use $\cos kt$ and $\sin kt/k$ as a fundamental set of solutions of the homogeneous equation.) Find an analogous formula in the case $k = 0$ and show that it can also be obtained by computing $\lim_{k \to 0} \phi(t)$.

8. Given the equation

$$y'' + 5y' + 4y = f(t).$$

Use the variation of constants formula to prove that:
  (a) If f is bounded on $0 \leq t < \infty$ (that is, there exists a constant $M \geq 0$ such that $|f(t)| \leq M$ on $0 \leq t < \infty$), then every solution of $y'' + 5y' + 4y = f(t)$ is bounded on $0 \leq t < \infty$.
  (b) If also $f(t) \to 0$ as $t \to \infty$, then every solution ϕ of $y'' + 5y' + 4y = f(t)$ satisfies $\phi(t) \to 0$ as $t \to \infty$.
9. Can you formulate Exercise 7 for the general equation

$$y'' + a_1 y' + a_2 y = f(t) \qquad a_1, a_2 \text{ constant}$$

with a_1, a_2 suitably restricted?

10. Formulate the analog of the corollary to Theorem 1 for the third-order equation

$$y''' + p_1(t)y'' + p_2(t)y' + p_3(t)y = f(t).$$

[*Hint:* Let $\psi_1(t)$, $\psi_2(t)$, $\psi_3(t)$ be linearly independent solutions of the corresponding homogeneous equation. Then proceed as in the corollary.]

11. Find the general solution of each of the following differential equations.
  (a) $y''' - 8y = e^{2t}$.
  (b) $y^{(4)} + 16y = f(t)$, f continuous on $-\infty < t < \infty$.

12. Solve the initial value problem (1.12), (1.13) for $\omega^2 \neq k/m$ in Example 5, Chapter 1, and then answer the questions (ii), (iii) formulated in that example.

13. Answer the questions (i) and (ii) formulated in Example 6, Chapter 1. [*Hint:* Solve (1.14) subject to zero initial conditions.]

• MISCELLANEOUS EXERCISES

1. For each of the following differential equations, determine the largest intervals on which a unique solution is certain to exist as an application of Corollary 1, Theorem 1, Section 6.2. In each problem it is assumed that you are given initial conditions of the form $\phi(t_0) = y_0$, $\phi'(t_0) = z_0$.
  (a) $ty'' + y = t^2$.
  (b) $t^2(t-3)y'' + y' = 0$.
  (c) $y'' + \sqrt{t}y = 0$.
  (d) $(1 + t^2)y'' - y' + ty = \cos t$.
  (e) $e^t y'' - \sin t y' + y = t^3$.
  (f) $y'' - \log|t| y = 0$.

2. In each of the following, let $\phi_1(t)$ and $\phi_2(t)$ be solutions of the differential equation

$$L(y) = y'' + p(t)y' + q(t)y = 0$$

on some interval $\mathcal{I}$, where p and q are continuous on $\mathcal{I}$.

(a) If $\phi_1(t_0) = \phi_2(t_0) = 0$ for some t_0 in $\mathscr{I}$, show that the solutions ϕ_1 and ϕ_2 cannot form a fundamental set of solutions on $\mathscr{I}$.

(b) If the solutions ϕ_1 and ϕ_2 both have a maximum or minimum at some point t_1 in $\mathscr{I}$, show that ϕ_1 and ϕ_2 cannot form a fundamental set of solutions on $\mathscr{I}$.

(c) Let ϕ_1 and ϕ_2 form a fundamental set of solutions on $\mathscr{I}$ which both have an inflection point at some point t_2 in $\mathscr{I}$. Show that $p(t_2) = q(t_2) = 0$.

(d) Let ϕ_1 and ϕ_2 form a fundamental set of solutions on $\mathscr{I}$. Show that $\psi_1 = \phi_1 + \phi_2$, $\psi_2 = \phi_2 - 2\phi_2$ also form a fundamental set of solutions on $\mathscr{I}$.

3. (a) Let ϕ_1 and ϕ_2 be solutions of $L(y) = y'' - 4ty' + (4t^2 - 2)y = 0$ on the interval $-\infty < t < \infty$, satisfying the initial conditions $\phi_1(1) = 1$, $\phi_1'(1) = \frac{1}{3}$, $\phi_2(1) = 3$, $\phi_2'(1) = 1$. Are these solutions linearly independent on $-\infty < t < \infty$? Justify your answer.

(b) Show that $\psi_1(t) = \exp(t^2)$ is a solution of the equation $L(y) = 0$ and find a second linearly independent solution on $-\infty < t < \infty$. [Hint: Look for a solution ψ_2 of the form $\psi_2(t) = u(t)\psi_1(t)$; substitute and find $u(t)$ to make ψ_1, ψ_2 linearly independent solutions.]

(c) Find the solutions ϕ_1 and ϕ_2 in part (a).

4. Given that the equation

$$ty'' - (2t + 1)y' + 2y = 0 \qquad (t > 0)$$

has a solution of the form e^{ct} for some c, find the general solution. [Hint: First find what c must be; then find a second linearly independent solution as in Exercise 3(b).

5. Find the general solution of each of the following differential equations.
(a) $y'' + y = \operatorname{cosec} t \cot t$ $\quad (0 < t < \pi)$.
(b) $y'' - 6y' + 9y = e^t$. $\qquad$ (c) $y'' - 6y' + 9y = t^2 e^{3t}$.
(d) $y^{(8)} + 8y^{(4)} + 16y = 0$. $\qquad$ (e) $y^{(4)} - 2y'' + y = e^t + \sin t$.
(f) $y''' + y' = \tan t$ $\quad (0 < t < \pi/2)$.
(g) $y^{(4)} + y = g(t)$ $\quad g$ continuous.
(h) $y'' + y = h(t)$, where

$$h(t) = t \quad (0 \le t \le \pi), \qquad h(t) = \pi \cos(\pi - t) \quad (\pi < t \le 2\pi)$$

and h is periodic with period 2π.

6. (a) One solution of the equation

$$L(y) = t^2 y'' + ty' + \left(t^2 - \frac{1}{4}\right)y = 0 \qquad (t > 0)$$

is $t^{-1/2} \sin t$. Find the general solution of the equation $L(y) = 3t^{1/2} \sin t$, where $t > 0$. [Hint: Use the method suggested in Exercise 3(b) previously.]

(b) Repeat part (a) for the equation

$$2ty'' + (1 - 4t)y' + (2t - 1)y = e^t$$

given that e^t is one solution of the homogeneous equation.

7. The current i in amperes in an electrical circuit with resistance R ohms, inductance L henrys, and capacitance C farads in series (see Figure 6.1), is governed by the equation

$$Li'' + Ri' + \frac{1}{C}i = E'(t),$$

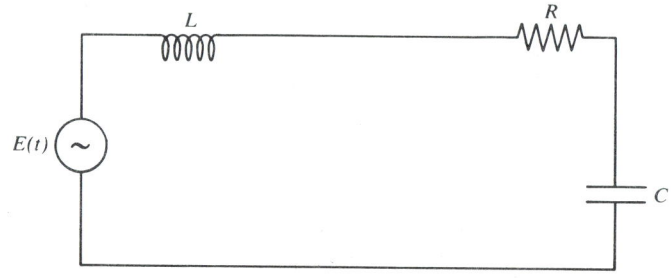

Figure 6.1

where $E(t)$ is the applied voltage. Suppose the applied voltage is a constant E_0.
(a) Show that the current decreases exponentially if $CR^2 > 4L$.
(b) Find the current if $CR^2 = 4L$ and $i(0) = 0$, $i'(0) = E_0/L$.
(c) Find the current if $CR^2 < 4L$ and $i(0) = 0$, $i'(0) = E_0/L$.

8. Consider the circuit of Exercise 7. Find the current if

$$E(t) = \begin{cases} 0 & t \leq 0 \\ E_0 \sin \alpha t & t > 0 \end{cases}$$

assuming $i(0) = 0$, $i'(0) = 0$.

9. Find the current in an electrical circuit with inductance and capacitance in series, but no resistance, and an applied voltage $E(t)$ given by

$$E(t) = \begin{cases} t \sin \frac{\pi}{2} t & (0 \leq t < 1) \\ 1 & (t > 1) \end{cases}$$

with $i(0) = 0$, $i'(0) = 0$.

10. Show that the solutions of the differential equation

$$y'' + py' + qy = 0$$

(p and q positive constants) are oscillations with amplitudes which decrease exponentially when $p^2 < 4q$ (light damping) and decrease exponentially without oscillating if $p^2 > 4q$ (over damping). How do they behave if $p^2 = 4q$ (critical damping)?

11. Determine a particular solution of the differential equation

$$y'' + py' + qy = A \cos kt$$

of the form $\phi(t) = B \cos(kt - \alpha)$. Show that the amplitude of the oscillation B is a maximum if $k = (q - p^2/2)^{1/2}$ (called the resonant frequency) provided $p^2 < 4q$. What happens in the case $p^2 \geq 2q$? Sketch the graph of the amplitude as a function of k for (i) $p^2 < 2q$, (ii) $p^2 \geq 2q$. [*Hint:* Use undetermined coefficients to solve the differential equation.]

Chapter 7

EIGENVALUES, EIGENVECTORS, AND LINEAR SYSTEMS OF DIFFERENTIAL EQUATIONS WITH CONSTANT COEFFICIENTS

We have seen how to solve the scalar equation $y' = ay$, and we know that every solution is of the form $e^{at}c$, where c is a constant. In this chapter we will learn how to find a fundamental matrix of the system $\mathbf{y}' = A\mathbf{y}$, where A is a constant $n \times n$ matrix. The explicit calculation of such a fundamental matrix will lead us naturally to the study of eigenvalues and eigenvectors of matrices.

7.1 The Exponential of a Matrix

In order to find a fundamental matrix of the system

$$\mathbf{y}' = A\mathbf{y}, \tag{7.1}$$

we first need to define **the exponential of a matrix.** If $M \in \mathscr{F}_{nn}$, we define the matrix $\exp M$ (or e^M) to be the sum of the series

$$\exp M = I + M + \frac{M^2}{2!} + \frac{M^3}{3!} + \cdots + \frac{M^k}{k!} + \cdots = \sum_{k=0}^{\infty} \frac{M^k}{k!}, \tag{7.2}$$

where I is the $n \times n$ identity matrix. (Note that $M^0 = I$ and $0! = 1$.) To justify this definition, we must show that the right-hand side of (7.2) makes

7.1 The Exponential of a Matrix

sense. It is not difficult to define a suitable notion of convergence of a series of matrices and to show, using this definition, that exp M is well defined for every matrix $M \in \mathscr{F}_{nn}$. This is done in Appendix 1.

An important property of the exponential matrix is that if $P, M \in \mathscr{F}_{nn}$ and if P and M commute ($MP = PM$), then

$$\exp(M + P) = \exp M \cdot \exp P. \tag{7.3}$$

To prove this, we apply the definition (7.2) to the left-hand side of (7.3) We obtain

$$\exp(M + P) = \sum_{k=0}^{\infty} \frac{(M + P)^k}{k!}. \tag{7.4}$$

By the binomial theorem and $MP = PM$,

$$(M + P)^k = \sum_{l=0}^{k} \frac{k!}{l!(k-l)!} M^l P^{k-l}.$$

(If x, y are real or complex numbers and $k > 0$ is an integer, the binomial theorem states that $(x + y)^k = \sum_{l=0}^{k} k!/[l!(k-l)!] \, x^l y^{k-l}$. If x and y are matrices in $\mathscr{F}_{nn}$ which commute, the same result holds.) Therefore, canceling $k!$, we obtain

$$\exp(M + P) = \sum_{k=0}^{\infty} \left[\sum_{l=0}^{k} \frac{M^l}{l!} \frac{P^{k-l}}{(k-l)!} \right]. \tag{7.5}$$

On the other hand,

$$\exp M \cdot \exp P = \sum_{i=0}^{\infty} \frac{M^i}{i!} \cdot \sum_{j=0}^{\infty} \frac{P^j}{j!}.$$

By multiplication of absolutely convergent series, we have

$$\exp M \cdot \exp P = \sum_{k=0}^{\infty} C_k$$

$$C_k = \sum_{l=0}^{k} \frac{M^l}{l!} \frac{P^{k-l}}{(k-l)!}. \tag{7.6}$$

Comparison of (7.6) with (7.5) proves (7.3).

A useful property is that if T is a nonsingular $n \times n$ matrix,

$$T^{-1}(\exp M)T = \exp(T^{-1}MT). \tag{7.7}$$

EXERCISES

1. Verify (7.7). [*Hint:* Use (7.2).]
2. If $M \in \mathscr{F}_{nn}$ show that:
   (a) $\exp(c_1 M + c_2 M) = \exp c_1 M \exp c_2 M$ for any $c_1, c_2 \in \mathscr{F}$.
   (b) $(e^M)^{-1} = e^{-M}$.
   (c) $(e^M)^k = e^{kM}$, where k is any integer.
   (d) $e^0 = I$, where 0 is the $n \times n$ zero matrix.

We are now ready to establish the basic result for linear systems with constant coefficients

$$\mathbf{y}' = A\mathbf{y}. \tag{7.1}$$

Theorem 1. *The matrix*

$$\Phi(t) = \exp At \tag{7.8}$$

is the fundamental matrix of (7.1) *with* $\Phi(0) = I$ *on* $-\infty < t < \infty$.

Proof. That $\Phi(0) = I$ is obvious from (7.2). Using (7.2) with $M = At$ (well defined for $-\infty < t < \infty$ and every $n \times n$ matrix A), we have by differentiation[1]

$$(\exp At)' = A + \frac{A^2 t}{1!} + \frac{A^3 t^2}{2!} + \cdots + \frac{A^k t^{k-1}}{(k-1)!} + \cdots = A \exp At,$$

$-\infty < t < \infty$. Therefore, $\exp At$ is a solution matrix of (7.1) (its columns are solutions of (7.11)). Since $\det \Phi(0) = \det I = 1$, Theorem 2, Section 6.3, shows that $\Phi(t)$ is a fundamental matrix of (7.1). ∎

It follows from Theorem 1 and Eq. 6.23 (Section 6.3) that every solution ϕ of the system (7.1) has the form

$$\phi(t) = (\exp At)\mathbf{c} \quad (-\infty < t < \infty) \tag{7.8}$$

for a suitably chosen constant vector $\mathbf{c}$.

---

[1] It is easy to prove that the familiar theorems on differentiation of power series (see, for example, F. Brauer and J. A. Nohel, *Ordinary Differential Equations*, W. A. Benjamin, Inc., New York, 1967, p. 122), with real or complex coefficients hold essentially without change for power series having $n \times n$ matrices as coefficients.

7.1 The Exponential of a Matrix

● **EXERCISES**

3. Show that if ϕ is that solution of (7.1) satisfying $\phi(t_0) = \eta$, then
$$\phi(t) = [\exp A(t - t_0)]\eta \qquad (-\infty < t < \infty). \tag{7.9}$$
4. Show that if $\Phi(t) = e^{tA}$, then $\Phi^{-1}(t) = e^{-tA}$.

We now proceed to find some fundamental matrices in certain special cases; that is, we evaluate $\exp tA$ for certain matrices A.

Example 1. Find a fundamental matrix of the system $\mathbf{y}' = A\mathbf{y}$ if A is a diagonal matrix,

$$A = \begin{bmatrix} d_1 & & 0 \\ & d_2 & \\ & & \ddots \\ 0 & & d_n \end{bmatrix}.$$

From (7.2),

$$\exp At = I + \begin{bmatrix} d_1 & & 0 \\ & \ddots & \\ 0 & & d_n \end{bmatrix} \frac{t}{1!} + \begin{bmatrix} d_1^2 & & \\ & \ddots & \\ & & d_n^2 \end{bmatrix} \frac{t^2}{2!} + \cdots$$

$$+ \begin{bmatrix} d_1^k & & 0 \\ & \ddots & \\ 0 & & d_n^k \end{bmatrix} \frac{t^k}{k!} + \cdots$$

$$= \begin{bmatrix} \exp d_1 t & & & 0 \\ & \exp d_2 t & & \\ & & \ddots & \\ 0 & & & \exp d_n t \end{bmatrix}$$

and by Theorem 1 this is a fundamental matrix. This result is, of course, obvious since in the present case each equation of the system is $y_k' = d_k y_k$ ($k = 1, \ldots, n$) and can be integrated.

Example 2. Find a fundamental matrix of $\mathbf{y}' = A\mathbf{y}$ if

$$A = \begin{bmatrix} 3 & 1 \\ 0 & 3 \end{bmatrix}.$$

Eigenvalues, Eigenvectors, and Systems of Differential Equations

Since

$$A = \begin{bmatrix} 3 & 0 \\ 0 & 3 \end{bmatrix} + \begin{bmatrix} 0 & 1 \\ 0 & 0 \end{bmatrix}$$

and since these two matrices commute, we have

$$\exp At = \exp \begin{bmatrix} 3 & 0 \\ 0 & 3 \end{bmatrix} t \cdot \exp \begin{bmatrix} 0 & 1 \\ 0 & 0 \end{bmatrix} t$$

$$= \begin{bmatrix} e^{3t} & 0 \\ 0 & e^{3t} \end{bmatrix} \left\{ I + \begin{bmatrix} 0 & 1 \\ 0 & 0 \end{bmatrix} t + \begin{bmatrix} 0 & 1 \\ 0 & 0 \end{bmatrix}^2 \frac{t^2}{2!} + \cdots \right\}.$$

But

$$\begin{bmatrix} 0 & 1 \\ 0 & 0 \end{bmatrix}^2 = \begin{bmatrix} 0 & 0 \\ 0 & 0 \end{bmatrix}$$

and the infinite series terminates after two terms. Therefore,

$$\exp At = e^{3t} \begin{bmatrix} 1 & t \\ 0 & 1 \end{bmatrix}$$

and by Theorem 1 this is a fundamental matrix.

• **EXERCISES**

5. Find a fundamental matrix of the system $y' = Ay$ if

$$A = \begin{bmatrix} -2 & 1 & 0 \\ 0 & -2 & 1 \\ 0 & 0 & -2 \end{bmatrix}$$

and check your answer by direct integration of the given system.

6. Find a fundamental matrix of the system $y' = Ay$ if

$$A = \begin{bmatrix} 0 & 1 & 0 \\ & \ddots & 1 \\ 0 & & 0 \end{bmatrix},$$

where A is an $n \times n$ matrix.

7. Find a fundamental matrix of the system $\mathbf{y}' = A\mathbf{y}$, where A is the $n \times n$ matrix

$$A = \begin{bmatrix} 2 & 1 & & 0 \\ & 2 & 1 & \\ & & \ddots & \ddots & 1 \\ 0 & & & & 2 \end{bmatrix}$$

7.2 Eigenvalues and Eigenvectors of Matrices

The reader will have noticed that the examples and exercises presented so far, all of which involve the calculation of e^{tA}, are of a rather special form. In order to be able to handle more complicated problems and in order to obtain a general representation of solutions of (7.1) (that is, if we want to evaluate explicitly the entries of the matrix exp tA), we will need to introduce the notions of eigenvalue and eigenvector of a matrix.

To motivate these concepts, consider the system $\mathbf{y}' = A\mathbf{y}$, and look for a solution of the form

$$\boldsymbol{\phi}(t) = e^{\lambda t}\mathbf{c} \qquad (\mathbf{c} \neq \mathbf{0}).$$

where the constant λ and the vector $\mathbf{c}$ are to be determined. Such a form is suggested by the above examples. Substitution shows that $e^{\lambda t}\mathbf{c}$ is a solution if and only if

$$\lambda e^{\lambda t}\mathbf{c} = A e^{\lambda t}\mathbf{c}.$$

Since $e^{\lambda t} \neq 0$, this condition becomes

$$(\lambda I - A)\mathbf{c} = \mathbf{0}$$

which can be regarded as a linear homogeneous algebraic system for the vector $\mathbf{c}$. This system has a nontrivial solution if and only if λ is chosen in such a way that

$$\det(\lambda I - A) = 0.$$

(See Theorem 1, Section 3.8 and Theorem 2, Section 4.5). This suggests the following definitions. Unless otherwise stated we shall assume that our field of scalars is the complex numbers.

Definition 1. *Let $A \in \mathscr{C}_{nn}$. An eigenvalue of A is a scalar λ such that the algebraic system*

$$(\lambda I - A)\mathbf{x} = \mathbf{0} \tag{7.10}$$

has a nontrivial solution. Any such nontrivial solution of (7.10) *is called an eigenvector of A corresponding to the eigenvalue λ.*[2]

Definition 2. *Let $A \in \mathscr{C}_{nn}$. The polynomial of degree n,*

$$p(\lambda) = \det(\lambda I - A)$$

is called the characteristic polynomial of A.[3]

Therefore, the calculation preceding Definition 1 shows that $e^{\lambda t}\mathbf{c}$ is a solution of the linear system $\mathbf{y}' = A\mathbf{y}$ if and only if λ is an eigenvalue of A and $\mathbf{c}$ is a corresponding eigenvector. We will return to a discussion of the system $\mathbf{y}' = A\mathbf{y}$ in Section 7.3 after we have become familiar with properties of eigenvalues and eigenvectors.

In view of the remarks immediately preceding Definition 1, the eigenvalues of A are the roots of the polynomial equation $p(\lambda) = 0$. As $p(\lambda)$ is a polynomial of degree n, there are exactly n eigenvalues, not necessarily distinct. In particular, there is at least one eigenvalue and one eigenvector for every matrix A. If $\lambda = \lambda_0$ is a simple root of the equation $p(\lambda) = 0$, then λ_0 is called a **simple eigenvalue**. If $\lambda = \lambda_0$ is a k-fold root of the equation $p(\lambda) = 0$ (that is, $p(\lambda)$ has $(\lambda - \lambda_0)^k$, but not $(\lambda - \lambda_0)^{k+1}$, as a factor), then λ_0 is an **eigenvalue of multiplicity** k. Since the constant term in $p(\lambda)$ is $p(0) = \det(-A)$, if $\lambda = 0$ is not an eigenvalue of A, then $p(0) \neq 0$, and in this case A is nonsingular.

Example 1. Find the eigenvalues and corresponding eigenvectors of the matrix

$$A = \begin{bmatrix} 3 & 5 \\ -5 & 3 \end{bmatrix}.$$

---

[2] Since every real number is a complex number we may have $A \in \mathscr{R}_{nn}$, but the scalar λ may be complex even in such a case (see Example 1 following).

[3] The reader should note that for each $\lambda \in \mathscr{C}$, $\det(\lambda I - A)$ is well defined as a scalar in $\mathscr{C}$. The function p defined by the expression $p(\lambda) = \det(\lambda I - A)$ is a polynomial of degree n. We shall tacitly assume that such determinantal polynomials obey the rules of determinants discussed in Chapter 4.

The eigenvalues of A are roots of the equation

$$\det(A - \lambda I) = \det\begin{bmatrix} 3-\lambda & 5 \\ -5 & 3-\lambda \end{bmatrix} = \lambda^2 - 6\lambda + 34 = 0.$$

Thus, $\lambda_{1,\,2} = 3 \pm 5i$. The eigenvector

$$\mathbf{u} = \begin{bmatrix} u_1 \\ u_2 \end{bmatrix}$$

corresponding to the eigenvalue $\lambda_1 = 3 + 5i$ must satisfy the linear homogeneous algebraic system

$$(A - \lambda_1 I)\mathbf{u} = \begin{bmatrix} -5i & 5 \\ -5 & -5i \end{bmatrix} \begin{bmatrix} u_1 \\ u_2 \end{bmatrix} = 0.$$

Thus, u_1, u_2 satisfy the system of equations

$$-iu_1 + u_2 = 0$$
$$-u_1 - iu_2 = 0$$

and, therefore,

$$\mathbf{u} = \alpha \begin{bmatrix} 1 \\ i \end{bmatrix}$$

is an eigenvector for any constant α. Similarly, the eigenvector

$$\mathbf{v} = \begin{bmatrix} v_1 \\ v_2 \end{bmatrix}$$

corresponding to the eigenvalue $\lambda_2 = 3 - 5i$ is found to be

$$\mathbf{v} = \beta \begin{bmatrix} i \\ 1 \end{bmatrix}$$

for any constant β.

Example 2. Find the eigenvalues of the matrix

$$A = \begin{bmatrix} 2 & 1 \\ -1 & 4 \end{bmatrix}.$$

Consider the equation $\det(\lambda I - A) = 0$.

$$\det\begin{bmatrix} \lambda - 2 & -1 \\ 1 & \lambda - 4 \end{bmatrix} = (\lambda - 2)(\lambda - 4) + 1 = \lambda^2 - 6\lambda + 9 = 0.$$

Thus, $\lambda = 3$ is an eigenvalue of A of multiplicity two. To find a corresponding eigenvector we consider the system

$$(3I - A)\mathbf{c} = 0$$

or

$$\begin{bmatrix} 1 & -1 \\ 1 & -1 \end{bmatrix}\begin{bmatrix} c_1 \\ c_2 \end{bmatrix} = \begin{bmatrix} 0 \\ 0 \end{bmatrix} \quad \text{or} \quad \begin{array}{l} c_1 - c_2 = 0 \\ c_1 - c_2 = 0. \end{array}$$

Any vector $\mathbf{c}$ with components $c_1 = c_2$ is an eigenvector. Thus, the vector

$$\mathbf{c} = \alpha \begin{bmatrix} 1 \\ 1 \end{bmatrix},$$

where α is any scalar, is an eigenvector corresponding to the eigenvalue $\lambda = 3$.

● **EXERCISES**

1. Compute the eigenvalues and corresponding eigenvectors of each of the following matrices.

(a) $\begin{bmatrix} -1 & 2 \\ -2 & -1 \end{bmatrix}$.   (b) $\begin{bmatrix} -3 & 1 & 7 \\ 0 & 4 & -1 \\ 0 & 0 & 2 \end{bmatrix}$.   (c) $\begin{bmatrix} 1 & 2 \\ 4 & 3 \end{bmatrix}$.

(d) $\begin{bmatrix} 2 & 1 \\ -1 & 1 \end{bmatrix}$.   (e) $\begin{bmatrix} 1 & 0 & 3 \\ 8 & 1 & -1 \\ 5 & 1 & -1 \end{bmatrix}$.   (f) $\begin{bmatrix} 3 & -1 & 1 \\ 2 & 0 & 1 \\ 1 & -1 & 2 \end{bmatrix}$.

(g) $\begin{bmatrix} -1 & 0 & 0 \\ 0 & -1 & 0 \\ 0 & 0 & 5 \end{bmatrix}$.   (h) $\begin{bmatrix} -1 & 1 & 0 \\ 1 & -1 & 0 \\ 0 & 0 & 5 \end{bmatrix}$.

(i) $\begin{bmatrix} 2 & -3 & 3 \\ 4 & -5 & 3 \\ 4 & -4 & 2 \end{bmatrix}$ (eigenvalues are $-1, 2, -2$).

(j) $\begin{bmatrix} 1 & 2 & 1 \\ 1 & -1 & 1 \\ 2 & 0 & 1 \end{bmatrix}$ (eigenvalues are $-1, -1, 3$).

(k) $\begin{bmatrix} 0 & 1 & 0 \\ 0 & 0 & 1 \\ -6 & -11 & -6 \end{bmatrix}$ (eigenvalues are $-1, -2, -3$).

(l) $\begin{bmatrix} 0 & 1 & 0 \\ 0 & 0 & 1 \\ -2 & -5 & -4 \end{bmatrix}$ (eigenvalues are $-1, -1, -2$).

(m) $\begin{bmatrix} 4 & 1 & 0 & 0 & 0 \\ 0 & 4 & 1 & 0 & 0 \\ 0 & 0 & 4 & 0 & 0 \\ 0 & 0 & 0 & 4 & 0 \\ 0 & 0 & 0 & 0 & 0 \end{bmatrix}$.

(n) $\begin{bmatrix} 3 & -1 & -4 & 2 \\ 2 & 3 & -2 & -4 \\ 2 & -1 & -3 & 2 \\ 1 & 2 & -1 & -3 \end{bmatrix}$ [*Hint:* characteristic polynomial is $(\lambda - 1)^2 (\lambda + 1)^2$.]

2. Show that if A is a triangular matrix of the form

$$A = \begin{bmatrix} a_{11} & a_{12} & \cdots & a_{1n} \\ 0 & a_{22} & \cdots & \\ \vdots & 0 & \ddots & \vdots \\ 0 & \cdots & 0 & a_{nn} \end{bmatrix}$$

the eigenvalues of A are $\lambda = a_{ii}$, where $i = 1, \ldots, n$.

The reader will note that in Example 1 preceeding, the two eigenvectors **u** and **v** are linearly independent in $\mathscr{C}_2$ if $\alpha \neq 0$ and $\beta \neq 0$, since

$$\det [\mathbf{u}, \mathbf{v}] = \det \begin{bmatrix} \alpha & \beta i \\ \alpha i & \beta \end{bmatrix} = 2\alpha\beta \neq 0.$$

Therefore, by Theorem 1, Section 5.5, interpreted for $\mathscr{C}_2$, the vectors **u** and **v** form a basis of $\mathscr{C}_2$. However, in Example 2, the eigenvectors form a one-dimensional subspace of $\mathscr{C}_2$. In applications to differential equations as well as in matrix theory it is important to know whether the set of all eigenvectors (corresponding to the various eigenvalues) of a given matrix $A \in \mathscr{C}_{nn}$ form a basis of $\mathscr{C}_n$. As Example 1 shows, even if the matrix A is real, the eigenvectors may have complex components. Thus, we consider the eigenvectors as vectors in $\mathscr{C}_n$. If the matrix $A \in \mathscr{C}_{nn}$ has n distinct eigenvalues, the corresponding eigenvectors form a basis for $\mathscr{C}_n$. This is a consequence of the following result.

Theorem 1. *A set of k eigenvectors in $\mathscr{C}_n$ corresponding to any k distinct eigenvalues is linearly independent.*

Proof. We shall prove the theorem by induction on the number k of eigenvectors. For $k = 1$, the result is trivial. Now, assume that every set of $(p - 1)$ eigenvectors corresponding to $(p - 1)$ distinct eigenvalues of a given matrix A is linearly independent. Let $\mathbf{v}_1, \ldots, \mathbf{v}_p$ be eigenvectors of A corresponding to the eigenvalues $\lambda_1, \ldots, \lambda_p$, respectively, with $\lambda_i \neq \lambda_j$ for $i \neq j$. Suppose that there exist constants $c_1, c_2, \ldots, c_p$, not all zero, such that

$$c_1\mathbf{v}_1 + c_2\mathbf{v}_2 + \cdots + c_p\mathbf{v}_p = \mathbf{0}. \tag{7.11}$$

We may assume $c_1 \neq 0$. Applying $A - \lambda_1 I$ to both sides of this equation, and using $(A - \lambda_1 I)\mathbf{v}_j = (\lambda_j - \lambda_1)\mathbf{v}_j$ (with $j = 1, 2, \ldots, n$), we obtain

$$c_2(\lambda_2 - \lambda_1)\mathbf{v}_2 + c_3(\lambda_3 - \lambda_1)\mathbf{v}_3 + \cdots + c_p(\lambda_p - \lambda_1)\mathbf{v}_p = \mathbf{0}.$$

But $\mathbf{v}_2, \mathbf{v}_3, \ldots, \mathbf{v}_p$ are linearly independent by the inductive hypothesis, and therefore $c_j(\lambda_j - \lambda_1) = 0$, where $j = 2, 3, \ldots, p$. Since $\lambda_j \neq \lambda_1$, where $j = 2, 3, \ldots, p$, we have $c_j = 0$, where $j = 2, 3, \ldots, p$, and (7.11) becomes $c_1\mathbf{v}_1 = \mathbf{0}$. Since $\mathbf{v}_1 \neq \mathbf{0}$, $c_1 = 0$ as well, which shows that $\mathbf{v}_1, \ldots, \mathbf{v}_p$ are linearly independent. This proves the theorem by induction. ∎

We remind the reader that since the characteristic polynomial of an $n \times n$ matrix is a polynomial of degree n there are at most n distinct eigenvalues. Since the eigenvectors span a subspace of $\mathscr{C}_n$, and since $\mathscr{C}_n$ is n dimensional, there are at most n linearly independent eigenvectors. Of course, in any case, there exists at least one eigenvector, since there is at least one (distinct) eigenvalue.

One reason that it is important to know whether the eigenvectors of a matrix $A \in \mathscr{C}_{nn}$ form a basis for $\mathscr{C}_n$ is that if they do, then the matrix A is similar to a diagonal matrix. (See Definition 3, Section 5.7):

Theorem 2. *Let $A \in \mathscr{C}_{nn}$. Let $\mathbf{v}_1, \ldots, \mathbf{v}_n$ be n linearly independent eigenvectors of A corresponding to the eigenvalues $\lambda_1, \ldots, \lambda_n$ (not necessarily distinct). Then there exists a nonsingular matrix $P \in \mathscr{C}_{nn}$ such that*

7.2 Eigenvalues and Eigenvectors of Matrices

$$P^{-1}AP = \begin{bmatrix} \lambda_1 & 0 & \cdots & 0 \\ 0 & \lambda_2 & \ddots & \vdots \\ \vdots & \ddots & \ddots & 0 \\ 0 & \cdots & 0 & \lambda_n \end{bmatrix} = B;$$

that is, A is similar to the diagonal matrix B.

Proof. Define the matrix

$$P = [\mathbf{v}_1, \ldots, \mathbf{v}_n].$$

By hypothesis, the eigenvectors $\mathbf{v}_1, \mathbf{v}_2, \ldots, \mathbf{v}_n$ are linearly independent; thus, P is nonsingular, by Theorem 1, Section 5.5 (interpreted for $\mathscr{C}_{nn}$). Now,

$$\begin{aligned} P^{-1}AP &= P^{-1}A[\mathbf{v}_1, \ldots, \mathbf{v}_n] \\ &= P^{-1}[A\mathbf{v}_1, \ldots, A\mathbf{v}_n] \\ &= P^{-1}[\lambda_1 \mathbf{v}_1, \ldots, \lambda_n \mathbf{v}_n], \end{aligned}$$

where in the last step we have used that $\mathbf{v}_k$ is an eigenvector corresponding to λ_k, so that $A\mathbf{v}_k = \lambda_k \mathbf{v}_k$, where $k = 1, \ldots, n$. But, by matrix multiplication, since $P^{-1}P = I$ and since $\mathbf{v}_k$ is the kth column of P, $P^{-1}\mathbf{v}_k$ is the kth column of I, $k = 1, 2, \ldots, n$. Therefore,

$$P^{-1}AP = \begin{bmatrix} \lambda_1 & 0 & \cdots & 0 \\ 0 & \lambda_2 & \ddots & \vdots \\ \vdots & \ddots & \ddots & 0 \\ 0 & \cdots & 0 & \lambda_n \end{bmatrix}. \blacksquare$$

The converse of Theorem 2 is also true:

Theorem 3. *If $A \in \mathscr{C}_{nn}$ is similar to a diagonal matrix D, then A possesses n linearly independent eigenvectors.*

Proof. Since A is similar to a diagonal matrix D, there exists a nonsingular matrix $P \in \mathscr{C}_{nn}$ such that

$$P^{-1}AP = D = \begin{bmatrix} d_1 & & & \\ & d_2 & & \\ & & \ddots & \\ & & & d_n \end{bmatrix}.$$

220 Eigenvalues, Eigenvectors, and Systems of Differential Equations

Since P is nonsingular, the columns of P are linearly independent (Theorem 1, Section 5.5). Suppose that

$$P = [\mathbf{v}_1, \mathbf{v}_2, \ldots, \mathbf{v}_n].$$

We will show that $\mathbf{v}_k$ is an eigenvector of A corresponding to the eigenvalue d_k. From $P^{-1}AP = D$ we have $AP = PD$. Writing this out, we obtain, by matrix multiplication,

$$AP = A[\mathbf{v}_1, \mathbf{v}_2, \ldots, \mathbf{v}_n] = [A\mathbf{v}_1, \ldots, A\mathbf{v}_n],$$

$$PD = [\mathbf{v}_1, \mathbf{v}_2, \ldots, \mathbf{v}_n] \begin{bmatrix} d_1 & & & \\ & d_2 & & \\ & & \ddots & \\ & & & d_n \end{bmatrix}$$

$$= [d_1\mathbf{v}_1, d_2\mathbf{v}_2, \ldots, d_n\mathbf{v}_n].$$

Therefore

$$[A\mathbf{v}_1, A\mathbf{v}_2, \ldots, A\mathbf{v}_n] = [d_1\mathbf{v}_1, d_2\mathbf{v}_2, \ldots, d_n\mathbf{v}_n],$$

and so $A\mathbf{v}_k = d_k\mathbf{v}_k$, where $k = 1, \ldots, n$. Thus, $\mathbf{v}_k$ is an eigenvector corresponding to the eigenvalue d_k. As already remarked the eigenvectors $\mathbf{v}_1, \mathbf{v}_2, \ldots, \mathbf{v}_n$ are linearly independent. ∎

If we combine Theorems 1 and 2, we obtain the following useful result.

Corollary to Theorem 2. *If a matrix A has distinct eigenvalues $\lambda_1, \ldots, \lambda_n$, then the corresponding eigenvectors form a basis of $\mathscr{C}_n$, and A is similar to the diagonal matrix*

$$B = \begin{bmatrix} \lambda_1 & & & \\ & \lambda_2 & & \\ & & \ddots & \\ & & & \lambda_n \end{bmatrix}$$

The reader will observe that both A and $P^{-1}AP$ in Theorem 2 have the same eigenvalues $\lambda_1, \ldots, \lambda_n$. This suggests the following result.

Theorem 4. *If $A, B \in \mathscr{C}_{nn}$ and if A and B are similar, then A and B have the same characteristic polynomials and hence also the same eigenvalues.*

Proof. Since A and B are similar, there exists a nonsingular matrix P such that $B = P^{-1}AP$. We compute,

$$\det(\lambda I - B) = \det(\lambda I - P^{-1}AP) = \det(\lambda P^{-1}P - P^{-1}AP)$$
$$= \det P^{-1}(\lambda I - A)P \qquad (7.12)$$
$$= \det P^{-1} \det(\lambda I - A) \det P = \det(\lambda I - A),$$

where we have used several properties of determinants. Since the eigenvalues of A are the roots of the equation $\det(\lambda I - A) = 0$ and the eigenvalues of B are the roots of $\det(\lambda I - B) = 0$, the result follows from (7.12). ∎

Corollary to Theorem 4. *Let $\mathbf{v}$ be an eigenvector of the matrix $A \in \mathscr{C}_{nn}$ corresponding to the eigenvalue λ. Let $P \in \mathscr{F}_{nn}$ be nonsingular. Then $P^{-1}\mathbf{v}$ is an eigenvector of $P^{-1}AP$ corresponding to the eigenvalue λ.*

Proof. Since $A\mathbf{v} = \lambda \mathbf{v}$, we have

$$(P^{-1}AP)(P^{-1}\mathbf{v}) = P^{-1}A\mathbf{v} = P^{-1}\lambda\mathbf{v} = \lambda P^{-1}\mathbf{v}. \quad \blacksquare$$

● **EXERCISE**

3. Justify each step in (7.12).

The converse of Theorem 4 is false, since, for example, the matrices

$$A = \begin{bmatrix} 1 & 0 \\ 0 & 1 \end{bmatrix} \quad \text{and} \quad B = \begin{bmatrix} 1 & 1 \\ 0 & 1 \end{bmatrix}$$

both have a double eigenvalue 1, but are not similar.

Example 3. For the matrix A in Example 1, find a matrix P such that the matrix $P^{-1}AP$ is diagonal.

By Example 1 and the remark following Exercise 2 with $\alpha = \beta = 1$,

$$\mathbf{u} = \begin{bmatrix} 1 \\ i \end{bmatrix}, \quad \mathbf{v} = \begin{bmatrix} i \\ 1 \end{bmatrix}$$

are linearly independent eigenvectors of A corresponding to the eigenvalues $3 + 5i$ and $3 - 5i$, respectively. As in Theorem 2, define

$$P = [\mathbf{u}, \mathbf{v}] = \begin{bmatrix} 1 & i \\ i & 1 \end{bmatrix}.$$

Since det $P \neq 0$, P is nonsingular; further

$$P^{-1} = \frac{1}{2}\begin{bmatrix} 1 & -i \\ -i & 1 \end{bmatrix}.$$

Hence,

$$\begin{aligned} P^{-1}AP &= \frac{1}{2}\begin{bmatrix} 1 & -i \\ -i & 1 \end{bmatrix}\begin{bmatrix} 3 & 5 \\ -5 & 3 \end{bmatrix}\begin{bmatrix} 1 & i \\ i & 1 \end{bmatrix} \\ &= \frac{1}{2}\begin{bmatrix} 3+5i & 5-3i \\ -5-3i & 3-5i \end{bmatrix}\begin{bmatrix} 1 & i \\ i & 1 \end{bmatrix} \\ &= \frac{1}{2}\begin{bmatrix} 6+10i & 0 \\ 0 & 6-10i \end{bmatrix} = \begin{bmatrix} 3+5i & 0 \\ 0 & 3-5i \end{bmatrix}. \end{aligned}$$

The reader will note that the diagonal elements of $P^{-1}AP$ are precisely the eigenvalues of A in accordance with Theorem 2.

Example 4. Determine the subspace of $\mathscr{C}_2$ spanned by the eigenvectors of the matrix A in Example 2.

As we saw in Example 2

$$\mathbf{u} = \alpha \begin{bmatrix} 1 \\ 1 \end{bmatrix} \quad (\alpha \neq 0)$$

is an eigenvector corresponding to the eigenvalue $\lambda = 3$ of A of multiplicity 2 for any value $\alpha \neq 0$. Since $\lambda = 3$ is the only eigenvalue of A, every eigenvector of A is of this form for some $\alpha \neq 0$. Thus, the set of all eigenvectors of A is the subspace of $\mathscr{C}_2$ spanned by the vector

$$\begin{bmatrix} 1 \\ 1 \end{bmatrix}.$$

Clearly, this subspace is the line in $\mathscr{C}_2$ passing through the point (1, 1) and the origin.

Example 5. Compute the eigenvalues, eigenvectors of the matrix

$$A = \begin{bmatrix} 4 & 3 \\ -1 & 0 \end{bmatrix}$$

7.2 Eigenvalues and Eigenvectors of Matrices

and find a matrix P such that $P^{-1}AP$ is diagonal. We have

$$\det [\lambda I - A] = \det \begin{bmatrix} \lambda - 4 & -3 \\ 1 & \lambda \end{bmatrix}$$
$$= \lambda^2 - 4\lambda + 3 = (\lambda - 3)(\lambda - 1).$$

Thus, $\lambda = 3$ and $\lambda = 1$ are eigenvalues. Corresponding eigenvectors are

$$\mathbf{u} = \begin{bmatrix} 3 \\ -1 \end{bmatrix} \quad \text{and} \quad \mathbf{v} = \begin{bmatrix} 1 \\ -1 \end{bmatrix},$$

respectively. Observe that both the eigenvalues and the corresponding eigenvectors are real. Moreover, the vectors $\mathbf{u}$ and $\mathbf{v}$ form a basis for $\mathcal{R}_2$. Clearly,

$$P = [\mathbf{u}, \mathbf{v}] = \begin{bmatrix} 3 & 1 \\ -1 & -1 \end{bmatrix}$$

is a matrix such that $P^{-1}AP$ is diagonal.

As Example 5 suggests, if **both** A and all eigenvalues $\lambda_1, \ldots, \lambda_n$ are real then we may replace $\mathcal{C}$ by $\mathcal{R}$ in Theorems 1, 2, and Corollary to Theorem 2 (in fact, as in Example 5 there exists a nonsingular matrix in $\mathcal{R}_{nn}$ such that $P^{-1}AP = B$) and Theorem 4. We have the following real analog to Theorem 3.

If A in $\mathcal{R}_{nn}$ is similar to a real diagonal matrix D, then A possesses n linearly independent eigenvectors in $\mathcal{R}_n$.

- **EXERCISES**

    4. For each of the matrices A in Exercises 1(a), (b), (c), (e), (g), (i), and (k), determine a nonsingular matrix P such that $P^{-1}AP$ is a diagonal matrix.
    5. Prove that the matrices

    $$A = \begin{bmatrix} 1 & 0 \\ 0 & 1 \end{bmatrix}, \quad B = \begin{bmatrix} 1 & 1 \\ 0 & 1 \end{bmatrix}$$

    are not similar.

    6. Determine the subspace, and its dimension, of $\mathcal{C}_n$ spanned by the eigenvectors of each matrix in Exercise 1.
    7. In the matrix A of Exercise 2 assume the diagonal elements a_{ii}, where $i = 1, \ldots, n$, are all distinct. Find the dimension of the subspace of $\mathcal{C}_n$ spanned by the eigenvectors of A.

7.3 Calculation of a Fundamental Matrix

We have seen in Theorem 1, Section 7.1, that $\exp tA$ is a fundamental matrix of the linear system with constant coefficients, $\mathbf{y}' = A\mathbf{y}$. We have also seen in Examples 1 and 2, Section 7.1, how to compute $\exp tA$ in certain special cases; in particular, we have seen how to compute $\exp tA$ when A is diagonal. We will now show how to compute a fundamental matrix Φ of the system $\mathbf{y}' = A\mathbf{y}$ when A has n linearly independent eigenvectors, which by Theorem 2, Section 7.2, is precisely the case when A is similar to a diagonal matrix. We postpone to Section 7.5 consideration of the completely general case of an arbitrary matrix A.

Suppose the matrix A has n linearly independent eigenvectors $\mathbf{v}_1, \mathbf{v}_2, \ldots, \mathbf{v}_n$ in $\mathscr{R}_n$ or $\mathscr{C}_n$ corresponding to the (not necessarily distinct) eigenvalues $\lambda_1, \lambda_2, \ldots, \lambda_n$. Motivated by the discussion at the beginning of Section 7.2, we claim that each vector function

$$\boldsymbol{\phi}_j(t) = \exp(\lambda_j t)\mathbf{v}_j \qquad (j = 1, \ldots, n)$$

is a solution of $\mathbf{y}' = A\mathbf{y}$, on $-\infty < t < \infty$. For,

$$\begin{aligned}\boldsymbol{\phi}_j'(t) &= \exp(\lambda_j t)\lambda_j \mathbf{v}_j \\ &= \exp(\lambda_j t)A\mathbf{v}_j \\ &= A\exp(\lambda_j t)\mathbf{v}_j \\ &= A\boldsymbol{\phi}_j(t) \qquad (j = 1, \ldots, n),\end{aligned}$$

where we have used the fact that $A\mathbf{v}_j = \lambda_j \mathbf{v}_j$, $j = 1, \ldots, n$. Define

$$\Phi(t) = [\boldsymbol{\phi}_1(t), \boldsymbol{\phi}_2(t), \ldots, \boldsymbol{\phi}_n(t)].$$

Since each column of Φ is a solution of $\mathbf{y}' = A\mathbf{y}$, Φ is a solution matrix of $\mathbf{y}' = A\mathbf{y}$ on $-\infty < t < \infty$. We have

$$\det \Phi(0) = \det[\mathbf{v}_1, \ldots, \mathbf{v}_n] \neq 0,$$

because the vectors $\mathbf{v}_1, \mathbf{v}_2, \ldots, \mathbf{v}_n$ are linearly independent (see Theorem 1, Section 5.5). It now follows from Theorem 2, Section 6.3, that $\det \Phi(t) \neq 0$ for $-\infty < t < \infty$ and that $\Phi(t)$ is a fundamental matrix of $\mathbf{y}' = A\mathbf{y}$ on $-\infty < t < \infty$. We have therefore proved the following result.

Theorem 1. *Let A be a constant matrix in $\mathscr{F}_{nn}$. Suppose $\mathbf{v}_1, \mathbf{v}_2, \ldots, \mathbf{v}_n$ are n linearly independent eigenvectors corresponding respectively to the eigenvalues $\lambda_1, \lambda_2, \ldots, \lambda_n$. Then*

7.3 Calculation of a Fundamental Matrix

$$\Phi(t) = [\exp(\lambda_1 t)\mathbf{v}_1, \exp(\lambda_2 t)\mathbf{v}_2, \ldots, \exp(\lambda_n t)\mathbf{v}_n]$$

is a fundamental matrix of the linear system with constant coefficients $\mathbf{y}' = A\mathbf{y}$ *on* $-\infty < t < \infty$. *In particular this is the case if the eigenvalues* $\lambda_1, \lambda_2, \ldots, \lambda_n$ *are distinct.*

Example 1. Find a fundamental matrix of the system $\mathbf{y}' = A\mathbf{y}$ if

$$A = \begin{bmatrix} 3 & 5 \\ -5 & 3 \end{bmatrix}.$$

By Example 1, Section 7.2, $\lambda_1 = 3 + 5i$ and $\lambda_2 = 3 - 5i$ are eigenvalues of A and

$$\mathbf{v}_1 = \begin{bmatrix} 1 \\ i \end{bmatrix}, \quad \mathbf{v}_2 = \begin{bmatrix} i \\ 1 \end{bmatrix}$$

are (linearly independent) eigenvectors corresponding to λ_1, λ_2, respectively. By Theorem 1

$$\Phi(t) = \begin{bmatrix} e^{(3+5i)t} & ie^{(3-5i)t} \\ ie^{(3+5i)t} & e^{(3-5i)t} \end{bmatrix}$$

is a fundamental matrix on $-\infty < t < \infty$.

In general, Theorem 1 does not yield $\exp tA$, even though it does yield a fundamental matrix $\Phi(t)$ of $\mathbf{y}' = A\mathbf{y}$. By Corollary 2 to Theorem 2, Section 6.3, since $\exp tA$ and $\Phi(t)$ are both fundamental matrices of $\mathbf{y}' = A\mathbf{y}$ on $-\infty < t < \infty$, there exists a nonsingular matrix C such that

$$\exp tA = \Phi(t)C. \tag{7.13}$$

Setting $t = 0$ in (7.13), we obtain $C = \Phi^{-1}(0)$. Thus,

$$\exp tA = \Phi(t)\Phi^{-1}(0). \tag{7.14}$$

Example 2. Find $\exp tA$ if A is the matrix in Example 1.
By (7.14), Example 1 and Example 3, Section 7.2, we have successively

$$\exp tA = \begin{bmatrix} e^{(3+5i)t} & ie^{(3-5i)t} \\ ie^{(3+5i)t} & e^{(3-5i)t} \end{bmatrix} \begin{bmatrix} 1 & i \\ i & 1 \end{bmatrix}^{-1}$$

$$= \frac{1}{2}\begin{bmatrix} e^{(3+5i)t} & ie^{(3-5i)t} \\ ie^{(3+5i)t} & e^{(3-5i)t} \end{bmatrix} \begin{bmatrix} 1 & -i \\ -i & 1 \end{bmatrix}$$

$$= \frac{1}{2}\begin{bmatrix} e^{(3+5i)t} + e^{(3-5i)t} & -i(e^{(3+5i)t} - e^{(3-5i)t}) \\ i(e^{(3+5i)t} - e^{(3-5i)t}) & e^{(3+5i)t} + e^{(3-5i)t} \end{bmatrix}$$

$$= e^{3t}\begin{bmatrix} \cos 5t & \sin 5t \\ -\sin 5t & \cos 5t \end{bmatrix}.$$

If A is real, $\exp tA$ is real from the definition (7.2). Thus, Eq. 7.14 gives at the same time a way of constructing a real fundamental matrix, whenever A is real. Example 2 is a special case of this remark.

● **EXERCISES**

1. Find a fundamental matrix of the system $y' = Ay$; also find $\exp tA$ for each of the following coefficient matrices.

(a) $A = \begin{bmatrix} 1 & 2 \\ 4 & 3 \end{bmatrix}$ (see Exercise 1(c), Section 7.2).

(b) $A = \begin{bmatrix} -2 & 1 \\ -1 & -2 \end{bmatrix}$.

(c) $A = \begin{bmatrix} 1 & 0 & 3 \\ 8 & 1 & -1 \\ 5 & 1 & -1 \end{bmatrix}$ (see Exercise 1(e), Section 7.2).

(d) $A = \begin{bmatrix} 2 & -3 & 3 \\ 4 & -5 & 3 \\ 4 & -4 & 2 \end{bmatrix}$ (see Exercise 1(i), Section 7.2).

(e) $A = \begin{bmatrix} 0 & 1 & 0 \\ 0 & 0 & 1 \\ -6 & -11 & -6 \end{bmatrix}$ (see Exercise 1(k), Section 7.2).

2. Show that the scalar second-order differential equation $u'' + pu' + qu = 0$ is equivalent to the system $y' = Ay$ with

$$A = \begin{bmatrix} 0 & 1 \\ -q & -p \end{bmatrix}$$

and compute the eigenvalues λ_1, λ_2 of A.

3. Compute a fundamental matrix for the system in Exercise 2 if $\lambda_1 \neq \lambda_2$, that is, if $p^2 \neq 4q$, and construct the general solution of the scalar second-order equation (compare with Theorem 2, Section 6.4).

An alternative way of producing a real fundamental matrix if A is a real 2×2 matrix is contained in the following exercises.

● EXERCISES

4. Given the matrix

$$A = \begin{bmatrix} 0 & 1 \\ -1 & 0 \end{bmatrix}.$$

Show that $A^2 = -I$, $A^3 = -A$, $A^4 = I$, and compute A^m, where m is an arbitrary positive integer.

5. Use the result of Exercise 4 and the definition (7.2) to show that

$$e^{tA} = \begin{bmatrix} \cos t & \sin t \\ -\sin t & \cos t \end{bmatrix}.$$

[*Hint:* $\cos t = 1 - \dfrac{t^2}{2!} + \dfrac{t^4}{4!} + \cdots$,

$\sin t = t - \dfrac{t^3}{3!} + \dfrac{t^5}{5!} + \cdots$].

6. Compute e^{tA}, if

$$A = \begin{bmatrix} 2 & 1 \\ -1 & 2 \end{bmatrix}.$$

[*Hint:* Use Exercise 4 and 5.]

We close this section with the solution of the nonhomogeneous system

$$\mathbf{y}' = A\mathbf{y} + \mathbf{g}(t), \tag{7.15}$$

where A is a constant matrix and g is a given continuous function on $-\infty < t < \infty$. The variation of constants formula (Theorem 1, Section 6.5) with $\Phi(t) = \exp tA$ as a fundamental matrix of the homogeneous system now becomes particularly simple in appearance. We have $\Phi^{-1}(s) = \exp(-sA)$, $\Phi(t)\Phi^{-1}(s) = \exp[(t-s)A]$; if the initial condition is $\phi(t_0) = \eta$, $\phi_h(t) = \exp[(t-t_0)A]\eta$ and the solution of (7.15) is

$$\phi(t) = \exp[(t-t_0)A]\eta + \int_{t_0}^{t} \exp[(t-s)A]\mathbf{g}(s)\,ds$$

$$(-\infty < t < \infty), \tag{7.16}$$

where e^{tA} is the fundamental matrix of the homogeneous system that we can construct by the method shown in this section. Note how easy it is to compute the inverse of Φ and also $\Phi(t)\Phi^{-1}(s)$ in this case. However, it may not be possible to evaluate explicitly the integral in (7.15) except in special cases.

Example 3. Find the solution Φ of the system $\mathbf{y}' = A\mathbf{y} + \mathbf{g}(t)$ satisfying the initial condition

$$\Phi(0) = \begin{bmatrix} 0 \\ 1 \end{bmatrix} \quad \text{if} \quad A = \begin{bmatrix} 3 & 5 \\ -5 & 3 \end{bmatrix} \quad \text{and} \quad \mathbf{g}(t) = \begin{bmatrix} e^{-t} \\ 0 \end{bmatrix}$$

From Example 2 preceding, we have

$$\exp tA = e^{3t} \begin{bmatrix} \cos 5t & \sin 5t \\ -\sin 5t & \cos 5t \end{bmatrix}.$$

Substituting in (7.16), we obtain (using $t_0 = 0$)

$$\Phi(t) = e^{3t} \begin{bmatrix} \cos 5t & \sin 5t \\ -\sin 5t & \cos 5t \end{bmatrix} \begin{bmatrix} 0 \\ 1 \end{bmatrix}$$

$$+ \int_0^t e^{3(t-s)} \begin{bmatrix} \cos 5(t-s) & \sin 5(t-s) \\ -\sin 5t(t-s) & \cos 5(t-s) \end{bmatrix} \begin{bmatrix} e^{-s} \\ 0 \end{bmatrix} ds$$

$$= e^{3t} \begin{bmatrix} \sin 5t \\ \cos 5t \end{bmatrix} + \int_0^t e^{3(t-s)} e^{-s} \begin{bmatrix} \cos 5(t-s) \\ -\sin 5(t-s) \end{bmatrix} ds.$$

In this case, we can evaluate the integrals as follows

$$\Phi(t) = e^{3t} \begin{bmatrix} \sin 5t \\ \cos 5t \end{bmatrix} + e^{3t} \int_0^t e^{-4s} \begin{bmatrix} \cos 5t \cos 5s + \sin 5t \sin 5s \\ -\sin 5t \cos 5s + \cos 5t \sin 5s \end{bmatrix} ds.$$

Using the formulas (these can be found by integration by parts)

$$\int_0^t e^{-4s} \cos 5s \, ds = \frac{e^{-4s}}{16 + 25} (-4 \cos 5s + 5 \sin 5s) \Big|_{s=0}^{s=t}$$

$$\int_0^t e^{-4s} \sin 5s \, ds = \frac{e^{-4s}}{16 + 25} (-4 \sin 5s - 5 \cos 5s) \Big|_{s=0}^{s=t}$$

we obtain

7.3 Calculation of a Fundamental Matrix

$$\phi(t) = e^{3t}\begin{bmatrix} \sin 5t \\ \cos 5t \end{bmatrix} + e^{3t}\begin{bmatrix} \cos 5t\left(\dfrac{e^{-4t}}{41}(-4\cos 5t + 5\sin 5t) + \dfrac{4}{41}\right) \\ + \sin 5t\left(\dfrac{e^{-4t}}{41}(-4\sin 5t - 5\cos 5t) + \dfrac{5}{41}\right) \\ -\sin 5t\left(\dfrac{e^{-4t}}{41}(-4\cos 5t + 5\sin 5t) + \dfrac{4}{41}\right) \\ +\cos 5t\left(\dfrac{e^{-4t}}{41}(-4\sin 5t - 5\cos 5t) + \dfrac{5}{41}\right) \end{bmatrix}.$$

Further simplification seems pointless. The reader will note that even such a simple example as above leads to a rather complicated answer.

● **EXERCISES**

7. Find the solution ϕ of the system

$$\mathbf{y}' = A\mathbf{y} + \mathbf{g}(t)$$

in each of the following cases:

(a) $\phi(0) = \begin{bmatrix} -1 \\ 1 \end{bmatrix}$, $A = \begin{bmatrix} 1 & 2 \\ 4 & 3 \end{bmatrix}$, $\mathbf{g}(t) = \begin{bmatrix} e^t \\ 1 \end{bmatrix}$ (see Exercise 1(a)).

(b) $\phi(0) = 0$, $A = \begin{bmatrix} 0 & 1 & -0 \\ 0 & 0 & 1 \\ -6 & -11 & -6 \end{bmatrix}$, $\mathbf{g}(t) = \begin{bmatrix} 0 \\ 0 \\ e^{-t} \end{bmatrix}$

(see Exercise 1(e)).

(c) $\phi(1) = \begin{bmatrix} 1 \\ 0 \\ 0 \end{bmatrix}$, $A = \begin{bmatrix} 2 & -3 & 3 \\ 4 & -5 & 3 \\ 4 & -4 & 2 \end{bmatrix}$, $\mathbf{g}(t)$ arbitrary

(see Exercise 1(d)).

8. By converting to an equivalent system, find the general solution of the scalar equation

$$y'' - y = f(t),$$

where f is continuous, by using the theory of this section.

9. Use the results of this section and Exercise 5 to find the general solution of the scalar equation

$$y'' + y = f(t),$$

where f is continuous.

7.4 Two-Dimensional Linear Systems

In this section we will solve completely two-dimensional systems of the form

$$\mathbf{y}' = A\mathbf{y}, \tag{7.17}$$

where A is a 2×2 matrix. More precisely, if we choose a nonsingular constant 2×2 matrix T, then the change of variable

$$\mathbf{y} = T\mathbf{z} \tag{7.18}$$

transforms (7.17) to the system

$$\mathbf{y}' = T\mathbf{z}' = A(T\mathbf{z}),$$

so that

$$\mathbf{z}' = (T^{-1}AT)\mathbf{z}. \tag{7.19}$$

We now show that it is always possible to choose the matrix T in such a way that the resulting system (7.19) can be solved easily. If $\psi(t)$ is a solution of (7.19), then $\phi(t) = T\psi(t)$ is a solution of (7.17), using (7.18). Conversely, since T is nonsingular, every solution of (7.17) gives a solution of (7.19). In other words, to solve (7.17) completely, it suffices to construct the matrix T in such a way that (7.19) can be solved.

Our first result concerns the case where the coefficient matrix A is complex.

Theorem 1. *Let $A \in \mathscr{C}_{22}$. There then exists a nonsingular matrix $T \in \mathscr{C}_{22}$ such that $T^{-1}AT$ has one of the following forms:*

(i) $\begin{bmatrix} \lambda & 0 \\ 0 & \mu \end{bmatrix}$ $(\lambda \neq \mu)$;

(ii) $\begin{bmatrix} \lambda & 0 \\ 0 & \lambda \end{bmatrix}$;

(iii) $\begin{bmatrix} \lambda & 1 \\ 0 & \lambda \end{bmatrix}$.

Proof. Case (i) arises when A has two distinct eigenvalues λ, μ. In this case the result is an immediate consequence of Theorems 1 and 2, Section 7.2.

Case (ii) arises when A has an eigenvalue λ of multiplicity two for which there are two linearly independent eigenvectors. In this case, the result is the special case $n = 2$ of Theorem 2, Section 7.2.

Case (iii) arises when A has an eigenvalue λ of multiplicity two but the subspace of $\mathscr{C}_2$ consisting of all eigenvectors of A has dimension one; that is, any two eigenvectors of A are linearly dependent. Then there exist nonzero vectors in $\mathscr{C}_2$ which are not eigenvectors of A; at the same time we recall that there exists at least one eigenvector of A. Let $\mathbf{v}$ be a nonzero vector in $\mathscr{C}_2$ which is not an eigenvector of A and let

$$\mathbf{u} = (A - \lambda I)\mathbf{v}. \tag{7.20}$$

Since $\mathbf{v}$ is not an eigenvector, and $A - \lambda I$ is not the zero matrix, $\mathbf{u} \neq \mathbf{0}$; we will show that $\mathbf{u}$ is an eigenvector.

We first assert that the vectors $\mathbf{u}$ and $\mathbf{v}$ are linearly independent. Suppose then there exist constants c_1, c_2 such that

$$c_1 \mathbf{u} + c_2 \mathbf{v} = \mathbf{0}. \tag{7.21}$$

Since $\mathbf{u}$ and $\mathbf{v}$ are both different from zero, either c_1 and c_2 are both zero (in which case there is nothing to prove) or they are both different from zero. Using the definition of $\mathbf{u}$ and the fact that $c_1 \neq 0$, we may rewrite (7.21) as

$$c_1(A - \lambda I)\mathbf{v} + c_2 \mathbf{v} = \mathbf{0}$$

or

$$c_1\left[\left(A - \lambda I + \frac{c_2}{c_1}I\right)\mathbf{v}\right] = c_1\left[A - \left(\lambda - \frac{c_2}{c_1}\right)I\right]\mathbf{v} = \mathbf{0}.$$

This says that $\lambda - c_2/c_1$ is an eigenvalue of A. Since λ is the only eigenvalue, $c_2 = 0$, which implies $c_1 = 0$. Thus, $\mathbf{u}$ and $\mathbf{v}$ are linearly independent and span the space $\mathscr{C}_2$.

Let $\mathbf{x}$ be any eigenvector of A; we may therefore write

$$\mathbf{x} = a\mathbf{u} + b\mathbf{v}.$$

If $a = 0$, then $\mathbf{x}$ is a multiple of $\mathbf{v}$, and therefore $\mathbf{v}$ is an eigenvector (but by hypothesis it is not); thus $a \neq 0$. Now,

$$\begin{aligned}\mathbf{0} &= (A - \lambda I)\mathbf{x} = (A - \lambda I)(a\mathbf{u} + b\mathbf{v}) \\ &= a(A - \lambda I)\mathbf{u} + b(A - \lambda I)\mathbf{v} = a(A - \lambda I)\mathbf{u} + b\mathbf{u} \\ &= a[(A - \lambda I + \frac{b}{a}I)\mathbf{u}] \\ &= a[A - (\lambda - \frac{b}{a})I]\mathbf{u}.\end{aligned}$$

This says that $\lambda - b/a$ is an eigenvalue, and since λ is the only eigenvalue, $b = 0$. Therefore, $\mathbf{x}$ is a nonzero multiple of $\mathbf{u}$, and $\mathbf{u}$ must be an eigenvector.

Now, we define

$$T = [\mathbf{u}, \mathbf{v}] = \begin{bmatrix} u_1 & v_1 \\ u_2 & v_2 \end{bmatrix}. \tag{7.22}$$

As in case (i),

$$T^{-1} = \frac{1}{u_1 v_2 - u_2 v_1} \begin{bmatrix} v_2 & -v_1 \\ -u_2 & u_1 \end{bmatrix}.$$

We have, using $\mathbf{u} = (A - \lambda I)\mathbf{v} = A\mathbf{v} - \lambda\mathbf{v}$ and (7.20), (7.22)

$$\begin{aligned}T^{-1}AT &= T^{-1}[A\mathbf{u}, A\mathbf{v}] \\ &= T^{-1}[\lambda\mathbf{u}, \mathbf{u} + \lambda\mathbf{v}] = T^{-1}\{[\mathbf{0}, \mathbf{u}] + [\lambda\mathbf{u}, \lambda\mathbf{v}]\} \\ &= T^{-1}[\lambda\mathbf{u}, \lambda\mathbf{v}] + T^{-1}[\mathbf{0}, \mathbf{u}] \\ &= \lambda T^{-1}T + \frac{1}{u_1 v_2 - u_2 v_1}\begin{bmatrix} v_2 & -v_1 \\ -u_2 & u_1 \end{bmatrix}\begin{bmatrix} 0 & u_1 \\ 0 & u_2 \end{bmatrix} \\ &= \lambda I + \begin{bmatrix} 0 & 1 \\ 0 & 0 \end{bmatrix} = \begin{bmatrix} \lambda & 1 \\ 0 & \lambda \end{bmatrix}. \blacksquare\end{aligned}$$

With reference to the above proof, the reader should note that in showing that $\mathbf{u} = (A - \lambda I)\mathbf{v}$ is an eigenvector, we have actually shown that $(A - \lambda I)^2\mathbf{w} = \mathbf{0}$ for every vector $\mathbf{w}$ in $\mathscr{C}_2$. For, if $\mathbf{w} \in \mathscr{C}_2$ we may write $\mathbf{w} = a\mathbf{u} + b\mathbf{v}$ because $\mathbf{u}, \mathbf{v}$ form a basis for $\mathscr{C}_2$. Since $\mathbf{u}$ is an eigenvector, we obtain

$$\begin{aligned}(A - \lambda I)\mathbf{w} &= a(A - \lambda I)\mathbf{u} + b(A - \lambda I)\mathbf{v} \\ &= b(A - \lambda I)\mathbf{v}.\end{aligned}$$

Hence,

$$(A - \lambda I)^2 \mathbf{w} = (A - \lambda I)b(A - \lambda I)\mathbf{v} = b(A - \lambda I)\mathbf{u} = \mathbf{0}$$

because of (7.20) and the fact that $\mathbf{u}$ is an eigenvector of A.

We remark that if A is a real matrix and if its eigenvalues are real, then the matrix T constructed in each of the three cases above is real. However, if A is real but has complex eigenvalues (necessarily complex conjugates), the matrix T will not be real and it is of interest to learn the simplest form of $T^{-1}AT$ which can be achieved with a real matrix T. The answer lies in the following result.

Theorem 2. *Let A be a real 2×2 matrix with complex conjugate eigenvalues $\alpha \pm i\beta$. There then exists a real constant nonsingular matrix T such that*

$$T^{-1}AT = \begin{bmatrix} \alpha & \beta \\ -\beta & \alpha \end{bmatrix}.$$

Proof. Let $\mathbf{u} + i\mathbf{v}$ be an eigenvector corresponding to the eigenvalue $\alpha + i\beta$, where $\mathbf{u}$ and $\mathbf{v}$ are real. If $\mathbf{v} = \mathbf{0}$, so that this eigenvector is real,

$$A\mathbf{u} = (\alpha + i\beta)\mathbf{u}$$

the left side of which is real and the right side of which is not real. Thus, $\mathbf{v} \neq \mathbf{0}$, and a similar argument shows that $\mathbf{u} \neq \mathbf{0}$.

We define the matrix T with columns $\mathbf{u}$ and $\mathbf{v}$,

$$T = [\mathbf{u}, \mathbf{v}] = \begin{bmatrix} u_1 & v_1 \\ u_2 & v_2 \end{bmatrix}.$$

In order to show that T is nonsingular, we must show that $\mathbf{u}$ and $\mathbf{v}$ are linearly independent. Suppose not; then there exist real constants c_1, c_2 both different from zero such that

$$c_1 \mathbf{u} + c_2 \mathbf{v} = \mathbf{0}.$$

Since $c_1 \neq 0$, and $A(\mathbf{u} + i\mathbf{v}) = (\alpha + i\beta)(\mathbf{u} + i\mathbf{v})$, we have

$$A(\mathbf{u} + i\mathbf{v}) = A(-\frac{c_2}{c_1}\mathbf{v} + i\mathbf{v}) = (\alpha + i\beta)(-\frac{c_2}{c_1} + i)\mathbf{v};$$

thus,
$$A\mathbf{v} = \left(\alpha + \frac{c_2}{c_1}\beta\right)\mathbf{v}.$$

Then $\mathbf{v}$ is a real eigenvector. But, as remarked above, $\mathbf{u} \neq \mathbf{0}$ and therefore this is impossible. Thus, $\mathbf{u}$ and $\mathbf{v}$ are linearly independent and T is non-singular.

Taking real and imaginary parts in the equation
$$A(\mathbf{u} + i\mathbf{v}) = (\alpha + i\beta)(\mathbf{u} + i\mathbf{v}),$$

we obtain
$$A\mathbf{u} = \alpha\mathbf{u} - \beta\mathbf{v}, \qquad A\mathbf{v} = \beta\mathbf{u} + \alpha\mathbf{v}.$$

Therefore,
$$AT = [A\mathbf{u}, A\mathbf{v}] = [\alpha\mathbf{u} - \beta\mathbf{v}, \beta\mathbf{u} + \alpha\mathbf{v}]$$

and
$$T^{-1}AT = \frac{1}{u_1 v_2 - u_2 v_1}\begin{bmatrix} v_2 & -v_1 \\ -u_2 & u_1 \end{bmatrix}\begin{bmatrix} \alpha u_1 - \beta v_1 & \beta u_1 + \alpha v_1 \\ \alpha u_2 - \beta v_2 & \beta u_2 + \alpha v_2 \end{bmatrix}$$
$$= \begin{bmatrix} \alpha & \beta \\ -\beta & \alpha \end{bmatrix}. \blacksquare$$

Consider now the system (7.19), where $T^{-1}AT = B$ has one of the forms in Theorems 1 and 2. We can write down the fundamental matrix $\exp Bt$ in each case. We have

CASE (i). *If*
$$B = \begin{bmatrix} \lambda & 0 \\ 0 & \mu \end{bmatrix},$$

then
$$\exp Bt = \begin{bmatrix} e^{\lambda t} & 0 \\ 0 & e^{\mu t} \end{bmatrix}.$$

7.4 Two-Dimensional Linear Systems

CASE (ii). *If*

$$B = \begin{bmatrix} \lambda & 0 \\ 0 & \lambda \end{bmatrix},$$

then

$$\exp Bt = \begin{bmatrix} e^{\lambda t} & 0 \\ 0 & e^{\lambda t} \end{bmatrix}.$$

CASE (iii). *If*

$$B = \begin{bmatrix} \lambda & 1 \\ 0 & \lambda \end{bmatrix},$$

then

$$\exp Bt = \begin{bmatrix} e^{\lambda t} & te^{\lambda t} \\ 0 & e^{\lambda t} \end{bmatrix}.$$

Further, if

$$B = \begin{bmatrix} \alpha & \beta \\ -\beta & \alpha \end{bmatrix},$$

then

$$\exp Bt = e^{\alpha t} \begin{bmatrix} \cos \beta t & \sin \beta t \\ -\sin \beta t & \cos \beta t \end{bmatrix}.$$

● **EXERCISE**

1. Show that $\exp Bt$ has the form indicated in each of these cases. [*Hint:* for special cases of the last one, see Examples 1 and 2, Section 7.3, or Exercises 4 and 5, Section 7.3]

In order to describe the behavior of many physical systems, it is convenient to picture all solutions of the two-dimensional systems geometrically. For the system

$$\mathbf{y}' = A\mathbf{y}, \qquad \mathbf{y} = \begin{bmatrix} y_1 \\ y_2 \end{bmatrix},$$

Eigenvalues, Eigenvectors, and Systems of Differential Equations

this is conveniently done by regarding t as a parameter and then picturing the solutions in the (y_1, y_2) plane. This plane is called the **phase plane**, and the graphical representation of all solutions is called the **phase portrait** of the system. Each curve in the phase portrait is called an **orbit**.

Definition 1. *A point* **p** *with coordinates* (p_1, p_2) *is called a critical point of the system* $\mathbf{y}' = A\mathbf{y}$ *if and only if* $A\mathbf{p} = \mathbf{0}$.

The origin is always a critical point. If A is nonsingular, the origin is clearly the only critical point since the system $A\mathbf{x} = \mathbf{0}$ has only the trivial solution.

Example 1. Construct the phase portrait for the system $\mathbf{y}' = A\mathbf{y}$, where

$$A = \begin{bmatrix} -2 & 0 \\ 0 & -3 \end{bmatrix}, \quad \mathbf{y} = \begin{bmatrix} y_1 \\ y_2 \end{bmatrix}.$$

Since $\det A \neq 0$, the origin is the only critical point. A fundamental matrix is

$$\exp At = \begin{bmatrix} e^{-2t} & 0 \\ 0 & e^{-3t} \end{bmatrix}.$$

Let $\boldsymbol{\phi}(t, \boldsymbol{\eta})$ be that solution of $\mathbf{y}' = A\mathbf{y}$ for which $\boldsymbol{\phi}(0, \boldsymbol{\eta}) = \boldsymbol{\eta}$. Then

$$\boldsymbol{\phi}(t, \boldsymbol{\eta}) = (\exp At)\boldsymbol{\eta} = \begin{bmatrix} e^{-2t}\eta_1 \\ e^{-3t}\eta_2 \end{bmatrix}.$$

Here we have arbitrarily chosen $t_0 = 0$. Notice that $\boldsymbol{\phi}(t - t_0, \boldsymbol{\eta})$ is that solution passing through the point $\boldsymbol{\eta}$ at $t = t_0$. Let $\boldsymbol{\eta} = (\eta_1, \eta_2)$ be any point in the (y_1, y_2) plane. Then the solution $\boldsymbol{\phi}(t, \boldsymbol{\eta})$ for $t > 0$ is represented by the parametric equations $y_1 = \phi_1(t) = e^{-2t}\eta_1$, $y_2 = \phi_2(t) = e^{-3t}\eta_2$ for $t > 0$ and this represents the portion of the curve shown in Figure 7.1 between $\boldsymbol{\eta}$ and the origin, as is verified by elementary calculus; the arrow indicates the direction of increasing t. Notice that the slope of the tangent to this curve, dy_2/dy_1, also tends to zero as $t \to +\infty$, because

$$\frac{dy_2}{dy_1} = \frac{y_2'}{y_1'} = \frac{-3\eta_2 e^{-3t}}{-2\eta_1 e^{-2t}} = \frac{3}{2}\frac{\eta_2}{\eta_1} e^{-t}.$$

7.4 Two-Dimensional Linear Systems 237

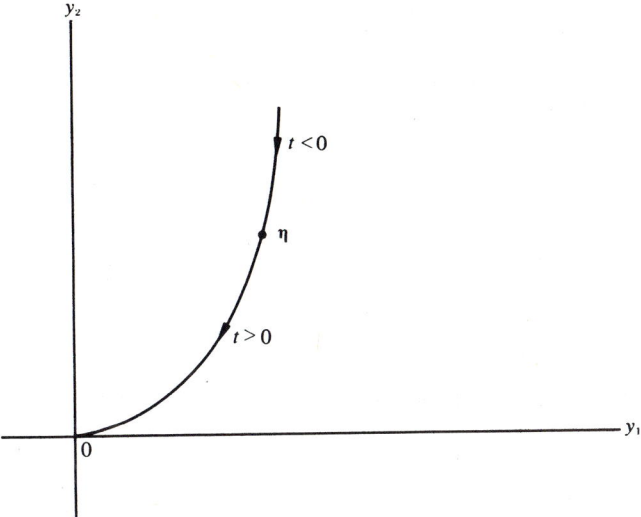

Figure 7.1

Similarly, $t < 0$ represents the portion of the curve in Figure 7.1 above the point $\mathbf{\eta}$. It should be noted that $\lim_{t \to +\infty} \boldsymbol{\phi}(t, \mathbf{\eta}) = \mathbf{0}$; that is, both the solution and the orbit approach the origin as $t \to +\infty$. Proceeding in this way by choosing various points of the phase plane as initial points, we obtain the **phase portrait** of the system, shown in Figure 7.2. Notice that every orbit approaches the origin (as $t \to +\infty$).

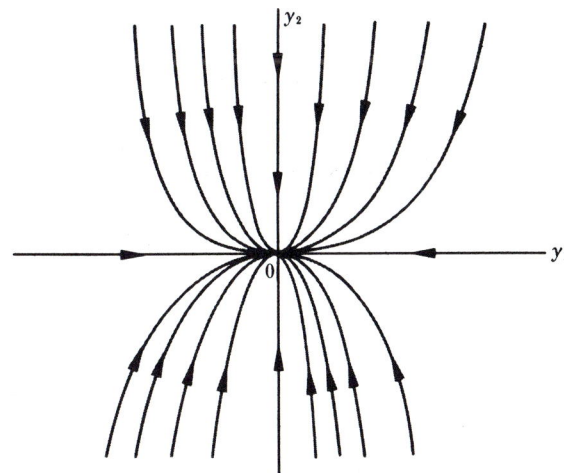

Figure 7.2

• EXERCISES

2. Obtain the phase portrait of the system $\mathbf{y}' = A\mathbf{y}$, where

$$A = \begin{bmatrix} 2 & 0 \\ 0 & 3 \end{bmatrix}.$$

3. Obtain the phase portrait of the system $\mathbf{y}' = A\mathbf{y}$, where

$$A = \begin{bmatrix} -2 & 1 \\ 0 & -2 \end{bmatrix}.$$

4. Obtain the phase portrait of the system $\mathbf{y}' = A\mathbf{y}$, where

$$A = \begin{bmatrix} 2 & 0 \\ 0 & -3 \end{bmatrix}.$$

5. Obtain the phase portrait for the scalar equation $y'' + 4y = 0$. [*Hint:* Use the system $y_1' = y_2'$, $y_2' = -4y_1$. By a phase portrait of a scalar second-order equation we mean the phase portrait of the equivalent first-order system.]

We shall now analyze the general **real** two-dimensional linear system $\mathbf{y}' = A\mathbf{y}$ with constant coefficients. As at the beginning of this section, let us make the change of variable $\mathbf{y} = T\mathbf{z}$, where T is a nonsingular constant matrix (to be determined), and substitute, obtaining the system

$$\mathbf{z}' = (T^{-1}AT)\mathbf{z} \tag{7.19}$$

whose coefficient $T^{-1}AT$ is similar to A. For simplicity, and because this is the case that arises in applications most frequently, we consider only the case $\det A \neq 0$. This means that zero is not an eigenvalue of A, and that the origin is the only critical point. For the case when $\det A = 0$, the reader is referred to Exercises 19 and 20.

As is done in Theorems 1 and 2, we may show that there is a **real** nonsingular matrix T such that $T^{-1}AT$ is equal to one of the following six matrices:

(i) $\begin{bmatrix} \lambda & 0 \\ 0 & \mu \end{bmatrix}$, where $\mu < \lambda < 0$ or $0 < \mu < \lambda$.

(ii) $\begin{bmatrix} \lambda & 0 \\ 0 & \lambda \end{bmatrix}$, where $\lambda > 0$ or $\lambda < 0$.

(iii) $\begin{bmatrix} \lambda & 0 \\ 0 & \mu \end{bmatrix}$, $\mu < 0 < \lambda$.

7.4 Two-Dimensional Linear Systems

(iv) $\begin{bmatrix} \lambda & 1 \\ 0 & \lambda \end{bmatrix}$, where $\lambda > 0$ or $\lambda < 0$.

(v) $\begin{bmatrix} \sigma & v \\ -v & \sigma \end{bmatrix}$, where $v \neq 0$ and $\sigma > 0$, or $\sigma < 0$.

(vi) $\begin{bmatrix} 0 & v \\ -v & 0 \end{bmatrix}$, where $v \neq 0$.

The cases (v) and (vi) correspond to complex conjugate eigenvalues of A, $\sigma \pm iv$ and $\pm iv$, respectively. In the four remaining cases the eigenvalues λ, μ are real. We obtain the possible phase portraits of (7.19) by assuming that $T^{-1}AT$ is one of the forms (i)–(vi).

CASE (i). (This is essentially Example 1 and Exercise 2 above.) The solution of (7.19) through the point $(\eta_1, \eta_2) \neq (0, 0)$ at $t = 0$ is

$$\phi(t) = \begin{bmatrix} e^{\lambda t}\eta_1 \\ e^{\mu t}\eta_2 \end{bmatrix}.$$

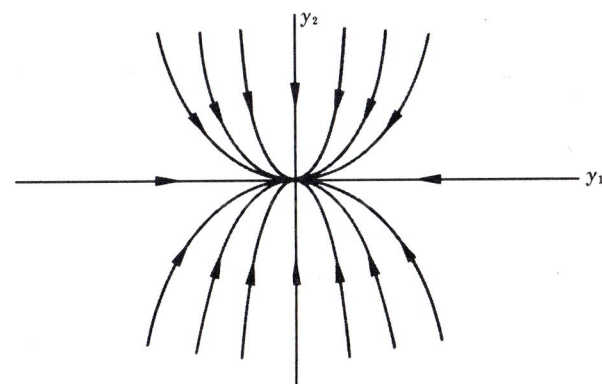

Figure 7.3

If $\mu < \lambda < 0$, we have $\phi(t) \to 0$ as $t \to +\infty$ and we obtain the phase portrait in Figure 7.3 with every orbit tending to the origin as $t \to +\infty$. If $0 < \mu < \lambda$, we obtain the phase portrait in Figure 7.4 with every orbit tending away from the origin as $t \to +\infty$. Arrows indicate the section of increasing t. The origin in Figures 7.3 and 7.4 corresponding to Case (i) is called an **improper node**.

240   *Eigenvalues, Eigenvectors, and Systems of Differential Equations*

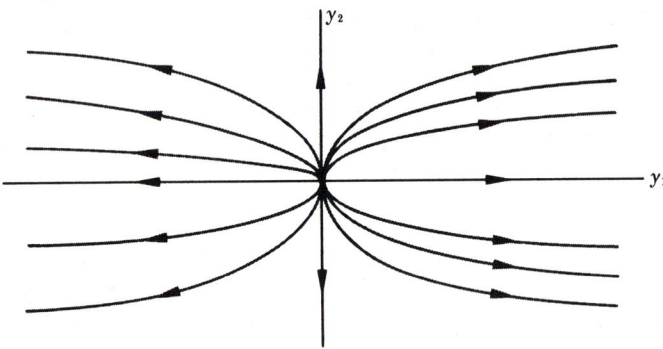

Figure 7.4

● **EXERCISE**

6. Justify the phase portrait for Case (i) with $0 < \mu < \lambda$.

CASE (ii).   Here the solution of (7.19) through $(\eta_1, \eta_2) \neq (0, 0)$ at $t = 0$ is

$$\phi(t) = \begin{bmatrix} e^{\lambda t} \eta_1 \\ e^{\lambda t} \eta_2 \end{bmatrix}$$

and if $\lambda > 0$, we obtain the phase portrait in Figure 7.5, whereas the case $\lambda < 0$ corresponds to Figure 7.6. Note that all orbits are straight lines tending away from the origin if $\lambda > 0$ and toward the origin if $\lambda < 0$.

The ratio $\phi_2(t)/\phi_1(t)$ if $\eta_1 \neq 0$ is constant, as is $\phi_1(t)/\phi_2(t)$ if $\eta_2 = 0$. The origin in Case (ii) is called a **proper node**.

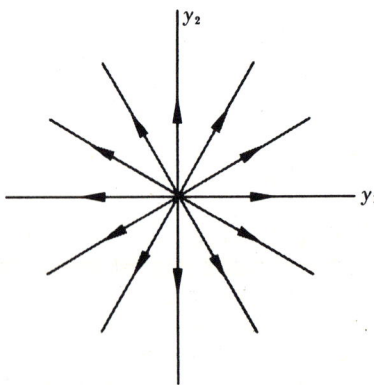

Figure 7.5

7.4 Two-Dimensional Linear Systems   241

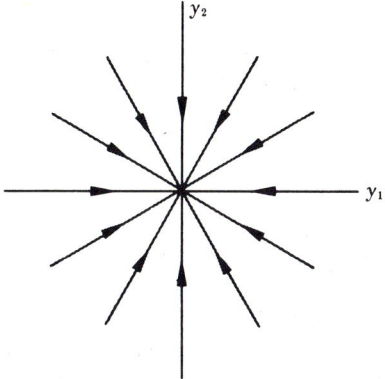

Figure 7.6

CASE (iii). Here

$$\phi(t) = \begin{bmatrix} e^{\lambda t}\eta_1 \\ e^{\mu t}\eta_2 \end{bmatrix},$$

with $\mu < 0$ and $\lambda > 0$, is the solution through (η_1, η_2) at $t = 0$. Now, as $t \to +\infty$, $\phi_1(t) \to \pm\infty$ according as $\eta_1 > 0$ or $\eta_1 < 0$ and $\phi_2(t) \to 0$ as $t \to +\infty$. It is easy to see that if $|\lambda| = |\mu|$, the orbits would be rectangular hyperbolas; for arbitrary $\lambda > 0$, $\mu < 0$ they resemble these curves as shown in Figure 7.7. Quite naturally, the origin in Case (iii) is called a **saddle point**.

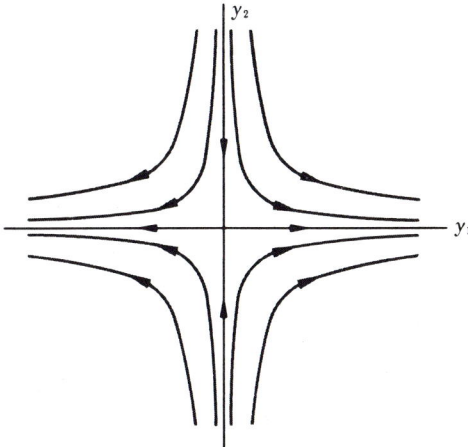

Figure 7.7

Eigenvalues, Eigenvectors, and Systems of Differential Equations

● **EXERCISE**

7. Construct the phase portrait in Case (iii) if $\lambda < 0$ and $\mu > 0$.

CASE (iv). Here

$$\phi(t) = \begin{bmatrix} \eta_1 + \eta_2 t \\ \eta_2 \end{bmatrix} e^{\lambda t}$$

is that solution passing through (η_1, η_2) at $t = 0$ and if $\lambda < 0$ the phase portrait is easily characterized by the fact that every orbit tends to the origin as $t \to +\infty$ and has the same limiting direction at $(0, 0)$. For $dy_2/dy_1 = \phi_2'/\phi_1' = (\lambda \phi_2/(\lambda \phi_1 + \phi_2)) \to 0$ as $t \to +\infty$ (see Figure 7.8). The origin in Case (iv) is called (as in Case (i)) an **improper node**.

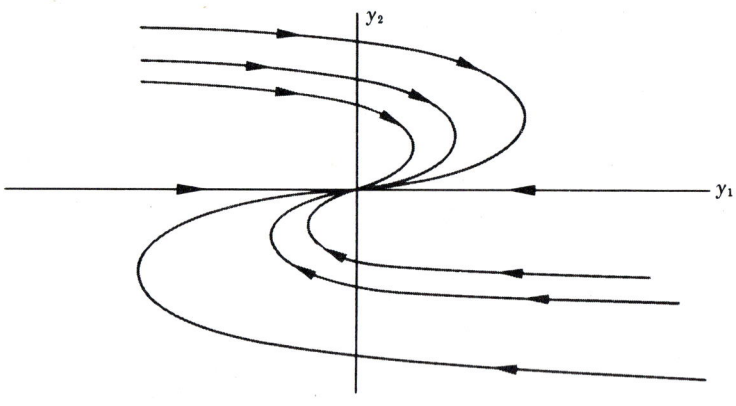

Figure 7.8

● **EXERCISE**

8. Construct the phase portrait in Case (iv) with $\lambda > 0$.

CASE (v). Here the solution, for the case $\sigma > 0$, passing through the point (η_1, η_2) at $t = 0$ is

$$\phi(t) = e^{\sigma t} \begin{bmatrix} \eta_1 \cos vt + \eta_2 \sin vt \\ -\eta_1 \sin vt + \eta_2 \cos vt \end{bmatrix}.$$

Let $\rho = (\eta_1^2 + \eta_2^2)^{1/2}$, $\cos \alpha = \eta_1/\rho$, $\sin \alpha = \eta_2/\rho$. Then

$$\phi(t) = e^{\sigma t} \begin{bmatrix} \rho \cos(vt - \alpha) \\ -\rho \sin(vt - \alpha) \end{bmatrix}.$$

7.4 Two-Dimensional Linear Systems

Letting r, θ be the polar coordinates, $y_1 = r \cos \theta$, $y_2 = r \sin \theta$, we may write the solution in polar form $r(t) = \rho e^{\sigma t}$, $\theta(t) = -(vt - \alpha)$. Eliminating the parameter t, we have $r = C \exp[(-\sigma/v)\theta]$, where $C = \rho \exp[(\sigma/v)\alpha]$. Thus, the phase portrait is a family of spirals, as shown in Figure 7.9, for the case $\sigma > 0$, $v > 0$ and the origin is called a **spiral point**. In this case, the orbits tend away from zero as $t \to +\infty$ (or, equivalently, approach zero as $t \to -\infty$).

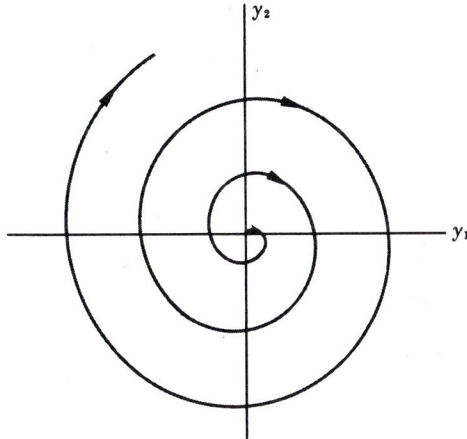

Figure 7.9

● **EXERCISE**

9. Sketch the phase portrait for the Case (v) in case $\sigma < 0$, $v < 0$.

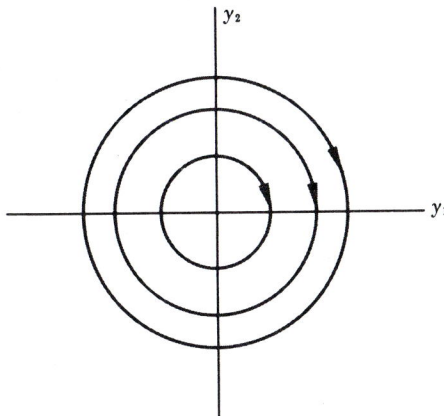

Figure 7.10

CASE (vi). This is a special case of Case (v) with $\sigma = 0$. From the above formulas we see that the orbits are concentric circles of radius ρ oriented as shown for $v > 0$ in Figure 7.10. The origin is called a **center**.

● **EXERCISE**

10. Sketch the phase portrait in Case (vi) when $v < 0$.

We observe from the possible cases considered above that all solutions of (7.19) and also their orbits tend to the origin as $t \to +\infty$ if and only if both eigenvalues of A have negative real parts; in this case we say that the origin is an **attractor of the linear system.** (7.19).

● **EXERCISES**

Sketch the phase portrait of each of the following scalar equations by converting to an equivalent system. Identify the origin and decide whether it is an attractor.

11. $x'' + x = 0$.
12. $x'' - 3x' + x = 0$.
13. $x'' + 3x' + x = 0$.
14. $x'' + 3x' - x = 0$.
15. $x'' - 3x' + 2x = 0$.
16. $x'' + 3x' + 2x = 0$.
17. $x'' - 2x' + x = 0$.
18. $x'' - x' - 6x = 0$.

19. To illustrate the complexity of the case when the origin is not the only critical point of a linear system consider the system

$$y_1' = y_1 - y_2,$$
$$y_2' = 2y - 2y_2.$$

(a) Show that there is a line of critical points.
(b) Sketch the phase portrait.
[*Hint:* $y_2' = 2y_1'$, and the eigenvalues of the coefficient matrix are 0 and -1.]

20. Repeat as much as you can of Exercise 19 for the system

$$y_1' = a_{11}y_1 + a_{12}y_2,$$
$$y_2' = a_{21}y_1 + a_{22}y_2,$$

where $a_{11}a_{22} - a_{12}a_{21} = 0$, but not all of $a_{11}, a_{12}, a_{21}, a_{22}$ are zero.

7.5 The General Case

In Section 7.3, we learned how to find a fundamental matrix of the system $\mathbf{y}' = A\mathbf{y}$ in the case when the matrix A possesses n linearly independent eigenvectors. (In particular, this covers the case when the eigenvalues of A are all distinct.) We then learned how to determine the matrix $\exp tA$ in this case, and we found that each element of $\exp tA$ is a linear combination of exponential functions. In Section 7.4 we treated the general 2×2 case, and we observed that if the matrix A has fewer than two linearly independent eigenvectors, then $\exp tA$ contains at least one term of the form $te^{\lambda t}$ (see Case (iii) of Theorem 1, Section 7.4). This shows that the solution may have a more complicated form than that suggested in Section 7.3.

In this section we will determine the form of $\exp tA$ when A is an arbitrary $n \times n$ matrix. We note that the following method, which leads to the formula (7.26), is always applicable and contains the results of Sections 7.3 and 7.4 as special cases. However, we warn the reader that this section is somewhat more difficult.

We will need the following result from linear algebra; its proof may be found in Appendix 3.

Let A be a (complex) $n \times n$ matrix. Compute $\lambda_1, \lambda_2, \ldots, \lambda_k$, the distinct eigenvalues of A with respective multiplicities $n_1, n_2, \ldots, n_k$, where $n_1 + n_2 + \cdots + n_k = n$. Corresponding to each eigenvalue λ_j of multiplicity n_j consider the system of linear equations

$$(A - \lambda_j I)^{n_j} \mathbf{x} = 0 \qquad (j = 1, 2, \ldots, k). \tag{7.23}$$

The solutions of each such linear system obviously span a subspace of $\mathscr{C}_n$ which we call X_j, where $j = 1, 2, \ldots, k$. The result from linear algebra needed tells us that *for every* $\mathbf{x} \in \mathscr{C}_n$ *there exist unique vectors* $\mathbf{x}_1, \mathbf{x}_2, \ldots, \mathbf{x}_k$, *where* $\mathbf{x}_j \in X_j$ $(j = 1, \ldots, k)$, *such that*

$$\mathbf{x} = \mathbf{x}_1 + \mathbf{x}_2 + \cdots + \mathbf{x}_k. \tag{7.24}$$

It is important to know that the linear algebraic system (7.23) has n_j linearly independent solutions so that the dimension of X_j is n_j. We note that if all the eigenvalues of A are distinct, that is, if each $n_j = 1$, where $j = 1, \ldots, k$ and $k = n$, then the vectors $\mathbf{x}_1, \mathbf{x}_2, \ldots, \mathbf{x}_n$ are suitable multiples of fixed eigenvectors that are linearly independent and span $\mathscr{C}_n$. Thus, if

$\mathbf{v}_1, \ldots, \mathbf{v}_n$ is a fixed set of linearly independent eigenvectors of A and if $\mathbf{x}$ is an arbitrary vector, the vectors $\mathbf{x}_j$ are given by $\mathbf{x}_j = c_j \mathbf{v}_j$ for some scalars c_j, where $j = 1, \ldots, n$.

It may, in fact, happen that in (7.23), $(A - \lambda_j I)^q \mathbf{x} = 0$ for all $\mathbf{x}$ in X_j and $q < n_j$, as the following example demonstrates.

Example 1. Consider the matrix

$$A = \begin{bmatrix} 4 & 1 & 0 & 0 & 0 \\ 0 & 4 & 1 & 0 & 0 \\ 0 & 0 & 4 & 0 & 0 \\ 0 & 0 & 0 & 4 & 0 \\ 0 & 0 & 0 & 0 & 4 \end{bmatrix}.$$

Here $\mathscr{C}_n$ is $\mathscr{C}_5$, and $\lambda = 4$ is the only eigenvalue, with multiplicity 5; that is, $k = 1$. Since there is only one eigenvalue, no decomposition into subspaces is necessary, and $\mathscr{C}_5 = X_1$. According to the theorem, $n_1 = 5$, so that certainly $(A - 4I)^5 = 0$. However, as the reader can easily verify $(A - 4I)^3 = 0$ but $(A - 4I)^2 \neq 0$. Therefore, $q = 3 < n_1 = 5$.

To apply this theory to the linear system $\mathbf{y}' = A\mathbf{y}$, we look for that solution $\boldsymbol{\phi}(t)$ satisfying the initial condition $\boldsymbol{\phi}(0) = \boldsymbol{\eta}$. By Theorem 1, Section 7.1, $\boldsymbol{\phi}(t) = e^{tA}\boldsymbol{\eta}$ and our object is to evaluate $e^{tA}\boldsymbol{\eta}$ explicitly, that is, to see exactly what the components of $\boldsymbol{\phi}$ are. We compute $\lambda_1, \lambda_2, \ldots, \lambda_k$, the distinct eigenvalues of A of multiplicities $n_1, n_2, \ldots, n_k$, respectively. We apply the theorem to the initial vector $\boldsymbol{\eta}$ and in accordance with (7.24), we have

$$\boldsymbol{\eta} = \mathbf{v}_1 + \mathbf{v}_2 + \cdots + \mathbf{v}_k, \tag{7.25}$$

where $\mathbf{v}_j$ is some suitable vector in the subspace X_j, with $j = 1, \ldots, k$. Since the subspace X_j is generated by the system (7.23), $\mathbf{v}_j$ must be some solution of (7.23). Now, by (7.25), $e^{tA}\boldsymbol{\eta} = \sum_{j=1}^{k} e^{tA}\mathbf{v}_j$, and we may write

$$\begin{aligned} e^{tA}\mathbf{v}_j &= \exp(\lambda_j t) \exp[(A - \lambda_j I)t]\mathbf{v}_j \\ &= \exp(\lambda_j t)[I + t(A - \lambda_j I) + \frac{t^2}{2!}(A - \lambda_j I)^2 + \cdots \\ &\quad + \frac{t^{n_j - 1}}{(n_j - 1)!}(A - \lambda_j I)^{n_j - 1}]\mathbf{v}_j \end{aligned}$$

for $-\infty < t < +\infty$, where the series in parentheses terminates because $\mathbf{v}_j$ is a solution of (7.23); (thus the term $(A - \lambda_j I)^{n_j}\mathbf{v}_j = 0$ and all subsequent terms in the expansion of the matrix $\exp[(A-\lambda_j I)t]$ are zero). Observe that the vectors $\mathbf{w}_j = (A - \lambda_j I)^p \mathbf{v}_j$, for $p = 0, 1, \ldots, n_j - 1$, belong to the subspace X_j because

$$(A - \lambda_j I)^{n_j} \mathbf{w}_j = (A - \lambda_j I)^{n_j}[(A - \lambda_j I)^p \mathbf{v}_j] = (A - \lambda_j I)^{n_j + p} \mathbf{v}_j = \mathbf{0}.$$

Thus, the vector $e^{tA}\mathbf{v}_j$ remains in X_j for each t, $-\infty < t < \infty$. Applying the above calculations to the solution $\boldsymbol{\phi}(t) = e^{tA}\boldsymbol{\eta}$ of $\mathbf{y}' = A\mathbf{y}$, we have

$$\boldsymbol{\phi}(t) = e^{tA}\boldsymbol{\eta} = e^{tA}\sum_{j=1}^{k}\mathbf{v}_j = \sum_{j=1}^{k}e^{tA}\mathbf{v}_j$$

$$= \sum_{j=1}^{k} \exp(\lambda_j t)\left[I + t(A - \lambda_j I) + \cdots + \frac{t^{n_j-1}}{(n_j - 1)!}(A - \lambda_j I)^{n_j - 1}\right]\mathbf{v}_j$$

or, finally, the solution $\boldsymbol{\phi}$ satisfying $\boldsymbol{\phi}(0) = \boldsymbol{\eta}$ is

$$\boldsymbol{\phi}(t) = \sum_{j=1}^{k} \exp(\lambda_j t)\left[\sum_{i=0}^{n_j-1} \frac{t^i}{i!}(A - \lambda_j I)^i\right]\mathbf{v}_j, \quad (-\infty < t < \infty). \quad (7.26)$$

We point out again that if $(A - \lambda_j I)^{q_j} = 0$, where $q_j < n_j$, then the sum on i in Eq. 7.26 will contain only q_j, rather than n_j, terms. This formula also tells us precisely how the components of the solution behave as functions of t for any given coefficient matrix A.

If A has only one distinct eigenvalue, there is no need to decompose as in (7.25). In this case we know that $(A - \lambda I)^n \mathbf{x} = \mathbf{0}$ for every $\mathbf{x} \in \mathscr{C}_n$; that is, that $(A - \lambda I)^n$ is the zero matrix. Therefore, from the series definition of $\exp tA$, we have

$$\exp tA = e^{\lambda t}e^{(A-\lambda I)t} = e^{\lambda t}\sum_{i=0}^{n-1}\frac{t^i}{i!}(A - \lambda I)^i; \quad (7.27)$$

that is, the series terminates after at most n terms and $\exp tA$ is easily computed.

Example 2. Solve the initial problem $\mathbf{y}' = A\mathbf{y}$, $\mathbf{y}(0) = \boldsymbol{\eta}$, if A is the matrix in Example 2, Section 7.2. Also, obtain a fundamental matrix.

From Example 2, Section 7.2, we know that $\lambda_1 = 3$ is an eigenvalue of multiplicity 2. In the above notation, $n_1 = 2$. Therefore, only the subspace X_1 in this case ($X_1 = \mathscr{C}_2$) is relevant. We readily calculate

$$A - 3I = \begin{bmatrix} -1 & 1 \\ -1 & 1 \end{bmatrix}$$

and we also see that

$$(A - 3I)^2 = \begin{bmatrix} 0 & 0 \\ 0 & 0 \end{bmatrix},$$

so that (7.23) is satisfied for every vector in $\mathscr{C}_2$. Substituting in (7.26) with

$$n_1 = 2, \quad \boldsymbol{\eta} = \begin{bmatrix} \eta_1 \\ \eta_2 \end{bmatrix},$$

we find

$$\boldsymbol{\phi}(t) = e^{3t} [I + t(A - 3I)] \boldsymbol{\eta}$$

and therefore

$$\begin{aligned}\boldsymbol{\phi}(t) &= e^{3t} \left\{ I + t \begin{bmatrix} -1 & 1 \\ -1 & 1 \end{bmatrix} \right\} \begin{bmatrix} \eta_1 \\ \eta_2 \end{bmatrix} \\ &= e^{3t} \begin{bmatrix} \eta_1 + t(-\eta_1 + \eta_2) \\ \eta_2 + t(-\eta_1 + \eta_2) \end{bmatrix}\end{aligned} \quad (7.28)$$

is the solution with $\boldsymbol{\phi}(0) = \boldsymbol{\eta}$. To construct a fundamental matrix we may use the formula (7.27) and we obtain (since $(A - 3I)^2 = 0$),

$$\exp tA = e^{3t}[I + (A - 3I)t]$$

$$= e^{3t} \left\{ \begin{bmatrix} 1 & 0 \\ 0 & 1 \end{bmatrix} + \begin{bmatrix} -1 & 1 \\ -1 & 1 \end{bmatrix} t \right\}$$

$$= e^{3t} \begin{bmatrix} 1 - t & t \\ -t & 1 + t \end{bmatrix}.$$

Alternatively, to construct $\exp tA$, we may use (7.28) as follows.

$$\exp tA = \exp tA \begin{bmatrix} 1 & 0 \\ 0 & 1 \end{bmatrix} = \left[\exp tA \begin{bmatrix} 1 \\ 0 \end{bmatrix}, \exp tA \begin{bmatrix} 0 \\ 1 \end{bmatrix} \right],$$

where the solution vectors

$$\exp tA \begin{bmatrix} 1 \\ 0 \end{bmatrix} \quad \text{and} \quad \exp tA \begin{bmatrix} 0 \\ 1 \end{bmatrix}$$

are found from (7.28) by substituting first

$$\eta = \begin{bmatrix} 1 \\ 0 \end{bmatrix}$$

and then

$$\eta = \begin{bmatrix} 0 \\ 1 \end{bmatrix},$$

respectively. This gives the matrix already found above.

Example 3. Consider the system.

$$\begin{aligned} x_1' &= 3x_1 - x_2 + x_3 \\ x_2' &= 2x_1 \quad\quad\;\; + x_3 \\ x_3' &= x_1 - x_2 + 2x_3 \end{aligned}$$

which has coefficient matrix

$$A = \begin{bmatrix} 3 & -1 & 1 \\ 2 & 0 & 1 \\ 1 & -1 & 2 \end{bmatrix}.$$

Find that solution ϕ satisfying the initial condition

$$\phi(0) = \begin{bmatrix} \eta_1 \\ \eta_2 \\ \eta_3 \end{bmatrix} = \eta,$$

and also find $\exp tA$.

The characteristic polynomial of A is $\det(\lambda I - A) = (\lambda - 1)(\lambda - 2)^2$, and therefore the eigenvalues are $\lambda_1 = 1$, $\lambda_2 = 2$ with multiplicities $n_1 = 1$, $n_2 = 2$, respectively. In the notation of (7.23), we consider the systems of algebraic equations

$$(A - I)\mathbf{x} = 0 \quad \text{and} \quad (A - 2I)^2 \mathbf{x} = 0$$

250 Eigenvalues, Eigenvectors, and Systems of Differential Equations

in order to determine the subspaces X_1 and X_2 of $\mathscr{C}_3$. Taking these in succession, we have first

$$(A - I)\mathbf{x} = \begin{bmatrix} 2 & -1 & 1 \\ 2 & -1 & 1 \\ 1 & -1 & 1 \end{bmatrix} \mathbf{x} = \mathbf{0} \quad \text{or} \quad \begin{array}{l} 2x_1 - x_2 + x_3 = 0 \\ 2x_1 - x_2 + x_3 = 0 \\ x_1 - x_2 + x_3 = 0. \end{array}$$

Thus, X_1 is the subspace spanned by the vectors

$$\begin{bmatrix} x_1 \\ x_2 \\ x_3 \end{bmatrix}$$

with $x_1 = 0$, $x_2 = x_3$, and clearly dim $X_1 = 1$. Next,

$$(A - 2I)^2 \mathbf{x} = \begin{bmatrix} 0 & 0 & 0 \\ -1 & 1 & 0 \\ -1 & 1 & 0 \end{bmatrix} \mathbf{x} = \mathbf{0} \quad \text{or} \quad \begin{array}{l} -x_1 + x_2 = 0 \\ -x_1 + x_2 = 0. \end{array}$$

Thus, X_2 is the subspace spanned by vectors

$$\begin{bmatrix} x_1 \\ x_2 \\ x_3 \end{bmatrix}$$

with $x_1 = x_2$ and x_3 arbitrary; clearly, dim $X_2 = 2$. The reader is advised to picture these subspaces in $\mathscr{C}_3$. Observe that the rank of the matrix $A - I$ is 2. Thus, by Theorem 2, Section 5.6, dim $X_1 = 3 - 2 = 1$. Similarly, the rank of the matrix $(A - 2I)^2$ is clearly 1 and dim $X_2 = 3 - 1 = 2$.

We now wish to find vectors $\mathbf{v}_1 \in X_1$, $\mathbf{v}_2 \in X_2$ such that we can write the initial vector $\boldsymbol{\eta}$ as in (7.25):

$$\boldsymbol{\eta} = \mathbf{v}_1 + \mathbf{v}_2.$$

Since $\mathbf{v}_1 \in X_1$,

$$\mathbf{v}_1 = \begin{bmatrix} 0 \\ \alpha \\ \alpha \end{bmatrix}$$

7.5 The General Case

for some scalar α, and since $\mathbf{v}_2 \in X_2$,

$$\mathbf{v}_2 = \begin{bmatrix} \beta \\ \beta \\ \gamma \end{bmatrix}$$

for some scalars β, γ. Therefore,

$$\begin{bmatrix} \eta_1 \\ \eta_2 \\ \eta_3 \end{bmatrix} = \begin{bmatrix} 0 \\ \alpha \\ \alpha \end{bmatrix} + \begin{bmatrix} \beta \\ \beta \\ \gamma \end{bmatrix}$$

so that $\beta = \eta_1$, $\alpha + \beta = \eta_2$, $\alpha + \gamma = \eta_3$. Solving these equations for α, β, γ, we find that $\alpha = \eta_2 - \eta_1$, $\beta = \eta_1$, $\gamma = \eta_3 - \eta_2 + \eta_1$, and

$$\mathbf{v}_1 = \begin{bmatrix} 0 \\ \eta_2 - \eta_1 \\ \eta_2 - \eta_1 \end{bmatrix} \quad \mathbf{v}_2 = \begin{bmatrix} \eta_1 \\ \eta_1 \\ \eta - \eta_2 + \eta_1 \end{bmatrix}.$$

Thus, by the formula (7.26), we find that the solution $\boldsymbol{\phi}$ such that $\boldsymbol{\phi}(0) = \boldsymbol{\eta}$ is given by

$$\boldsymbol{\phi}(t) = e^t \mathbf{v}_1 + e^{2t}(I + t(A - 2I))\mathbf{v}_2$$

$$= e^t \begin{bmatrix} 0 \\ \eta_2 - \eta_1 \\ \eta_2 - \eta_1 \end{bmatrix} + e^{2t}\left(I + t\begin{bmatrix} 1 & -1 & 1 \\ 2 & -2 & 1 \\ 1 & -1 & 0 \end{bmatrix}\right)\begin{bmatrix} \eta_1 \\ \eta_1 \\ \eta - \eta_2 + \eta_1 \end{bmatrix} \quad (7.29)$$

$$= e^t \begin{bmatrix} 0 \\ \eta_2 - \eta_1 \\ \eta_2 - \eta_1 \end{bmatrix} + e^{2t} \begin{bmatrix} 1+t & -t & t \\ 2t & 1-2t & t \\ t & -t & 1 \end{bmatrix} \begin{bmatrix} \eta_1 \\ \eta_1 \\ \eta_3 - \eta_2 + \eta_1 \end{bmatrix}.$$

To find $\exp tA$, we put $\boldsymbol{\eta}$ successively equal to

$$\begin{bmatrix} 1 \\ 0 \\ 0 \end{bmatrix}, \quad \begin{bmatrix} 0 \\ 1 \\ 0 \end{bmatrix}, \quad \begin{bmatrix} 0 \\ 0 \\ 1 \end{bmatrix}$$

in (7.29). We obtain the three linearly independent solutions that we use as columns of the matrix

$$e^{tA} = \begin{bmatrix} (1+t)e^{2t} & -te^{2t} & te^{2t} \\ -e^t + (1+t)e^{2t} & e^t - te^{2t} & te^{2t} \\ -e^t + e^{2t} & e^t - e^{2t} & e^{2t} \end{bmatrix}.$$

252   Eigenvalues, Eigenvectors, and Systems of Differential Equations

Example 4. Find $\exp tA$ if

$$A = \begin{bmatrix} 4 & 1 & 0 & 0 & 0 \\ 0 & 4 & 1 & 0 & 0 \\ 0 & 0 & 4 & 0 & 0 \\ 0 & 0 & 0 & 4 & 0 \\ 0 & 0 & 0 & 0 & 4 \end{bmatrix}.$$

Using the results of Example 1, we have $(A - 4I)^3 = 0$, so that $(A - 4I)^3 \mathbf{x} = \mathbf{0}$ for any vector $\mathbf{x}$ in $\mathscr{C}_5$ and the initial vector $\boldsymbol{\eta}$ remains arbitrary. Since there is only one eigenvalue ($\lambda = 4$), only the subspace $X_1 = \mathscr{C}_5$ is relevant and we have, from (7.27) (since $(A - 4I)^3 = 0$),

$$\exp tA = e^{4t}[I + t(A - 4I) + \frac{t^2}{2!}(A - 4I)^2].$$

Therefore,

$$\exp tA = e^{4t}\left\{ I + t\begin{bmatrix} 0 & 1 & 0 & 0 & 0 \\ 0 & 0 & 1 & 0 & 0 \\ 0 & 0 & 0 & 0 & 0 \\ 0 & 0 & 0 & 0 & 0 \\ 0 & 0 & 0 & 0 & 0 \end{bmatrix} + \frac{t^2}{2!}\begin{bmatrix} 0 & 0 & 1 & 0 & 0 \\ 0 & 0 & 0 & 0 & 0 \\ 0 & 0 & 0 & 0 & 0 \\ 0 & 0 & 0 & 0 & 0 \\ 0 & 0 & 0 & 0 & 0 \end{bmatrix}\right\}$$

$$= e^{4t}\begin{bmatrix} 1 & t & \dfrac{t^2}{2!} & 0 & 0 \\ 0 & 1 & t & 0 & 0 \\ 0 & 0 & 1 & 0 & 0 \\ 0 & 0 & 0 & 1 & 0 \\ 0 & 0 & 0 & 0 & 1 \end{bmatrix}.$$

Example 5. Solve the initial value problem

$$\mathbf{y}' = A\mathbf{y} + \mathbf{g}(t),$$

where A is the constant matrix in Example 2 and where

$$\mathbf{g}(t) = \begin{bmatrix} e^{3t} \\ 1 \end{bmatrix}$$

with the initial condition $\boldsymbol{\phi}(0) = \boldsymbol{\eta}$.
From Example 2 we have

$$\Phi(t) = e^{tA} = e^{3t}\begin{bmatrix} 1-t & t \\ -t & 1+t \end{bmatrix}.$$

7.5 The General Case

$$\Phi(t)\Phi^{-1}(s) = \exp[(t-s)A] = \exp[3(t-s)]\begin{bmatrix} 1-(t-s) & t-s \\ -(t-s) & 1+(t-s) \end{bmatrix}$$

$$\exp[(t-s)A]\mathbf{g}(s) = e^{3t}\begin{bmatrix} 1-(t-s)+e^{-3s}(t-s) \\ -(t-s)+e^{-3s}(1+t-s) \end{bmatrix}$$

Therefore, using (7.16) (Section 7.3),

$$\boldsymbol{\phi}(t) = e^{3t}\begin{bmatrix} 1-t & t \\ -t & 1+t \end{bmatrix}\boldsymbol{\eta} + e^{3t}\int_0^t \begin{bmatrix} 1-(t-s)+e^{-3s}(t-s) \\ -(t-s)+e^{-3s}(1+t-s) \end{bmatrix} ds.$$

There seems to be little point in evaluating the integrals.

● **EXERCISES**

Find the fundamental matrix $\exp tA$ for each of the following systems $\mathbf{y}' = A\mathbf{y}$ having the coefficient matrix given. Also find a particular solution satisfying the given initial condition.

1. $A = \begin{bmatrix} 1 & 2 \\ 4 & 3 \end{bmatrix}$; $\boldsymbol{\eta} = \begin{bmatrix} 3 \\ 3 \end{bmatrix}$. (See Exercise 1(c), Section 7.2).

2. $A = \begin{bmatrix} 2 & -3 \\ 3 & -4 \end{bmatrix}$; $\boldsymbol{\eta} = \begin{bmatrix} 1 \\ 2 \end{bmatrix}$.

3. $A = \begin{bmatrix} 1 & 0 & 3 \\ 8 & 1 & -1 \\ 5 & 1 & -1 \end{bmatrix}$; $\boldsymbol{\eta} = \begin{bmatrix} 0 \\ -2 \\ -7 \end{bmatrix}$. (See Exercise 1(e), Section 7.2).

4. $A = \begin{bmatrix} 3 & -1 & -4 & 2 \\ 2 & 3 & -2 & -4 \\ 2 & -1 & -3 & 2 \\ 1 & 2 & -1 & -3 \end{bmatrix}$; $\boldsymbol{\eta} = \begin{bmatrix} 1 \\ 0 \\ -1 \\ 0 \end{bmatrix}$.

(See Exercise 1(n), Section 7.2).

[*Note:* The characteristic polynomial is $(\lambda - 1)^2 \cdot (\lambda + 1)^2$.]

5. Find that solution of the system

$$y_1' = y_1 + y_2 + \sin t, \qquad y_2' = 2y_1 + \cos t$$

such that $y_1(0) = 1$, $y_2(0) = 1$. [*Hint:* Find a fundamental matrix of the homogeneous system; then use the variation of constants formula (7.16) in Section 7.3.]

254   *Eigenvalues, Eigenvectors, and Systems of Differential Equations*

Consider the scalar linear differential equation of second order

$$y'' + py' + qy = 0, \tag{7.30}$$

where p and q are constants. We can solve this equation as a special case of the theory developed here as outlined in the following exercises. Note that we have already done this in a different way in Section 6.4.

● **EXERCISES**

6. Show that Eq. 7.30 is equivalent to the system $y' = Ay$ with

$$A = \begin{bmatrix} 0 & 1 \\ -q & -p \end{bmatrix}$$

and compute the eigenvalues λ_1, λ_2 of A.

7. Compute a fundamental matrix for the system in Exercise 6 if $\lambda_1 \neq \lambda_2$; that is, if $p^2 \neq 4q$, and construct the general solution of Eq. 7.30 in this case.

8. Compute a fundamental matrix for the system in Exercise 6 in the case $\lambda_1 = \lambda_2 = \lambda$; that is, $p^2 = 4q$, and construct the general solution of (7.30) in the case $p^2 = 4q$. Note that $A - \lambda I$ is never zero in this case, so that the fundamental matrix, as well as the general solution of (7.30) must necessarily contain a term in $te^{\lambda t}$.

9. Generalize the results of Exercises 6, 7, and 8 to the scalar equation

$$y''' + p_1 y'' + p_2 y' + p_3 y = 0,$$

where p_1, p_2, p_3 are constants. (Needless to say, you are not expected actually to solve a cubic equation.)

7.6 Solution of Example 8, Chapter 1. An Electric Circuit

The particular circuit in question led us to the initial value problem (1.23) which we write in the form

$$\begin{aligned} v_1' &= -\frac{5}{3} i_1 &&+ \frac{5}{3} i_s(t), & v_1(0) &= 0.6 \\ i_1' &= \frac{5}{3} v_1 - \frac{5}{3} v_2 &&, & i_1(0) &= 1 \\ v_2' &= 6 i_1 - 6 v_2 &&, & v_2(0) &= 1.2. \end{aligned} \tag{7.31}$$

7.6 Solution of Example 8, Chapter 1. An Electric Circuit

Define

$$\mathbf{y} = \begin{bmatrix} v_1 \\ i_1 \\ v_2 \end{bmatrix}, \quad A = \begin{bmatrix} 0 & -\frac{3}{5} & 0 \\ \frac{5}{3} & 0 & -\frac{5}{3} \\ 0 & 6 & -6 \end{bmatrix}, \quad \mathbf{g}(t) = \begin{bmatrix} \frac{5}{3} i_s(t) \\ 0 \\ 0 \end{bmatrix}.$$

Then (7.31) has the form

$$\mathbf{y}' = A\mathbf{y} + \mathbf{g}(t), \quad \mathbf{y}(0) = \begin{bmatrix} 0.6 \\ 1 \\ 1.2 \end{bmatrix}. \tag{7.32}$$

To solve this initial value problem we first solve the homogeneous system $\mathbf{y}' = A\mathbf{y}$ in (7.32). By Theorem 1, Section 7.1, $\exp tA$ is a fundamental matrix of the homogeneous system and to find it we first find the eigenvalues of A. The eigenvalues are the roots of the characteristic polynomial,

$$\det(\lambda I - A) = \begin{bmatrix} \lambda & \frac{3}{5} & 0 \\ -\frac{5}{3} & \lambda & \frac{5}{3} \\ 0 & -6 & \lambda + 6 \end{bmatrix} = \lambda^3 + 6\lambda^2 + 11\lambda + 6$$

$$= (\lambda + 1)(\lambda + 2)(\lambda + 3).$$

Hence, the eigenvalues are $\lambda_1 = -1$, $\lambda_2 = -2$, $\lambda_3 = -3$.

Since the eigenvalues of A are distinct, we may proceed by the method of Section 7.3, and we next compute an eigenvector corresponding to each eigenvalue. Corresponding to the eigenvalue $\lambda_1 = -1$, we consider the system

$$(-I - A)\mathbf{x} = \begin{bmatrix} -1 & \frac{3}{5} & 0 \\ -\frac{5}{3} & -1 & \frac{5}{3} \\ 0 & -6 & 5 \end{bmatrix} \mathbf{x} = \mathbf{0}.$$

256   Eigenvalues, Eigenvectors, and Systems of Differential Equations

By elementary row operations we have

$$\begin{bmatrix} -1 & \frac{3}{5} & 0 \\ -\frac{5}{3} & -1 & \frac{5}{3} \\ 0 & -6 & 5 \end{bmatrix} \underset{\sim}{R} \begin{bmatrix} 1 & -\frac{3}{5} & 0 \\ -\frac{5}{3} & -1 & \frac{5}{3} \\ 0 & -6 & 5 \end{bmatrix} \underset{\sim}{R} \begin{bmatrix} 1 & -\frac{3}{5} & 0 \\ 0 & -2 & \frac{5}{3} \\ 1 & -6 & 5 \end{bmatrix}$$

$$\underset{\sim}{R} \begin{bmatrix} 1 & -\frac{3}{5} & 0 \\ 0 & 1 & -\frac{5}{6} \\ 0 & -6 & 5 \end{bmatrix} \underset{\sim}{R} \begin{bmatrix} 1 & -\frac{3}{5} & 0 \\ 0 & 1 & -\frac{5}{6} \\ 0 & 0 & 0 \end{bmatrix}.$$

Thus, the vector

$$\mathbf{v}_1 = \begin{bmatrix} \frac{3}{5} \\ 1 \\ \frac{6}{5} \end{bmatrix}$$

is a solution of $(-I - A)\mathbf{x} = \mathbf{0}$. Notice that (miraculously) $\mathbf{v}_1 = \mathbf{y}(0)$ (it was intentionally given this way). Similarly,

$$\mathbf{v}_2 = \begin{bmatrix} \frac{3}{10} \\ 1 \\ \frac{3}{2} \end{bmatrix}, \quad \mathbf{v}_3 = \begin{bmatrix} \frac{1}{10} \\ \frac{1}{2} \\ 1 \end{bmatrix}$$

are eigenvectors corresponding to the eigenvalues $\lambda_2 = -2$ and $\lambda_2 = -3$ respectively. By Theorem 1, Section 7.3,

$$\Phi(t) = [e^{-t}\mathbf{v}_1, e^{-2t}\mathbf{v}_2, e^{-2t}\mathbf{v}_3]$$

7.6 Solution of Example 8, Chapter 1. An Electric Circuit

is a fundamental matrix of the system $\mathbf{y}' = A\mathbf{y}$. By formula (7.14), Section 7.3,

$$\exp tA = \Phi(t)\Phi^{-1}(0).$$

By Theorem 1, Section 4.7,

$$\Phi^{-1}(0) = \frac{\operatorname{adj} \Phi(0)}{\det \Phi(0)},$$

and we find, after an elementary but tedious calculation, that

$$\Phi^{-1}(0) = \begin{bmatrix} \dfrac{25}{6} & -2.5 & \dfrac{5}{6} \\ -\dfrac{20}{3} & 8 & -\dfrac{10}{3} \\ 5 & -9 & 5 \end{bmatrix}. \tag{7.33}$$

The reader can also verify that $\Phi(0)\Phi^{-1}(0) = \Phi^{-1}(0)\Phi(0) = I$. Hence,

$$\exp tA = \begin{bmatrix} \dfrac{3}{5}e^{-t} & \dfrac{3}{10}e^{-2t} & \dfrac{1}{10}e^{-3t} \\ e^{-t} & e^{-2t} & \dfrac{1}{2}e^{-3t} \\ \dfrac{6}{5}e^{-t} & \dfrac{3}{2}e^{-2t} & e^{-3t} \end{bmatrix} \begin{bmatrix} \dfrac{25}{6} & -2.5 & \dfrac{5}{6} \\ -\dfrac{20}{3} & 8 & -\dfrac{10}{3} \\ 5 & -9 & 5 \end{bmatrix}$$

$$\tag{7.34}$$

$$= \begin{bmatrix} \left(\dfrac{5}{2}e^{-t} - 2e^{-2t} + \dfrac{1}{2}e^{-3t}\right) & \left(-\dfrac{3}{2}e^{-t} + \dfrac{12}{5}e^{-2t} - \dfrac{9}{10}e^{-3t}\right) & \left(\dfrac{1}{2}e^{-t} - e^{-2t} + \dfrac{1}{2}e^{-3t}\right) \\ \left(\dfrac{25}{6}e^{-t} - \dfrac{20}{3}e^{-2t} + \dfrac{5}{2}e^{-3t}\right) & \left(-2.5e^{-t} + 8e^{-2t} - \dfrac{9}{2}e^{-3t}\right) & \left(\dfrac{5}{6}e^{-t} - \dfrac{10}{3}e^{-2t} + \dfrac{5}{2}e^{-3t}\right) \\ \left(5e^{-t} - 10e^{-2t} + 5e^{-3t}\right) & \left(-3e^{-t} + 12e^{-2t} - 9e^{-3t}\right) & \left(e^{-t} - 5e^{-2t} + 5e^{-3t}\right) \end{bmatrix}.$$

(Notice that at $t = 0$ this matrix reduces to the identity matrix as it should).

Thus, the solution ϕ_h of the homogeneous system $\mathbf{y}' = A\mathbf{y}$ satisfying the given initial condition at $t = 0$ is (by matrix vector multiplication)

$$\phi_h(t) = \exp tA \begin{bmatrix} 0.6 \\ 1 \\ 1.2 \end{bmatrix} = \exp(tA)\mathbf{v}_1 = \begin{bmatrix} 0.6e^{-t} \\ e^{-t} \\ 1.2e^{-t} \end{bmatrix}.$$

By the variation of constants formula applied to the system (7.32), we obtain as the solution ϕ to the given initial value problem

$$\phi(t) = \exp(tA)\mathbf{v}_1 + \int_0^t \exp(t-s)A\mathbf{g}(s)\,ds.$$

In view of the special form of $\mathbf{g}(s)$ we obtain from (7.33) (replacing t by $t - s$):

$$\phi(t) = \begin{bmatrix} 0.6 \\ 1 \\ 1.2 \end{bmatrix} e^{-t} + \frac{5}{3}\int_0^t \begin{bmatrix} \frac{5}{2}e^{-(t-\tau)} - 2e^{-2(t-\tau)} + \frac{1}{2}e^{-3(t-\tau)} \\ \frac{25}{6}e^{-(t-\tau)} - \frac{20}{3}e^{-2(t-\tau)} + \frac{5}{2}e^{-3(t-\tau)} \\ 5e^{-(t-\tau)} - 10e^{-2(t-\tau)} + 5e^{-3(t-\tau)} \end{bmatrix} i_s(\tau)\,d\tau.$$

(7.35)

Thus, if the source current $i_s(t)$ is given, the solution (7.35) can be simplified further and if $i_s(t)$ is sufficiently simple, the integrals may be evaluated explicitly (for example, if the source current is $3\sin \omega t$) and then the desired voltages v_1, v_2 and the current i_1 can be determined explicitly.

● **EXERCISES**

1. Find the voltages $v_1(t)$, $v_2(t)$, and the current $i_1(t)$, if $i_s(t) = 3\sin \omega t$.
2. Discuss the behavior as $t \to \pm\infty$ of the voltages $v_1(t)$, $v_2(t)$, and of the current $i_1(t)$, in the case that $i_s(t) = 3\sin \omega t$.
3. Repeat Exercise 2 if $i_s(t) = 3e^{-\alpha t}\sin \omega t$, > 0.

We close this section with several remarks. Define the matrix

$$T = \Phi(0) = [\mathbf{v}_1, \mathbf{v}_2, \mathbf{v}_2],$$

and make the change of variable

$$\mathbf{y} = T\mathbf{z}.$$

7.6 Solution of Example 8, Chapter 1. An Electric Circuit

Then $\mathbf{y}' = T\mathbf{z}'$, and the system (7.29) becomes

$$T\mathbf{z}' = AT\mathbf{z} + \mathbf{g}(t)$$

or

$$\mathbf{z}' = T^{-1}AT\mathbf{z} + T^{-1}\mathbf{g}(t), \qquad \mathbf{z}(0) = T^{-1}\mathbf{y}(0). \tag{7.36}$$

The matrix $T^{-1}AT$ is, of course, similar to the matrix A for any matrix T (see Definition 4, Section 5.7), but with the present choice of the matrix T (see Exercise 4) and the vector $\mathbf{g}(t)$, the system (7.36) becomes

$$\mathbf{z}' = \begin{bmatrix} -1 & 0 & 0 \\ 0 & -2 & 0 \\ 0 & 0 & -3 \end{bmatrix} \mathbf{z} + i_s(t)T^{-1}\begin{bmatrix} 1 \\ 0 \\ 0 \end{bmatrix}, \tag{7.37}$$

where $\mathbf{z}(0) = T^{-1}\mathbf{y}(0)$ and where T^{-1} is given by (7.33). Of course, (7.37) is readily solved (almost by inspection) because it is **uncoupled**. If we denote by $\psi(t)$ the solution of (7.37) satisfying the initial condition $\psi(0) = \mathbf{z}(0) = T^{-1}\mathbf{y}(0)$, then the solution $\phi(t)$ of the original problem (given by (7.35)) is $\phi(t) = T\psi(t)$.

● **EXERCISES**

  4. Find the solution $\psi(t)$ of the system (7.37) satisfying the original condition $\psi(0) = T^{-1}\mathbf{y}(0)$ with $\mathbf{y}(0)$ given in (7.32).
  5. Use the result of Exercise 4 to find the solution of the original initial value problem. (Compare your result with (7.35); they should be identical.)

The procedure outlined here, which involves the diagonalization of the coefficient matrix $A \in \mathcal{R}_{nn}$, in general, has the following interpretation. According to Theorems 2 and 3, Section 7.2 the coefficient matrix A is similar to a diagonal matrix if and only if A possesses n linearly independent eigenvectors. (This is always the case if the eigenvalues of A are distinct.) Then the motion of any system such as (7.32) driven by any input $\mathbf{g}(t)$ and starting from any set of initial conditions can be thought of as the superposition of the motions of the **uncoupled first-order systems** (7.37); for if ψ is a solution of the uncoupled system then $\phi = T\psi$ is a solution of the original coupled system (7.32).

Another remark of some physical importance is the following. A reader may prefer to think in terms of a mechanical rather than an electrical system. An electrical system such as the one in Figure 1.6, Chapter 1, which we have studied has an **equivalent mechanical analog**. This is obtained as follows.

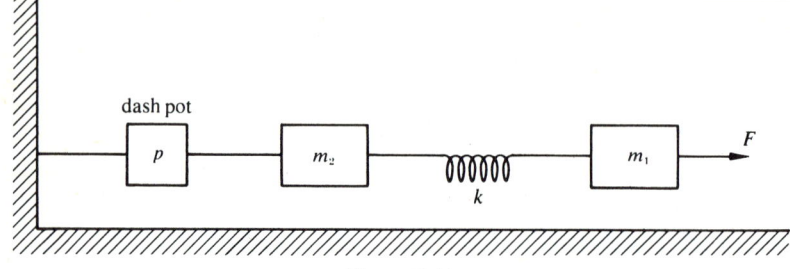

Figure 7.11

Consider the mechanical system sliding on a frictionless table as shown in Figure 7.11. Two masses m_1 and m_2 are connected by a spring (obeying Hooke's law), having spring constant k. A force f is applied to the mass m_1 and the mass m_2 is connected to a rigid wall by means of a dash pot which offers a resistance to motion proportional to the velocity, with proportionality constant p. The system moves in a straight line on a frictionless table. Let x_1, x_2 denote the displacements from equilibrium of the masses m_1 and m_2, respectively. At equilibrium $f = 0$ and the system is at rest. Let $v_i = x_i'$, with $i = 1, 2$, be the velocities of the masses and let $l = x_2 - x_1$ be the increase in length of the spring with respect to its unstretched length (length under no tension). By Newton's Second Law of Motion, we have

(i) for the mass m_1:

$$m_1 v_1' = \frac{l}{k} + f;$$

(ii) for the mass m_2:

$$m_2 v_2' = -p v_2 - \frac{l}{k}$$

and from the definition of l and v_i

$$l = v_2 - v_1.$$

Putting this together, we obtain

$$\begin{aligned} v_1' &= \frac{1}{m_1 k} l + \frac{1}{m_1} f \\ l' &= -v_1 + v_2 \\ v_2' &= -\frac{1}{m_2 k} l - \frac{p}{m_2} v_2. \end{aligned} \tag{7.38}$$

If we now identify

$$f = i_s, \quad -\frac{l}{k} = i_1, \quad m_1 = \frac{3}{5}, \quad m_2 = \frac{1}{6}, \quad k = \frac{3}{5},$$

the mechanical system (7.38) and the electrical system (7.31) already solved become identical.

There is a procedure for establishing mechanical analogs from electrical ones and vice versa. For our purposes it suffices to see one example, such as the one above. For more details the interested reader is referred to one of the standard references such as L. A. Zadeh and C. A. Desoer, *Linear Systems Theory* (McGraw-Hill, New York, 1963).

● EXERCISE

6. Solve the problem for the mechanical system posed in Example 7, Chapter 1. [*Hint:* Reduce the linear system (1.18) of two second-order equations to an equivalent linear system of four first-order equations and apply the methods of this chapter.]

● MISCELLANEOUS EXERCISES

1. What is wrong with the following calculation for arbitrary continuous matrix $A(t)$?

$$\frac{d}{dt}\left[\exp\int_{t_0}^{t} A(s)\,ds\right] = A(t)\exp\left[\int_{t_0}^{t} A(s)\,ds\right]$$

so that $\exp\left(\int_{t_0}^{t} A(s)\,ds\right)$ is a fundamental matrix of $y' = A(t)y$ for an arbitrary continuous matrix $A(t)$.

2. Find a fundamental matrix for the system $y' = Ay$, where A is the matrix.

(a) $A = \begin{bmatrix} 2 & 1 \\ 3 & 4 \end{bmatrix}.$ (b) $A = \begin{bmatrix} -1 & 8 \\ 1 & 1 \end{bmatrix}.$

(c) $A = \begin{bmatrix} 1 & 1 \\ 3 & -2 \end{bmatrix}.$ (d) $A = \begin{bmatrix} 2 & -3 \\ 1 & -2 \end{bmatrix}.$

(e) $A = \begin{bmatrix} 5 & 3 \\ -3 & -1 \end{bmatrix}.$ (f) $A = \begin{bmatrix} 1 & -1 & 1 \\ 1 & 1 & -1 \\ 2 & -1 & 0 \end{bmatrix}.$

(g) $A = \begin{bmatrix} -3 & 4 & -2 \\ 1 & 0 & 1 \\ 6 & -6 & 5 \end{bmatrix}$.

(h) $A = \begin{bmatrix} -2 & 1 & 0 \\ 1 & 3 & -1 \\ -1 & 2 & 3 \end{bmatrix}$.

(i) $A = \begin{bmatrix} 4 & -1 & -1 \\ 1 & 2 & -1 \\ 1 & -1 & 2 \end{bmatrix}$.

(j) $A = \begin{bmatrix} -1 & 1 & -2 \\ 4 & 1 & 0 \\ 2 & 1 & -1 \end{bmatrix}$.

(k) $A = \begin{bmatrix} 2 & 1 & 0 \\ 0 & 2 & 4 \\ 1 & 0 & -1 \end{bmatrix}$.

(l) $A = \begin{bmatrix} 2 & -1 & -1 \\ 2 & -1 & -2 \\ -1 & 1 & 2 \end{bmatrix}$.

(m) $A = \begin{bmatrix} 4 & -1 & 0 \\ 3 & 1 & -1 \\ 1 & 0 & 1 \end{bmatrix}$.

3. Sketch the phase portrait for each of the systems in Exercises 2(a)–2(e) and determine in each case whether the origin is a node, saddle point, spiral point, or center. For which of these is the origin an attractor?

4. Find the general solution of the system $y' = Ay + b(t)$ in each of the following cases

(a) $A = \begin{bmatrix} 0 & 1 \\ 1 & 0 \end{bmatrix}$, $b(t) = \begin{bmatrix} 2e^t \\ t^2 \end{bmatrix}$.

(b) $A = \begin{bmatrix} -1 & 2 \\ -2 & 3 \end{bmatrix}$, $b(t) = \begin{bmatrix} 1 \\ 0 \end{bmatrix}$.

(c) $A = \begin{bmatrix} 4 & -3 \\ 2 & -1 \end{bmatrix}$, $b(t) = \begin{bmatrix} \sin t \\ -2 \cos t \end{bmatrix}$.

(d) $A = \begin{bmatrix} 2 & 1 \\ 1 & 2 \end{bmatrix}$, $b(t) = \begin{bmatrix} 2e^t \\ -3e^{4t} \end{bmatrix}$.

(e) $A = \begin{bmatrix} 1 & -1 \\ 2 & -1 \end{bmatrix}$, $b(t) = \begin{bmatrix} 1 \\ \cos t \\ 0 \end{bmatrix}$.

5. Suppose m is not an eigenvalue of the matrix A. Show that the nonhomogeneous system

$$y' = Ay + ce^{mt}$$

has a solution of the form

$$\phi(t) = pe^{mt}$$

and calculate the vector p in terms of A and c.

Miscellaneous Exercises   263

6. Suppose m is not an eigenvalue of the matrix A. Show that the nonhomogeneous system

$$\mathbf{y}' = A\mathbf{y} + \sum_{j=0}^{k} \mathbf{c}_j t^j e^{mt}$$

has a solution of the form

$$\mathbf{\phi}(t) = \sum_{j=0}^{k} \mathbf{p}_j t^j e^{mt}.$$

[*Hint:* Show that $\mathbf{p}_j$ satisfies the algebraic system

$$(A - mI)\mathbf{p}_k = -\mathbf{c}_k$$
$$(A - mI)\mathbf{p}_j = (j+1)\mathbf{p}_{j+1} - \mathbf{c}_j \qquad (j = 0, 1, \ldots, k-1)$$

and that these systems can be solved recursively.]

7. Consider the system

$$t\mathbf{y}' = A\mathbf{y},$$

where A is a constant matrix. Show that $t^A = e^{A \log t}$ is a fundamental matrix for $t \neq 0$ in two ways: (i) by direct substitution; (ii) by making the change of variable $|t| = e^s$.

8. Find the general solution of the system

$$t\mathbf{y}' = A\mathbf{y} + \mathbf{b}(t).$$

9. Consider the system of differential equations

$$y_1'' - 3y_1' + 2y_1 + y_2' - y_2 = 0,$$
$$y_1' - 2y_1 + y_2' + y_2 = 0.$$

(a) Show that this system is equivalent to the system of first-order equations $\mathbf{u}' = A\mathbf{u}$, where

$$\mathbf{u} = \begin{bmatrix} u_1 \\ u_2 \\ u_3 \end{bmatrix} = \begin{bmatrix} y_1 \\ y_1' \\ y_2 \end{bmatrix}, \qquad A = \begin{bmatrix} 0 & 1 & 0 \\ -4 & 4 & 2 \\ 2 & -1 & -1 \end{bmatrix}.$$

(b) Find a fundamental matrix for the system in part (a).
(c) Find the general solution of the original system.
(d) Find the solution of the original system satisfying the initial conditions

$$y_1(0) = 0, \qquad y_1'(0) = 1, \qquad y_2(0) = 0.$$

10. Repeat the procedure of Exercise 9 for the system

$$y_1'' + y_2'' - y_2' + y_2 = 0,$$
$$y_1' + y_1 + y_2'' + y_2' = 0.$$

In part (d) find that solution satisfying the initial conditions $y_1(0) = 0$, $y_1'(0) = 1$, $y_2(0) = 0$, $y_2'(0) = 2$.

11. Consider the matrix differential equation

$$Y' = AY + YB,$$

where A, B, and Y are $n \times n$ matrices.

(a) Show that the solution satisfying the initial condition $Y(0) = C$, where C is a given $n \times n$ matrix, is given by

$$Y(t) = e^{At} C e^{Bt}.$$

(b) Show that

$$Z = -\int_0^\infty e^{At} C e^{Bt}\, dt$$

is the unique solution of the matrix equation

$$AX + XB = C$$

whenever the integral exists.

(c) Show that the integral for Z in part (b) exists if all eigenvalues of both A and B have negative real parts.

Chapter 8 | THE LAPLACE TRANSFORM

In this chapter we shall study a method of solving initial value problems for linear differential equations and linear systems of differential equations. This method does not enable us to solve any problems in differential equations which we could not solve by the methods already studied in Chapters 6 and 7. It does, however, provide a simple and often employed technique for solving many problems which arise in applications, particularly those involving linear equations with constant coefficients. One reason for its usefulness is that it enables us to find the particular solution of the differential equation which satisfies the initial conditions directly, rather than first finding the general solution and then using the initial conditions to determine constants.

8.1 Introduction

For every function $f(t)$ of a suitable class, we define the **Laplace transform** $F(s)$, also denoted by $\mathscr{L}(f)$, by

$$F(s) = \int_0^\infty e^{-st} f(t)\, dt = \lim_{A \to \infty} \int_0^A e^{-st} f(t)\, dt, \qquad (8.1)$$

where it is understood that we restrict our considerations to these values of the complex parameter s and to those functions f for which the above limit exists.

Example 1. To calculate the Laplace transform of the constant function 1, we evaluate

$$\int_0^\infty e^{-st}\,dt = \lim_{A\to\infty} \int_0^A e^{-st}\,dt = \lim_{A\to\infty} \left[\frac{1}{s} - \frac{e^{-sA}}{s}\right] = \frac{1}{s} \qquad (\mathscr{R}s > 0). \qquad (8.2)$$

Thus $\mathscr{L}(1) = 1/s$. Clearly, the integral does not converge for $\mathscr{R}s \leq 0$.

Example 2. We can calculate the Laplace transform of e^{zt} by almost exactly the same process as that used in Example 1.

$$\mathscr{L}(e^{zt}) = \int_0^\infty e^{-st}e^{zt}\,dt = \int_0^\infty e^{-(s-z)t}\,dt = \frac{1}{s-z} \qquad (\mathscr{R}s > \mathscr{R}z). \qquad (8.3)$$

Notice that the only difference between the integrals evaluated here and in Example 1 is that where there was an s in Example 1, there is an $(s - z)$ in Example 2. This is an example of a general principle to which we shall return later.

The operator defined by (8.1) is linear in the sense that if $f_1(t)$ and $f_2(t)$ have Laplace transforms $F_1(s)$ and $F_2(s)$, respectively, and if a and b are constants, then $af_1(t) + bf_2(t)$ has Laplace transform $aF_1(s) + bF_2(s)$.

The idea behind the Laplace transform method is very simple. It will be shown that every solution of any linear homogeneous differential equation with constant coefficients has a Laplace transform. Also, the Laplace transform of the derivatives of f can be expressed in terms of the Laplace transform of f and the values of f and its derivatives at $t = 0$. This then means that if ϕ is the solution of a linear differential equation with constant coefficients which satisfies some given initial conditions at $t = 0$, the Laplace transform of ϕ satisfies a **linear algebraic equation** rather than a differential equation. When we have solved this algebraic equation, we need only find the function whose Laplace transform is the solution of this algebraic equation. This may often be facilitated by tables of Laplace transforms. Of course, in order to be sure that the function found by this procedure is the same as the function ϕ, we need a uniqueness theorem for Laplace transforms, to the effect that two different functions cannot have the same Laplace transform.

8.2 Basic Properties of the Laplace Transform

In defining the Laplace transform by Eq. 8.1, we must impose some conditions on the function $f(t)$ to assure the convergence of the integral. We consider functions $f(t)$ defined for $0 < t < \infty$ which grow sufficiently slowly near $t = 0$ to assure the convergence of the integral at zero (the integral may

8.2 Basic Properties of the Laplace Transform

be improper); we also require that the function $f(t)$ grows sufficiently slowly for large t to assure the convergence of the infinite integral for some values of the complex parameter s; finally, we require $f(t)$ to be integrable over every closed subinterval of $0 < t < \infty$. This leads us to the following definition.

Definition 1. *A function f on $0 < t < \infty$ is said to be of exponential growth at infinity if it satisfies an inequality of the form*

$$|f(t)| \le Me^{ct}, \tag{8.4}$$

for some real constants $M > 0$ and c, for all sufficiently large t.

Now, we define *the class Λ of functions on $0 < t < \infty$* which are
 (i) absolutely integrable at zero (that is, $\lim_{\delta \to 0+} \int_\delta^a |f(t)|\, dt$ exists for sufficiently small $a > 0$);
 (ii) piecewise continuous on $0 < t < \infty$;
 (iii) of exponential growth at infinity.

This is the class of functions for which we wish to define the Laplace transform. Clearly, the functions 1, t, t^n (n a positive integer), $\sin t$, $\cos t$, e^{zt} for any complex z are in the class Λ, but $\exp t^2$ is not.

Theorem 1. *If f is a function in the class Λ, the integral $\int_0^\infty e^{-st} f(t)\, dt$ converges absolutely for all complex numbers s with sufficiently large real part.*

Proof. For small t, $|e^{-st}|$ is bounded, and therefore the assumption that f is absolutely integrable at zero implies the convergence of the integral $\int_0^\delta |e^{-st} f(t)|\, dt$ for every $\delta > 0$. If $\mathscr{R}s = \sigma$, $|e^{-st}| = e^{-\sigma t}$, and then (8.4) yields

$$|e^{-st} f(t)| \le Me^{-(\sigma - c)t},$$

for $t \ge T$, where T is some number greater than zero. Therefore, if $\sigma > c$, the infinite integral $\int_T^\infty |e^{-st} f(t)|\, dt$ converges because its integrand decreases exponentially to zero, and we have

$$\left| \int_T^\infty e^{-st} f(t)\, dt \right| \le \int_T^\infty |e^{-st} f(t)|\, dt \le M \int_T^\infty e^{-(\sigma - c)t}\, dt = \frac{M}{\sigma - c} e^{-(\sigma - c)T}.$$

Finally, because f is piecewise continuous on $\delta \le t \le T$, $\int_\delta^T |e^{-st} f(t)|\, dt$ exists, and this completes the proof of the theorem. ∎

The Laplace transform $F(s)$ of any function $f(t)$ in the class Λ is defined by Eq. 8.1 for all complex s with sufficiently large real part; we will usually be

268 The Laplace Transform

concerned only with real values of s. Sometimes we denote the Laplace transform by the symbol $\mathscr{L}$, to emphasize that the Laplace transform is an **operator**, which associates the function $F(s)$ with the function $f(t)$. Thus, we write

$$F(s) = \mathscr{L}\{f(t)\}.$$

As already remarked, this operator is linear.

Example 1. If $f(t)$ is a complex-valued function of the class Λ, $f(t) = u(t) + iv(t)$, where u, v are real, we have

$$\mathscr{L}\{f(t)\} = \mathscr{L}\{u(t) + iv(t)\} = \mathscr{L}\{u(t)\} + i\mathscr{L}\{v(t)\}.$$

From the definition, it is clear that the Laplace transform of a real-valued function is real for real s. Thus, $\mathscr{L}\{u(t)\}$ is the real part of $\mathscr{L}\{f(t)\}$ and $\mathscr{L}\{v(t)\}$ is the imaginary part of $\mathscr{L}\{f(t)\}$. If $z = \alpha + i\beta$, and if $f(t) = e^{zt}$, we have

$$\mathscr{L}(e^{zt}) = \mathscr{L}\{e^{\alpha t}e^{i\beta t}\} = \mathscr{L}\{e^{\alpha t}\cos\beta t + ie^{\alpha t}\sin\beta t\} = \frac{1}{s-z}$$

$$= \frac{1}{s-\alpha-i\beta} = \frac{1}{s-\alpha-i\beta}\cdot\frac{s-\alpha+i\beta}{s-\alpha+i\beta} = \frac{s-\alpha+i\beta}{(s-\alpha)^2+\beta^2}. \quad (8.5)$$

When we take real and imaginary parts of (8.5) with s real, and then let s be complex again, we obtain

$$\mathscr{L}\{e^{\alpha t}\cos\beta t\} = \frac{s-\alpha}{(s-\alpha)^2+\beta^2},$$
$$\mathscr{L}\{e^{\alpha t}\sin\beta t\} = \frac{\beta}{(s-\alpha)^2+\beta^2} \quad (\mathscr{R}s > \alpha). \quad (8.6)$$

In particular, taking $\alpha = 0$, we obtain

$$\mathscr{L}\{\cos\beta t\} = \frac{s}{s^2+\beta^2}, \quad \mathscr{L}\{\sin\beta t\} = \frac{\beta}{s^2+\beta^2} \quad (\mathscr{R}s > 0). \quad (8.7)$$

We could use the same direct approach to compute the Laplace transforms of other functions such as t^k and $t^k e^{zt}$, but we can calculate these transforms less laboriously by using some additional general properties of Laplace transforms.

8.2 Basic Properties of the Laplace Transform

Suppose f is in the class Λ. If we differentiate (8.1) with respect to s under the integral sign, we obtain formally

$$F'(s) = -\int_0^\infty t e^{-st} f(t)\, dt. \tag{8.8}$$

Since f is of exponential growth at infinity, satisfying the bound (8.4),

$$|te^{-st}f(t)| \le Mte^{-(\sigma-c)t},$$

where $\sigma = \mathcal{R}s$, and the integral (8.8) converges absolutely for $\sigma > c$. From this we can prove that the relation (8.8) is valid if the real part of s is sufficiently large, that is, if $\mathcal{R}s > c$. In fact, we can use the same argument to justify repeated differentiation under the integral sign, which gives the formula

$$F^{(k)}(s) = (-1)^k \int_0^\infty t^k e^{-st} f(t)\, dt \qquad (k = 1, 2, \ldots). \tag{8.9}$$

Comparing (8.9) with the definition of the Laplace transform, we see that we have sketched the proof of the following result.

Theorem 2. *The Laplace transform of a function f in the class Λ has derivatives of all orders and these derivatives are given by (8.9). If $f(t)$ belongs to the class Λ, then $t^k f(t)$ also belongs to the class Λ for every positive integer k, and its Laplace transform is given by*

$$\mathscr{L}\{t^k f(t)\} = (-1)^k \frac{d^k}{ds^k} \mathscr{L}\{f(t)\} \qquad (k = 1, 2, \ldots). \tag{8.10}$$

If we apply (8.10) with $f(t) = 1$ and use (8.2), we obtain

$$\mathscr{L}\{t^k\} = (-1)^k \frac{d^k}{ds^k}\left(\frac{1}{s}\right) = \frac{k!}{s^{k+1}} \qquad (\mathcal{R}s > 0;\; k = 1, 2, \ldots). \tag{8.11}$$

If we apply (8.10) with $f(t) = e^{zt}$ and use (8.3), we obtain

$$\mathscr{L}\{t^k e^{zt}\} = (-1)^k \frac{d^k}{ds^k}\left(\frac{1}{s-z}\right) = \frac{k!}{(s-z)^{k+1}}$$

$$(\mathcal{R}s > \mathcal{R}z;\; k = 1, 2, \ldots). \tag{8.12}$$

By letting $z = \alpha + i\beta$ and taking real and imaginary parts, we can obtain the formulas

$$\mathscr{L}\{t^k e^{\alpha t} \cos \beta t\} = \frac{k!\,\mathscr{R}[(s-\alpha)+i\beta]^{k+1}}{[(s-\alpha)^2+\beta^2]^{k+1}},$$

$$\mathscr{L}\{t^k e^{\alpha t} \sin \beta t\} = \frac{k!\,\mathscr{I}[(s-\alpha)+i\beta]^{k+1}}{[(s-\alpha)^2+\beta^2]^{k+1}},$$

(8.13)

$$\mathscr{L}\{t^k \cos \beta t\} = \frac{k!\,\mathscr{R}(s+i\beta)^{k+1}}{(s^2+\beta^2)^{k+1}}, \qquad \mathscr{L}\{t^k \sin \beta t\} = \frac{k!\,\mathscr{I}(s+i\beta)^{k+1}}{(s^2+\beta^2)^{k+1}}.$$

(8.14)

By evaluating the indicated real and imaginary parts in (8.13) and (8.14), we can obtain explicit expressions for these Laplace transforms.

● EXERCISES

1. Find the Laplace transforms of $t \cos \beta t$ and $t \sin \beta t$, using (8.14).

2. Find the Laplace transforms of $t \cos \beta t$ and $t \sin \beta t$ directly from the definition.

3. Calculate the Laplace transforms of
(a) $\cos^2 \beta t$. [*Hint:* Use the half-angle formula.]
(b) $\sin \beta t \cos \beta t$.

4. Calculate the Laplace transform of the function f given by

$$f(t) = \begin{cases} 0 & (0 < t < 1) \\ t & (1 < t < 2) \\ 0 & (t > 2). \end{cases}$$

5. Calculate the Laplace transform of the function f given by

$$f(t) = \begin{cases} \sin 2t & (0 < t < \pi) \\ 0 & (t > \pi). \end{cases}$$

The relation between (8.2) and (8.3) suggests the multiplication of a function by an exponential does not affect its Laplace transform except for causing a translation of the independent variable. This is in fact a general property.

Theorem 3. *If the function f in the class Λ has Laplace transform F, then the Laplace transform of $e^{at}f(t)$, for any constant a (real or complex) is $F(s-a)$.*

8.2 Basic Properties of the Laplace Transform

Proof. It is easy to verify that $e^{at}f(t)$ also belongs to the class Λ. The only part of the verification which is not completely obvious is that $e^{at}f(t)$ is of exponential growth at infinity. If f satisfies (8.4), and if $\alpha = \mathscr{R}a$, then

$$|e^{at}f(t)| \le e^{\alpha t}Me^{ct} = Me^{(\alpha+c)t},$$

and thus $e^{at}f(t)$ is of exponential growth at infinity and has a Laplace transform, which we may calculate directly. We obtain

$$\mathscr{L}\{e^{at}f(t)\} = \int_0^\infty e^{-st}e^{at}f(t)\,dt = \int_0^\infty e^{-(s-a)t}f(t)\,dt = F(s-a),$$

and the result is proved. ∎

● **EXERCISES**

6. Use Theorem 3 to derive (8.6) from (8.7).
7. Use Theorem 3 to derive (8.13) from (8.14).
8. Using the results of Exercise 1, find the Laplace transforms of $te^{at}\cos\beta t$ and $te^{at}\sin\beta t$.
9. Let $f(t) = \phi(t)$ for $0 < t < a$, and let f be periodic with period a, so that $f(t+a) = f(t)$ for $0 < t < \infty$. Show that

$$\mathscr{L}\{f(t)\} = \frac{\int_0^a e^{-st}\phi(t)\,dt}{1 - e^{-as}} \qquad (\mathscr{R}s > 0).$$

[*Hint:* Write

$$\mathscr{L}\{f(t)\} = \int_0^a e^{-st}f(t)\,dt + \int_a^{2a} e^{-st}f(t)\,dt + \cdots,$$

and transform each integral so that the range of integration is $[0, a]$.]

10. Let f be the "square-wave function" given by $f(t) = 1$ $(0 < t < a/2)$, $f(t) = -1$ $(a/2 < t < a)$ with f periodic of period a. Sketch the graph and find the Laplace transform of f.
11. Let f be the "square-wave function" of Exercise 10 and let $g(t) = \int_0^t f(s)\,ds$. Sketch the graph and find the Laplace transform of g.

The usefulness of the Laplace transform in finding solutions of a differential equation depends on the fact that the Laplace transform of the derivative of a function can be expressed easily in terms of the Laplace transform of the function.

Theorem 4. *Let f be a function in the class Λ whose derivative also belongs to the class Λ, and let the Laplace transform of f be F. Then*

$$\mathscr{L}\{f'(t)\} = sF(s) - f(0). \tag{8.15}$$

Proof. We simply apply the definition of the Laplace transform and integrate by parts. This gives

$$\begin{aligned}
\mathscr{L}\{f'(t)\} &= \int_0^\infty e^{-st} f'(t)\, dt = \lim_{A \to \infty} \int_0^A e^{-st} f'(t)\, dt \\
&= \lim_{A \to \infty} \left\{ [e^{-st} f(t)]_0^A + \int_0^A s e^{-st} f(t)\, dt \right\} \\
&= -f(0) + s \int_0^\infty e^{-st} f(t)\, dt = sF(s) - f(0),
\end{aligned}$$

where we have used the fact that $\lim_{A \to \infty} e^{-sA} f(A) = 0$ for $\mathscr{R}s$ sufficiently large. ∎

The result of Theorem 4 can easily be extended to derivatives of higher order. It is convenient to introduce some additional notation. For each positive integer k, *we define Λ^k to be the class of functions in Λ which have continuous derivatives up to order k on $(0, \infty)$ and whose derivatives up to order k also belong to Λ.* Thus, the hypothesis of Theorem 4 is that f belongs to the class Λ^1.

Theorem 5. *If f belongs to the class Λ^k for some positive integer k, and if F is the Laplace transform of f, then*

$$\begin{aligned}
\mathscr{L}\{f^{(j)}(t)\} &= s^j F(s) - s^{j-1} f(0) - s^{j-2} f'(0) - \cdots - s f^{(j-2)}(0) - f^{(j-1)}(0) \\
&(j = 1, 2, \ldots, k).
\end{aligned} \tag{8.16}$$

Proof. We can prove (8.16) most easily by induction on j for any fixed k, remembering that the induction procedure cannot be carried out for $j > k$ since the hypotheses do not guarantee the existence of the Laplace transforms of derivatives of order higher than k. The case $j = 1$ of (8.16) was proved in Theorem 4. If (8.16) has been established for $f^{(j)}(t)$, we can write $f^{(j+1)}(t)$ as the first derivative of $f^{(j)}(t)$ and apply Theorem 4. We obtain

$$\begin{aligned}
\mathscr{L}\{f^{(j+1)}(t)\} &= s\mathscr{L}\{f^{(j)}(t)\} - f^{(j)}(0) \\
&= s[s^j F(s) - s^{j-1} f(0) - \cdots - f^{(j-1)}(0)] - f^{(j)}(0) \\
&= s^{j+1} F(s) - s^j f(0) - \cdots - s f^{(j-1)}(0) - f^{(j)}(0),
\end{aligned}$$

and this is Eq. 8.16 with j replaced by $(j+1)$. Thus, Theorem 5 is proved by induction. ∎

8.2 Basic Properties of the Laplace Transform

The formula (8.16) for calculating the Laplace transforms of the derivatives of a function is the key to the solution of linear differential equations by means of Laplace transforms. We shall study this subject in more detail in Sections 8.4 and 8.5. However, we give some simple examples here which we have already solved in Chapter 6 to illustrate the main idea.

Example 2. Let us find the solution ϕ_0 of the familiar first-order differential equation

$$y' + ay = 0, \tag{8.17}$$

which satisfies the initial condition

$$\phi_0(0) = y_0, \tag{8.18}$$

where a and y_0 are given constants. From our previous solution in Section 6.1 we know that the solution ϕ_0 is in the class Λ^1. Thus, we should be able to find ϕ_0 by Laplace transforms. Let $Y_0(s) = \mathscr{L}(\phi_0)$. Using (8.15), we may take the Laplace transform of every term in the equation

$$\phi_0'(t) + a\phi_0(t) = 0,$$

satisfied by ϕ_0, and we obtain

$$sY_0(s) - \phi_0(0) + aY_0(s) = 0 \quad \text{or} \quad (s + a)Y_0(s) = y_0,$$

using (8.18). This yields

$$Y_0(s) = \frac{y_0}{s + a},$$

and the only remaining problem is to find a function which has this expression as its Laplace transform. As we have seen in Example 2, Section 8.1, $y_0 e^{-at}$ is such a function. By direct verification and application of Theorem 1, Section 6.2, we know that it is the only solution of Eq. 8.17 satisying (9.18). However, as motivation for more complicated problems, it is useful to look at this a little differently. At this stage, we do not know that it is the only such function, but if we accept the truth of the statement that two different continuous functions cannot have the same Laplace transform, then we conclude that $\phi_0(t) = y_0 e^{-at}$. Since $y_0 e^{-at}$ does belong to the class Λ^1, our reasoning is valid, except for the uniqueness statement which will be stated precisely in Section 8.3.

Example 3. Now, let us apply the same method to find the solution ϕ of the nonhomogeneous equation

$$y' + ay = f(t), \tag{8.19}$$

which satisfies the initial condition

$$\phi(0) = y_0. \tag{8.20}$$

Here f is a given function belonging to the class Λ. The problem in (8.19) and (8.20) can be solved by the method of Example 3, Section 6.1, and from the solution it is obvious that if f is in the class Λ, the solution ϕ is in the class Λ'. We proceed as in Example 2; we may assume that ϕ belongs to the class Λ' and we let $Y(s)$ be the Laplace transform of ϕ and $F(s)$ the Laplace transform of f. Using (8.15) we take the Laplace transform of every term in the equation

$$\phi'(t) + a\phi(t) = f(t),$$

satisfied by ϕ, and we obtain

$$sY(s) - \phi(0) + aY(s) = F(s) \quad \text{or} \quad (s+a)Y(s) = y_0 + F(s),$$

using (8.20). This yields

$$Y(s) = \frac{y_0}{s+a} + \frac{F(s)}{s+a}.$$

We must now find a function which has this expression as its Laplace transform. We have already seen in Example 2 that $y_0 e^{-at}$ has Laplace transform $y_0/(s+a)$, but we have no method as yet of finding a function whose Laplace transform is $(F(s)/(s+a))$. This suggests that we will need a method of finding a function whose Laplace transform is the product of two given functions. In this case the given functions are $F(s)$ and $1/(s+a)$, which are the Laplace transforms of $f(t)$ and e^{-at}, respectively. This problem will be studied in Section 8.3. For the present, we remark that the method of Section 6.1 gives

$$\phi(t) = y_0 e^{-at} + \int_0^t e^{-a(t-u)} f(u) \, du,$$

and this suggests that $\int_0^t e^{-a(t-u)} f(u) \, du$ should have the Laplace transform $F(s)/(s+a)$.

Example 4. In Examples 2 and 3 we have used the Laplace transform to solve linear differential equations with constant coefficients. To show that the method is less useful for linear equations with variable coefficients, let us attempt to find the solution ψ of the equation

$$y' + 2ty = \sin t, \tag{8.21}$$

which satisfies the initial condition

$$\psi(0) = y_0. \tag{8.22}$$

In Example 3, Section 6.1, we found that

$$\psi(t) = y_0 \exp(-t^2) + \exp(-t^2) \int_0^t \exp(s^2) \sin s \, ds.$$

When we try to obtain this solution by the use of Laplace transforms, we meet with serious difficulties. If we let $Z(s)$ be the Laplace transform of ψ and take the Laplace transform of every term in the equation

$$\psi'(t) + 2t\psi(t) = \sin t,$$

satisfied by ψ, we apply (8.10) and (8.15) and we obtain

$$sZ(s) - \psi(0) - 2Z'(s) = \frac{1}{s^2 + 1}. \tag{8.23}$$

While in Examples 2 and 3 we obtained an algebraic equation for the Laplace transform of the desired solution (and thus simplified the problem), here we have a differential equation for the transform of the solution which is no simpler than the original problem. **This example suggests that the usefulness of the Laplace transform method is limited mainly to equations with constant coefficients.**

8.3 The Inverse Transform

The examples of the previous section have indicated the need for finding a function with a given Laplace transform. The function $f(t)$ whose Laplace transform is $F(s)$ is called the **inverse Laplace transform** of $F(s)$. The inverse

The Laplace Transform

Laplace transform is, as we shall prove later in this section, linear. We must consider the following questions:

(i) If we know that $F(s)$ is the Laplace transform of a function $f(t)$, how can we compute the inverse transform f from a knowledge of F?

(ii) Is the inverse transform of a given function F unique?

An answer to the first question can be given in a theoretical way by means of the so-called complex inversion formula (see, for example, D. V. Widder, *The Laplace Transform* (Princeton University Press, Princeton, N.J., 1946), p. 66). However, the derivation and application of this formula require a knowledge of real and complex analysis. Therefore, we confine ourselves to a more elementary approach that will enable us to find the inverse transforms of some functions commonly arising in applications.

We have seen in Example 3, Section 8.2, that it would be useful to have a general method of calculating the inverse Laplace transform of a product of two functions each of whose inverse Laplace transforms are known. Let us now determine whether there is a means for doing this.

We assume that we are given $F(s)$ and $G(s)$, and that we can find functions $f(t)$ and $g(t)$ in the class Λ whose Laplace transforms are $F(s)$ and $G(s)$, respectively; thus $F(s) = \mathscr{L}(f(t))$ and $G(s) = \mathscr{L}(g(t))$. **Our problem is to determine the function $h(t)$ whose Laplace transform is the product $F(s)G(s)$, if such a function exists.** If there is such a function, then

$$\mathscr{L}\{h(t)\} = \int_0^\infty e^{-st} h(t)\, dt$$
$$= F(s)G(s) = \int_0^\infty e^{-su} f(u)\, du \int_0^\infty e^{-sv} g(v)\, dv, \tag{8.24}$$

where $\mathscr{R}s > \sigma$ for some real number σ. If $F(s) = \mathscr{L}\{f(t)\}$ for $\mathscr{R}s > \alpha$, and $G(s) = \mathscr{L}\{g(t)\}$ for $\mathscr{R}s > \beta$ (α, β real), then $\sigma = \max(\alpha, \beta)$. Since each integral converges absolutely for $\mathscr{R}s > \sigma$, we may write the product of the two integrals on the right side of (8.24) as a double integral, obtaining

$$\int_0^\infty e^{-st} h(t)\, dt = \int_0^\infty \int_0^\infty e^{-s(u+v)} f(u) g(v)\, du\, dv, \tag{8.25}$$

for $\mathscr{R}s > \sigma$.

We write this as an iterated integral,

$$\int_0^\infty e^{-st} h(t)\, dt = \int_0^\infty g(v) \left[\int_0^\infty e^{-s(u+v)} f(u)\, du \right] dv;$$

8.3 The Inverse Transform

this can be justified under our hypotheses because of the absolute convergence noted above. Making the change of variable $u + v = t$ in the inner integral, we obtain

$$\int_0^\infty e^{-st} h(t)\, dt = \int_0^\infty g(v) \left[\int_v^\infty e^{-st} f(t - v)\, dt \right] dv$$
$$= \int_0^\infty e^{-st} \left[\int_0^t f(t - v) g(v)\, dv \right] dt \qquad (\mathcal{R}s > \sigma), \tag{8.26}$$

where the interchange in the order of integration can again be justified.

● **EXERCISE**

1. Obtain the limits of integration in (8.26). [*Hint:* Draw a sketch of the region of integration.]

Equation 8.26 says $\mathcal{L}\{h(t)\} = \mathcal{L}[\int_0^t f(t - v) g(v)\, dv]$, which suggests that the solution to our problem is

$$h(t) = \int_0^t f(t - v) g(v)\, dv. \tag{8.27}$$

By reversing the argument we have just completed, we may prove that the Laplace transform of the function $h(t)$ defined by (8.27) is $F(s)G(s)$, as desired.

● **EXERCISE**

2. Show that if f and g belong to the class Λ and h is defined by (8.27), then

$$\int_0^\infty e^{-st} h(t)\, dt = \int_0^\infty f(u) e^{-su}\, du \int_0^\infty g(v) e^{-sv}\, dv.$$

We may summarize what we have sketched in the following theorem.

Theorem 1. *Let $f(t)$, $g(t)$ belong to the class Λ and let $F(s) = \mathcal{L}\{f(t)\}$, $G(s) = \mathcal{L}\{g(t)\}$. Then $\mathcal{L}[\int_0^t f(t - v) g(v)\, dv] = F(s)G(s)$.*

The function h defined by (8.27) is called the **convolution** of f and g, and is sometimes denoted by $h = f * g$ to indicate that the convolution of two functions resembles a product in many ways.

The Laplace Transform

● **EXERCISES**

   3. Show that $f * g = g * f$ for any two functions f, g in the class Λ.
   4. Show that $(f * g) * h = f * (g * h)$ for any three functions f, g, h in the class Λ.
   5. Show that $f * 0 = 0$ for any function f in the class Λ.
   6. Show that if f and g are defined for $-\infty < t < \infty$ but are both identically zero for $t < 0$, then their convolution $f * g$ can be written as $\int_{-\infty}^{\infty} f(t-v)g(v)\,dv$.

The convolution integral in Exercise 6 is sometimes used to define an operation analogous to multiplication for certain classes of functions. The reader should be careful not to carry the analogy too far; for example, it is not true that $f * 1 = f$, since for an arbitrary function f we have

$$\int_0^t f(t-s) \cdot 1 \, ds = \int_0^t f(\sigma)\, d\sigma \neq f(t).$$

The properties of the convolution make it possible for us to calculate some inverse Laplace transforms which would otherwise be difficult or impossible to find.

Example 1. Find the inverse Laplace transform of

$$1/(s^2 - 1) = [1/(s - 1)][1/(s + 1)].$$

Since $\mathscr{L}^{-1}\{1/(s-1)\} = e^t$, and $\mathscr{L}^{-1}\{1/(s+1)\} = e^{-t}$, we see that

$$\mathscr{L}^{-1}\left\{\frac{1}{s^2 - 1}\right\} = \int_0^t e^{t-u} e^{-u} \, du = e^t \int_0^t e^{-2u} \, du$$

$$= e^t \left[\frac{e^{-2u}}{-2}\right]_0^t = \frac{1}{2} e^t (1 - e^{-2t}) = \frac{1}{2}(e^t - e^{-t}).$$

There is another way to calculate this particular inverse transform. If we decompose $(1/(s^2 - 1))$ into partial fractions, we find

$$\frac{1}{s^2 - 1} = \frac{1}{(s-1)(s+1)} = \frac{1}{2}\left(\frac{1}{s-1} - \frac{1}{s+1}\right),$$

and thus, by the linearity of $\mathscr{L}^{-1}$ (this will be proved following the corollary to Theorem 2).

$$\mathscr{L}^{-1}\left\{\frac{1}{s^2 - 1}\right\} = \frac{1}{2}\left[\mathscr{L}^{-1}\left(\frac{1}{s-1}\right) - \mathscr{L}^{-1}\left(\frac{1}{s+1}\right)\right]$$

$$= \frac{1}{2}(e^t - e^{-t}).$$

8.3 The Inverse Transform

Example 2. Find the inverse Laplace transform of $(1/(s^2(s^2+1)))$. We have seen, in Eq. 8.11, that $\mathscr{L}^{-1}(1/(s^2)) = t$, and in Eq. 8.7, $\mathscr{L}^{-1}(1/(s^2+1)) = \sin t$. Now, by Theorem 1, $\mathscr{L}^{-1}(1/(s^2(s^2+1)))$ is the convolution of t and $\sin t$, which is

$$\int_0^t (t-u)\sin u\, du = t\int_0^t \sin u\, du - \int_0^t u \sin u\, du$$

$$= t - t\cos t + t\cos t - \sin t$$

$$= t - \sin t.$$

We may also obtain this result by partial fractions. For

$$\frac{1}{s^2(s^2+1)} = \frac{1}{s^2} - \frac{1}{s^2+1},$$

and hence, by either method

$$\mathscr{L}^{-1}\left(\frac{1}{s^2(s^2+1)}\right) = \mathscr{L}^{-1}\left(\frac{1}{s^2}\right) - \mathscr{L}^{-1}\left(\frac{1}{s^2+1}\right) = t - \sin t.$$

The above examples suggest that if we wish to find the inverse transform of a rational function $N(s)/D(s)$, where N and D are polynomials with the degree of N less than the degree of D, then we can take advantage of partial functions to decompose $N(s)/D(s)$ into a sum of terms each of whose inverse transform is easily found.

● **EXERCISES**

7. Find the inverse Laplace transform of each of the following functions. (We indicate the partial fraction decomposition for the convenience of the reader.)

(a) $\dfrac{s^2-6}{s^3+4s^2+3s} = \dfrac{A}{s} + \dfrac{B}{s+1} + \dfrac{C}{s+3}$.

(b) $\dfrac{1}{s^2(s^2+1)} = \dfrac{A}{s} + \dfrac{B}{s^2} + \dfrac{Cs+D}{s^2+1}$.

(c) $\dfrac{16}{s(s^2+4)^2} = \dfrac{A}{s} + \dfrac{Bs+C}{s^2+4} + \dfrac{Ds+E}{(s^2+4)^2}$.

(d) $\dfrac{F(s)}{s+a}$,

where $f(t)$ is in the class Λ and $\mathscr{L}\{f(t)\} = F(s)$.

(e) $\dfrac{F(s)}{s^2+1}$,

where $f(t)$ is in the class Λ and $\mathscr{L}\{f(t)\} = F(s)$.

8. Find the solution ϕ of the integral equation

$$\phi(t) + \int_0^t (t-u)\phi(u)\, du = 1.$$

[*Hint:* Take the Laplace transform of every term and use Theorem 1; then solve for $\mathscr{L}(\phi)$ and finally find ϕ.]

This is as far as we can go with systematic techniques for constructing inverse transforms, without using the inversion theorem. Beyond this the most useful aids in finding inverse transforms are a good table of Laplace transforms (a brief one may be found at the end of the chapter), experience, and luck.

Let us now turn to question (ii), namely, the uniqueness of the inverse transform. We state the following basic result without proof—a proof may be found in D. V. Widder, *The Laplace Transform* (Princeton University Press, Princeton, N.J., 1946).

Theorem 2. *Let f_1, f_2 be continuous on $0 < t < \infty$ and belong to the class Λ, and let their Laplace transforms $F_1(s)$ and $F_2(s)$, respectively, exist for $\mathscr{R}s \geq \sigma$. If $F_1(s) = F_2(s)$ for real $s \geq \sigma$, then $f_1(t) = f_2(t)$ on $0 \leq t < \infty$.*

Theorem 2 justifies the method used in Examples 2 and 3, Section 8.2, where we found the Laplace transform of an unknown function and then found this unknown function by taking the inverse transform. If the inverse transform were not unique, we would not know that the process of taking the inverse transform returns us to the original unknown function.

Another consequence of Theorem 2 is **the linearity of the inverse Laplace transform:** If the continuous functions $f_1(t)$ and $f_2(t)$ in the class Λ are the inverse transforms of $F_1(s)$ and $F_2(s)$, respectively, and if a and b are constants, then the inverse transform of $aF_1(s) + bF_2(s)$ is $af_1(t) + bf_2(t)$. In order to prove this, we need only observe that $af_1(t) + bf_2(t)$ has Laplace transform $aF_1(s) + bF_2(s)$ and use Theorem 2 to show that $af_1(t) + bf_2(t)$ is the only continuous function in the class Λ with Laplace transform $aF_1(s) + bF_2(s)$.

The problem of inversion of Laplace transforms presents many practical as well as mathematical problems. As a simple example, we give the following

8.4 Applications to Linear Equations with Constant Coefficients

theorem which shows that not all functions $F(s)$ can be the Laplace transform of a function $f(t)$ in the class Λ.

Theorem 3. *If f belongs to the class Λ and if $F(s)$ is its Laplace transform, then $\lim_{s \to \infty} F(s) = 0$.*

Proof. Since f belongs to the class Λ, it satisfies an inequality of the form (8.4) (see Section 8.2). Then, letting $\sigma = \mathcal{R}s$, we have

$$|F(s)| \leq M \int_0^\infty |e^{-st}| e^{ct}\, dt$$

$$= M \int_0^\infty e^{-\sigma t} e^{ct}\, dt = \frac{M}{\sigma - c} \qquad (\mathcal{R}s > c),$$

and it is clear that $\lim_{\mathcal{R}s \to \infty} F(s) = 0$. ∎

We have actually proved a little more than we claimed. Not only have we shown that $F(s)$ tends to zero as $s \to \infty$, but in fact that $|sF(s)|$ remains bounded as $\mathcal{R}s \to \infty$.

The question of which functions are Laplace transforms is a difficult one, and we cannot in this brief treatment give a more precise answer. In practice, we only apply Laplace transforms when we can prove the existence of an inverse transform by finding it explicitly, but the question remains an important one because it is often impossible to find the inverse transform explicitly.

8.4 Applications to Linear Equations with Constant Coefficients

Our main purpose in developing the Laplace transform has been to apply it to the solution of linear differential equations. We have suggested how this may be done by examples in Section 8.2, where we considered first-order differential equations. For equations of higher order, the general idea is the same, but there are some technical problems which arise when we try to find the inverse transform. In this section we shall discuss these technical problems and the means of dealing with them by a collection of examples. We shall concentrate on equations of the second order, but we shall also indicate the minor additional problems which arise for equations of higher order.

282 The Laplace Transform

In each of the examples involving a second-order differential equation we shall be seeking the solution ϕ of an equation of the form

$$a_0 y'' + a_1 y'' + a_2 y = f(t), \qquad (8.28)$$

where a_0, a_1, a_2 are constants, which satisfies the initial conditions

$$\phi(0) = y_0, \qquad \phi'(0) = y_1. \qquad (8.29)$$

Under suitable hypotheses on the function f, we may apply an existence theorem, such as the Corollary 1 to Theorem 1, Section 6.2, to conclude the existence of a unique solution ϕ of Eq. 8.28 which satisfies the initial conditions (8.29). When we attempt to find this solution by means of the Laplace transform, we can expect to obtain it only if it and its derivatives up to the second order are of exponential growth, that is, if it belongs to the class Λ^2. Thus, to conclude the existence of a solution which can be found by means of the Laplace transform, we need a theorem on the growth of solutions of linear differential equations with constant coefficients. Such a theorem is proved in Appendix 2. This theorem applied in the present context says that **every solution of a linear nonhomogeneous differential equation of order n with constant coefficients, whose nonhomogeneous term is in the class Λ, belongs to the class Λ^n**.

Example 1. Let us use the Laplace transform to find the solution ϕ of the equation $y'' + y' + 2y = 0$ which satisfies the initial conditions $\phi(0) = 1$, $\phi'(0) = 1$. We let Y be the Laplace transform of ϕ and take the Laplace transform of the equation

$$\phi''(t) + 3\phi'(t) + 2\phi(t) = 0,$$

satisfies by ϕ. Using (8.16), we obtain

$$s^2 Y(s) - s\phi(0) - \phi'(0) + 3[s Y(s) - \phi(0)] + 2 Y(s) = 0;$$

now, using $\phi(0) = \phi'(0) = 1$, we obtain

$$(s^2 + 3s + 2) Y(s) = (s + 3)\phi(0) + \phi'(0) = s + 4.$$

Thus,

$$Y(s) = \frac{s + 4}{s^2 + 3s + 2} = \frac{s + 4}{(s + 1)(s + 2)}. \qquad (8.30)$$

8.4 Applications to Linear Equations with Constant Coefficients

In order to take the inverse transform in (8.30), we must simplify the expression $(s+4)/((s+1)(s+2))$. We use the method of partial fractions to accomplish this. Letting

$$\frac{s+4}{(s+1)(s+2)} = \frac{A}{s+1} + \frac{B}{s+2},$$

for all s, we observe that

$$A = \lim_{s \to -1} \frac{s+4}{(s+1)(s+2)} \cdot (s+1) = \lim_{s \to -1} \frac{s+4}{s+2} = 3,$$

$$B = \lim_{s \to -2} \frac{s+4}{(s+1)(s+2)} \cdot (s+2) = \lim_{s \to -2} \frac{s+4}{s+1} = -2.$$

Thus,

$$Y(s) = \frac{s+4}{(s+1)(s+2)} = \frac{3}{s+1} - \frac{2}{s+2}.$$

Now, we may use (8.3) and the linearity of the inverse transform to find a function whose Laplace transform is $Y(s)$, and because of the uniqueness of the inverse transform (Theorem 2, Section 8.3), this function must be the desired solution ϕ. We see that $\phi(t) = 3e^{-t} - 2e^{-2t}$. We could, of course, have obtained this solution by using the methods of Section 6.4, but this would have involved first finding the general solution of (8.28) and then substituting the initial conditions (8.29) to determine the constants. **The Laplace transform does not solve problems which would otherwise be unsolvable, but it does provide an easy, practical method of solution for many problems.**

● EXERCISE

1. Verify the solution $\phi(t) = 3e^{-t} - 2e^{-2t}$ to Example 1 by direct substitution.

Example 2. Let us now use the Laplace transform to find the solution ϕ of the equation $y'' + 4y' + 4y = f(t)$ which satisfies the initial conditions $\phi(0) = 1$, $\phi'(0) = 2$, where $f(t)$ belongs to the class Λ. We let Y be the Laplace transform of ϕ, F the Laplace transform of f, and we take the Laplace transform of the equation

$$\phi''(t) + 4\phi'(t) + 4\phi(t) = f(t),$$

satisfied by ϕ. Much as in Example 1, we obtain

$$s^2 Y(s) - s\phi(0) - \phi'(0) + 4[sY(s) - \phi(0)] + 4Y(s) = F(s),$$

or

$$(s^2 + 4s + 4)Y(s) = (s + 4)\phi(0) + \phi'(0) + F(s)$$
$$= (s + 4) + 2 + F(s) = s + 6 + F(s).$$

Thus,

$$Y(s) = \frac{s + 6 + F(s)}{s^2 + 4s + 4} = \frac{s + 6}{(s + 2)^2} + \frac{F(s)}{(s + 2)^2}.$$

Again, we use partial fractions to simplify this expression. We must find constants A and B such that

$$\frac{s + 6}{(s + 2)^2} = \frac{A}{s + 2} + \frac{B}{(s + 2)^2}.$$

We observe that

$$B = \lim_{s \to -2} \frac{s + 6}{(s + 2)^2} \cdot (s + 2)^2 = \lim_{s \to -2} (s + 6) = 4,$$

$$\frac{A}{s + 2} = \frac{s + 6}{(s + 2)^2} - \frac{B}{(s + 2)^2} = \frac{s + 6}{(s + 2)^2} - \frac{4}{(s + 2)^2} = \frac{s + 2}{(s + 2)^2} = \frac{1}{s + 2}$$

which implies $A = 1$. Now,

$$Y(s) = \frac{s + 6}{(s + 2)^2} = \frac{1}{s + 2} + \frac{4}{(s + 2)^2} + \frac{F(s)}{(s + 2)^2},$$

and we may use (8.3), (8.12), and Theorem 1, Section 8.3, to find the inverse transform. This yields the solution.

$$\phi(t) = e^{-2t} + 4te^{-2t} + \int_0^t (t - u)e^{-2(t-u)} f(u)\, du$$

● **EXERCISE**

2. Verify the answer to Example 2 by direct substitution.

8.4 Applications to Linear Equations with Constant Coefficients

Examples 1 and 2 give us enough insight into the method to enable us to solve the general linear homogeneous equation of order n with constant coefficients. Let ϕ be the solution of the equation of order n,

$$L_n(y) = a_0 y^{(n)} + a_1 y^{(n-1)} + \cdots + a_n y = 0, \tag{8.31}$$

$(a_0, a_1, \ldots, a_n$ constants) which satisfies the initial conditions

$$\phi(0) = y_0, \phi'(0) = y, \ldots, \phi^{(n-1)}(0) = y_{n-1}. \tag{8.32}$$

We let Y be the Laplace transform of ϕ and take the Laplace transform of the equation

$$a_0 \phi^{(n)}(t) + a_1 \phi^{(n-1)}(t) + \cdots + a_n \phi(t) = 0.$$

We obtain

$$a_0[s^n Y(s) - s^{n-1} y_0 - s^{n-2} y_1 - \cdots - y_{n-1}]$$
$$+ a_1[s^{n-1} Y(s) - s^{n-2} y_0 - \cdots - y_{n-2}] + \cdots$$
$$+ a_{n-1}[s Y(s) - y_0] + a_n Y(s) = 0,$$

or

$$p(s)Y(s) = a_0(s^{n-1} y_0 + s^{n-2} y_1 + \cdots + y_{n-1})$$
$$+ a_1(s^{n-2} y_0 + \cdots + y_{n-2}) + \cdots + a_{n-1} y_0, \tag{8.33}$$

where p is the characteristic polynomial of the linear operator L_n (see Section 6.4).

$$p(s) = a_0 s^n + a_1 s^{n-1} + \cdots + a_n.$$

We may write (8.33) in the form

$$p(s) Y(s) = q(s),$$

where q is a polynomial of degree at most $(n-1)$ on the right-hand side of (8.33). (The degree could be less than $(n-1)$ if the given initial value y_0 is zero.) This polynomial q is linear in $y_0, y_1, \ldots, y_{n-1}$.

We now have

$$Y(s) = \frac{q(s)}{p(s)},$$

and the next step is to separate the rational function $q(s)/p(s)$ into partial fractions. If the roots of the polynomial p are $z_1, \ldots, z_k$ of multiplicities $m_1, \ldots, m_k$, respectively, then we can write

$$P(s) = a_0(s-z_1)^{m_1}(s-z_2)^{m_2} \cdots (s-z_k)^{m_k}.$$

The process of separating into partial fractions gives

$$\frac{q(s)}{p(s)} = \frac{a_{11}}{s-z_1} + \cdots + \frac{a_{1,m_1}}{(s-z_1)^{m_1}} + \frac{a_{21}}{s-z_2} + \cdots + \frac{a_{2,m_2}}{(s-z_2)^{m_2}}$$

$$+ \cdots + \frac{a_{k1}}{s-z_k} + \cdots + \frac{a_{k,m_k}}{(s-z_k)^{m_k}},$$

where the constants $a_{11}, \ldots, a_{k,m_k}$ may be calculated. We may now take the inverse transform of $Y(s)$ using (8.3) and (8.12). We obtain the solution

$$\phi(t) = a_{11}e^{z_1 t} + \cdots + \frac{a_{1,m_1}}{(m_1-1)!} t^{m_1-1} e^{z_1 t} + \cdots + a_{k1}e^{z_k t}$$

$$+ \cdots + \frac{a_{k,m_k}}{(m_k-1)!} t^{m_k-1} e^{z_k t},$$

which is, of course, the same as that obtained in Section 6.4. If some of the roots $z_1, \ldots, z_k$ are complex but the coefficients $a_0, \ldots, a_n$ of L are real, then we may express the solutions in terms of real functions just as in Section 6.4.

We remark if in place of Eq. 8.31 we consider the equation

$$L_n(y) = f(t),$$

where f belongs to the class Λ, we handle the additional term by using the convolution exactly as in Example 2 above.

The Laplace transform is used a great deal in the study of electrical circuits. We have already mentioned in Chapter 1 the linear electrical circuit consisting of a capacitance C, a resistance R, and an inductance L connected in series.

8.4 Applications to Linear Equations with Constant Coefficients

The voltage $v(t)$ across the capacitance may be described by the equation

$$Lv'' + Rv' + \frac{1}{C}v = 0.$$

Consider such a circuit with an external applied voltage $A \cos(kt + \alpha)$, such as might arise from connecting the circuit to a source of alternating current. Then the voltage $v(t)$ would be described by the equation

$$Lv'' + Rv'' + \frac{1}{C}v = A \cos(kt + \alpha). \tag{8.34}$$

In studying Eq. 8.34, we find it convenient to write the nonhomogeneous term as the real part of a complex exponential, solve the corresponding complex equation, and then take the real part of the solution, just as we suggested in Section 6.4. To do this, we define the complex number

$$b = Ae^{i\alpha}; \tag{8.35}$$

then

$$A \cos(kt + \alpha) = \mathscr{R}[Ae^{i(kt+\alpha)}] = \mathscr{R}(Ae^{i\alpha}e^{ikt})$$
$$= \mathscr{R}be^{ikt}.$$

Thus, instead of (8.34), we consider the complex equation

$$Ly'' + Ry' + \frac{1}{C}y = be^{ikt}. \tag{8.36}$$

Let Y be the Laplace transform of the solution ϕ of (8.36) which satisfies the initial conditions $\phi(0) = y_0$, $\phi'(0) = y_1$. Then, using (8.3) and (8.16), we see that

$$L[s^2 Y(s) - sy_0 - y_1] + R[sY(s) - y_0] + \frac{1}{C}Y(s) = \frac{b}{s - ik}$$

or

$$\left(Ls^2 + Rs + \frac{1}{C}\right)Y(s) = \frac{b}{s - ik} + L(sy_0 + y_1) + Ry_0, \tag{8.37}$$

The Laplace Transform

Let z_1 and z_2 be the roots of the polynomial $p(s) = Ls^2 + Rs + 1/C$, so that $Ls^2 + Rs + 1/C = L(s - z_1)(s - z_2)$. Then we can write (8.37) as

$$Y(s) = \frac{b}{L(s - z_1)(s - z_2)(s - ik)} + \frac{L(sy_0 + y_1) + Ry_0}{L(s - z_1)(s - z_2)}.$$

When we separate into partial fractions, we obtain

$$Y(s) = \frac{M}{s - ik} + \frac{N}{s - z_1} + \frac{P}{s - z_2},$$

with

$$M = \lim_{s \to ik} \frac{b}{L(s - z_1)(s - z_2)} = \frac{b}{p(ik)}.$$

When we take the inverse transform, we obtain the solution of (8.36) in the form $\phi(t) = Me^{ikt} + Ne^{z_1 t} + Pe^{z_2 t}$. If L, C, R are all positive constants, a reasonable hypothesis in applications, then the roots z_1 and z_2 of the polynomial $p(s)$ have negative real part. (Why?) We let $\phi_p(t) = Me^{ikt}$, $\phi_h(t) = Ne^{z_1 t} + Pe^{z_2 t}$. Then $\phi(t) = \phi_p(t) + \phi_h(t)$ and $\phi_h(t)$ tends to zero exponentially as $t \to +\infty$. The term $\phi_h(t)$ is called a **transient**, because its effect dies out, and the term $\phi_p(t)$ is called **the steady state**. As we wish to concentrate on this steady state, we do not examine the transient term further. (For this reason we did not actually compute the constants N and P.) The steady state is $(b/p(ik))e^{ikt}$. To calculate the corresponding voltage $v_1(t)$, we must take the real part of this expression. We define the **transfer function**

$$C(k) = \frac{1}{p(ik)}.$$

Then

$$\begin{aligned} v_1(t) &= \mathscr{R}[bC(k)e^{ikt}] = \mathscr{R}[Ae^{i\alpha}|C(k)|e^{i \arg C(k)}e^{ikt}] \\ &= \mathscr{R}[A|C(k)|e^{i(kt + \alpha + \arg C(k))}] \\ &= A|C(k)| \cos[kt + \alpha + \arg C)k)]. \end{aligned}$$

When we compare this steady-state output voltage $v_1(t)$ with the input voltage $A \cos(kt + \alpha)$, we see that the effect of the circuit has been to multiply the amplitude by the **gain function** $|C(k)|$ and to introduce a **phase lag** $\arg C(k)$.

8.4 Applications to Linear Equations with Constant Coefficients 289

Note that v_1 is independent of the initial conditions. The transfer function $C(k)$ is determined by the electrical circuit, but depends on the frequency of the input voltage. The essential principle in tuning a radio is to adjust the circuit, usually by varying the capacitance C, to maximize the gain function for a given k. In other electrical applications, it is necessary to vary k to maximize the gain function for given values of L, R, and C.

The reader is warned that we have assumed that L, R, and C are constant and that the circuit is linear in the above discussion. For time-dependent or nonlinear (vacuum tube) circuits, the resulting differential equations cannot, as a rule, be solved by means of the Laplace transform, and other methods must be developed. It may still be reasonable to define a gain function and a phase lag, but these will no longer be given by a transfer function.

The Laplace transform is sometimes used to solve linear nonhomogeneous differential equations whose nonhomogeneous terms do not belong to the class Λ. While the solutions obtained in those cases are then purely formal, it is possible to show that they are actual solutions in a more general sense. The justification requires a more sophisticated approach, such as the theory of distributions. Here, we shall only give an example to indicate the nature of the problem.

Let us again consider an electrical circuit consisting of a capacitance C, a resistance R, and an inductance L connected in series. However, now let us attempt to determine the behavior of the circuit if a large external voltage is applied over a very short time interval. Let this external voltage be defined by

$$\delta_\sigma(t) = \begin{cases} 1/\sigma & (0 < t < \sigma) \\ 0 & (t > \sigma). \end{cases}$$

Then the circuit is governed by the differential equation

$$Lv'' + Rv' + \frac{1}{C}v = \delta_\sigma(t). \tag{8.38}$$

Since $\delta_\sigma(t)$ is certainly in the class Λ, we can treat this equation by taking Laplace transforms, using

$$\begin{aligned}\mathscr{L}\{\delta_\sigma(t)\} &= \int_0^\infty \delta_\sigma(t) e^{-st}\, dt = \int_0^\sigma \frac{1}{\sigma} e^{-st}\, dt \\ &= \frac{1}{\sigma}\left[\frac{e^{-st}}{-s}\right]_{t=0}^{t=\sigma} = \frac{1}{\sigma s}(1 - e^{-\sigma s}).\end{aligned} \tag{8.39}$$

● EXERCISE

3. Find the steady-state solution ϕ_p of Eq. 8.38 satisfying the initial conditions $\phi(0) = y_0$, $\phi'(0) = y_1$. (By steady-state solution we mean, as above, the difference between the solution and those terms ϕ_h in the solution which (because L, R, C are positive constants) tend to zero as $t \to +\infty$.)

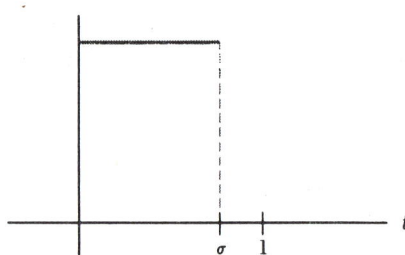

Figure 8.1

It is of interest to consider the "limiting behavior" of this electrical circuit as $\sigma \to 0$. The applied voltage $\delta_\sigma(t)$ does not tend to a limiting function in the usual sense (see Figure 8.1) when $\sigma \to 0$, but we can think of the limit as an impulse at $t = 0$ whose integral is 1, and call this a "generalized function" $\delta(t)$ (also the Dirac delta function). This would lead to the differential equation

$$Lv'' + Rv' + \frac{1}{C}v = \delta(t). \tag{8.40}$$

We treat this equation by Laplace transforms, proceeding as if we could write

$$\mathscr{L}\{\delta(t)\} = \lim_{\sigma \to 0} \mathscr{L}\{\delta_\sigma(t)\} = \lim_{\sigma \to 0} \frac{1 - e^{-\sigma s}}{\sigma s} = 1. \tag{8.41}$$

The reader should compare (8.41) with Theorem 3, Section 8.3, which says that for a function f in the class Λ, $\lim_{s \to \infty} \mathscr{L}f(t) = 0$, and should note that there is no contradiction here since δ does not belong to the class Λ, although δ_σ does.

If we let $Y(s)$ be the Laplace transform of the solution ϕ of (8.40) satisfying the initial conditions $\phi(0) = y_0$, $\phi'(0) = y_0$, we obtain, using (8.41) and writing $p(s) = Ls^2 + Rs + 1/C = L(s - z_1)(s - z_2)$ as before,

$$L[s^2 Y(s) - sy_0 - y_1] + R[sY(s) - y_0] + \frac{1}{C}Y(s) = 1,$$

8.4 Applications to Linear Equations with Constant Coefficients

$$\left(Ls^2 + Rs + \frac{1}{C}\right)Y(s) = 1 + L(sy_0 + y_1) + Ry_0, \tag{8.42}$$

$$Y(s) = \frac{1}{L(s - z_1)(s - z_2)} + \frac{L(sy_0 + y_1) + Ry_0}{L(s - z_1)(s - z_2)}.$$

This can be separated into partial fractions in the form

$$Y(s) = \frac{A}{s - z_1} + \frac{B}{s - z_2},$$

which leads to the solution $\phi(t) = Ae^{z_1 t} + Be^{z_2 t}$. Observe that if L, R, C are all positive, so that z_1 and z_2 have negative real part, $\lim_{t \to \infty} \phi(t) = 0$. Observe also that the output voltage ϕ is a continuous function, even though the input voltage δ is not.

Since the method used in this example has not been justified, the answer obtained must be verified by direct substitution. Such a verification requires an understanding of the meaning of $\delta(t)$. Let us also point out that we could obtain the same solution without using the Laplace transform by means of variation of constants (see Section 6.5). This would also involve the same nonrigorous calculations, and would also require verification by direct substitution.

● **EXERCISES**

Find, using Laplace transforms, the solution ϕ of each of the following differential equations which satisfies the given initial conditions.

4. $y'' - y = 0$, $\phi(0) = 0$, $\phi'(0) = 1$.
5. $y'' - 5y' + 6y = 0$, $\phi(0) = 0$, $\phi'(0) = 1$.
6. $y''' - 6y'' + 11y' - 6y = 0$, $\phi(0) = \phi'(0) = 0$, $\phi''(0) = 1$.
7. $y''' - 8y = e^{2t}$, $\phi(0) = \phi'(0) = 0$, $\phi''(0) = 1$.
8. $y'' - 9y = e^t$, $\phi(0) = 1$, $\phi'(0) = 0$.
9. $y'' - 9y = \sin t$, $\phi(0) = 1$, $\phi'(0) = 0$.
10. $y'' + 4y = \sin 2t$, $\phi(0) = 1$, $\phi'(0) = 1$.
11. $y'' - y = 0$, $\phi(1) = 0$, $\phi'(1) = 1$.

[*Hint:* Begin by making the change of independent variable $\tau = t - 1$ to move the initial time to zero.]

12. $y'' - 9y = f(t)$, $\phi(0) = 1$, $\phi'(0) = 0$, where f is in the class Λ. [*Hint:* Use the convolution.]

13. $y'' + 4y = f(t)$, $\phi(0) = 1$, $\phi'(0) = 1$, where f is in the class Λ.

14. Plot the gain and transfer functions of the linear differential operator defined by each of the following:
 (a) $L(y) = y'' + 4y' + 3y$.
 (b) $L(y) = y'' + 4y' + 4y$.
 (c) $L(y) = y'' + y$.

15. Find the steady-state solution of the differential equation $y'' + 9y = e^{2t}$.
16. Find the steady-state voltage in an electrical circuit with $L = 1$, $R = 10$, $C = \frac{1}{9}$ with an applied voltage $9\delta(t)$.

8.5 Applications to Linear Systems

The methods used in the previous section to obtain solutions of a linear differential equation with constant coefficients can be used with no essential change to obtain solutions of a linear system with constant coefficients. We shall not go into great detail in the solution of linear systems, but shall only give a few examples to illustrate the method. The applicability of the Laplace transform method to each of the systems considered follows from Theorem 2, Appendix 2.

Example 1. Find the solution $\phi = (\phi_1, \phi_2)$ of the system

$$y'_1 = 2y_1 + y_2$$
$$y'_2 = -y_1 + 4y_2,$$

which satisfies the initial conditions $\phi_1(0) = 0$, $\phi_2(0) = 1$. Also find a fundamental matrix. We let $Y_1(s) = \mathscr{L}\{\phi_1(t)\}$, $Y_2(s) = \mathscr{L}\{\phi_2(t)\}$. When we take Laplace transforms in the equations satisfied by ϕ_1, ϕ_2, we obtain

$$sY_1(s) - \phi_1(0) = 2Y_1(s) + Y_2(s)$$
$$sY_2(s) - \phi_2(0) = -Y_1(s) + 4Y_2(s)$$

or

$$(s-2)Y_1(s) - Y_2(s) = \phi_1(0) = 0$$
$$Y_1(s) + (s-4)Y_2(s) = \phi_2(0) = 1.$$

Solving for $Y_1(s)$ and $Y_2(s)$, and computing the inverse transform, we obtain successively

$$Y_1(s) = \frac{1}{(s-3)^2}, \quad Y_2(s) = \frac{s-2}{(s-3)^2} = \frac{1}{s-3} + \frac{1}{(s-3)^2};$$

8.5 Applications to Linear Systems

hence

$$\phi_1(t) = te^{3t}, \qquad \phi_2(t) = e^{3t} + te^{3t}.$$

To find a fundamental matrix we find the solution $\psi = (\psi_1, \psi_2)$ for which $\psi_1(0) = 1$, $\psi_2(0) = 0$. Proceeding as above we obtain

$$(s-2)Y_1(s) - Y_2(0) = \psi_1(0) = 1,$$
$$Y_1(s) + (s-4)Y_2(s) = \psi_2(0) = 0,$$

the solution of which is

$$Y_1(s) = \frac{s-4}{(s-3)^2} = \frac{1}{s-3} - \frac{1}{(s-3)^2},$$

$$Y_2(s) = \frac{-1}{(s-3)^2}.$$

Therefore, $\psi_1(t) = (1-t)e^{3t}$, $\psi_2(t) = te^{3t}$, and by Theorem 2, Section 6.3 a fundamental matrix is

$$\Phi(t) = (\phi(t), \psi(t)) = e^{3t}\begin{bmatrix} 1-t & t \\ -t & 1+t \end{bmatrix}.$$

The reader should compare this solution with that of Example 2, Section 7.5, for the initial condition imposed here.

● **EXERCISE**

1. Verify the solution obtained in Example 1 above by direct substitution.

By taking the Laplace transform in Example 1, we reduce the problem of solving a system of differential equations to the problem of solving a linear system of algebraic equations. This is a substantial simplification, as algebraic systems can be solved explicitly by an easily carried-out procedure if the order of the system is not too large.

Another type of problem for which the Laplace transform is very useful is a system of linear differential equations with constant coefficients of order higher than the first. While such a system can be reduced to a system of first-order equations by the methods of Chapter 6, it can also be solved directly by means of the Laplace transform without the need for this reduction.

294 The Laplace Transform

Example 2. Find the solution (ϕ_1, ϕ_2) of the system

$$y_1'' - 2y_1' - y_2' + 2y_2 = 0,$$
$$y_1' - 2y_1 + y_2' = -2e^{-t},$$

which satisfies the initial conditions $\phi_1(0) = 3$, $\phi_2'(0) = 2$, $\phi_2(0) = 0$. We let Y_1 and Y_2 be the Laplace transforms of ϕ_1 and ϕ_2, respectively. When we take Laplace transforms in the equations satisfied by ϕ_1 and ϕ_2, we obtain

$$[s^2 Y_1(s) - 3s - 2] - 2[sY_1(s) - 3] - sY_2(s) + 2Y_2(s) = 0$$

$$[sY_1(s) - 3] - 2Y_1(s) + sY_2(s) = -\frac{2}{s+1}$$

or

$$(s^2 - 2s)Y_1(s) - (s - 2)Y_2(s) = 3s + 2 - 6 = 3s - 4$$

$$(s - 2)Y_1(s) + sY_2(s) = -\frac{2}{s+1} + 3 = \frac{3s+1}{s+1}. \tag{8.43}$$

Solving the system (8.43), we obtain

$$Y_1(s) = \frac{3s^2 - 4s - 1}{(s+1)(s-1)(s-2)} = \frac{1}{s-1} + \frac{1}{s+1} + \frac{1}{s-2},$$

$$Y_2(s) = \frac{2}{(s+1)(s-1)} = \frac{1}{s-1} - \frac{1}{s+1}. \tag{8.44}$$

We take inverse transforms of (8.44) to obtain the solution

$$\phi_1(t) = e^t + e^{-t} + e^{2t}, \qquad \phi_2(t) = e^t - e^{-t}.$$

● **EXERCISE**

2. Verify the solution obtained in Example 2 by direct substitution.

To solve the problem of Example 2, we could proceed as follows. Transform the system to an equivalent system of three first-order equations, find a fundamental matrix for this system, and finally impose the initial conditions. It is clear that in most simple problems such as the one in Example 2, the use of Laplace transforms does give an answer more quickly.

8.5 Applications to Linear Systems

Example 3. Find the solution (ϕ_1, ϕ_2) of the system

$$y_1'' + 2y_1' + y_1 + y_2'' + y_2' = 0$$
$$y_1' + y_1 + y_2' = 0,$$

which satisfies the initial conditions $\phi_1(0) = 1$, $\phi_1'(0) = 0$, $\phi_2(0) = 1$, $\phi_2'(0) = 0$. We let Y_1 and Y_2 be the Laplace transforms of ϕ_1 and ϕ_2, respectively. When we take Laplace transforms in the equations satisfied by ϕ_1 and ϕ_2, we obtain

$$[s^2 Y_1(s) - s] + 2[s Y_1(s) - 1] + Y_1(s) + [s^2 Y_2(s) - s] +$$
$$+ [s Y_2(s) - 1] = 0$$
$$[s Y_1(s) - 1] + Y_1(s) + [s Y_2(s) - 1] = 0$$

or

$$(s^2 + 2s + 1) Y_1(s) + (s^2 + s) Y_2(s) = 2s + 3$$
$$(s + 1) Y_1(s) + s Y_2(s) = 2.$$
(8.45)

When we attempt to solve the algebraic system (8.45), we find that it is inconsistent! Hence, there is no solution to the given system of equations which satisfies the initial conditions.

The Laplace transform provides an alternate means, independent of the development of Chapter 7, for constructing a fundamental matrix for the system

$$\mathbf{y}' = A\mathbf{y},$$
(8.46)

where A is an arbitrary $n \times n$ constant matrix. This is accomplished in the following way.

We will say that if $\mathbf{f}(t)$ is a vector function with n components defined on $0 \leq t < \infty$, then $\mathbf{f} \in \Lambda$ if and only if each component of $\mathbf{f}$ is in the class Λ; we write $\mathscr{L}\{\mathbf{f}\} = \int_0^\infty \exp(-st)\mathbf{f}(t)\, dt$. The analog of formula (8.16) holds for vector functions (the proof is exactly the same as in the scalar case). It follows from Theorem 2, Appendix 2, that if $\boldsymbol{\phi}$ is a solution of (8.46) with $\boldsymbol{\phi}(0) = \boldsymbol{\eta}$, then $\boldsymbol{\phi} \in \Lambda^1$ for any initial vector $\boldsymbol{\eta}$. Let $\mathbf{Y}(s) = \mathscr{L}\{\boldsymbol{\phi}\}$. Taking Laplace transforms of both sides of (8.46) and using the initial condition, we obtain

$$s\mathbf{Y}(s) - \boldsymbol{\eta} = A\mathbf{Y}(s).$$

Thus,

$$(sI - A)\mathbf{Y}(s) = \boldsymbol{\eta}. \tag{8.47}$$

The system (8.47) is a linear nonhomogeneous system of n algebraic equations in n unknowns, namely the components $(Y_1(s), Y_2(s), \ldots, Y_n(s))$ of the vector $\mathbf{Y}(s)$. Clearly, if s is not equal to an eigenvalue of A, $\det(sI - A) \neq 0$, then (8.47) can be solved uniquely for $\mathbf{Y}(s)$ in terms of $\boldsymbol{\eta}$ and s by Cramer's rule (Theorem 4, Section 4.5). From it, since $\det(sI - A)$ is a polynomial of degree n, it is clear that $\mathbf{Y}(s)$ is a vector whose components are **rational functions of** s and linear in $(\eta_1, \eta_2, \ldots, \eta_n)$, the components of $\boldsymbol{\eta}$. Hence, each component of $\mathbf{Y}(s)$ can be expanded in partial fractions (the denominators will be integral powers of $(s - \lambda_j)$, where λ_j is an eigenvalue of A). Doing this we can then invert $\mathbf{Y}(s)$ to find the solution $\boldsymbol{\phi}(t)$ corresponding **to any initial vector** $\boldsymbol{\eta}$. Letting $\boldsymbol{\eta}$ successively take on the values

$$\boldsymbol{\eta}_1 = \begin{bmatrix} 1 \\ 0 \\ 1 \\ 0 \end{bmatrix}, \quad \boldsymbol{\eta}_2 = \begin{bmatrix} 0 \\ 1 \\ 0 \\ \vdots \\ 0 \end{bmatrix}, \ldots, \quad \boldsymbol{\eta}_n = \begin{bmatrix} 0 \\ \vdots \\ 0 \\ 1 \end{bmatrix}$$

(or any other n linearly independent constant vectors in $\mathscr{C}_n$ which form a basis), the solutions $\boldsymbol{\phi}_1, \boldsymbol{\phi}_2, \ldots, \boldsymbol{\phi}_n$ used as columns of the matrix Φ generate a fundamental matrix of $\mathbf{y}' = A\mathbf{y}$, such that $\Phi(0) = I$.

Example 4. Construct a fundamental matrix $\Phi(t)$ (for example, the one with $\Phi(0) = I$) for the system $\mathbf{y}' = A\mathbf{y}$, where

$$A = \begin{bmatrix} 3 & -1 & 1 \\ 2 & 0 & 1 \\ 1 & -1 & 2 \end{bmatrix}.$$

Compare your result with Example 3, Section 7.5.

We use the method outlined in the preceding paragraph. In that notation with $n = 3$, we have

$$s\mathbf{Y}(s) - \boldsymbol{\eta} = A\mathbf{Y}(s)$$

or

$$(sI - A)\mathbf{Y}(s) = \boldsymbol{\eta},$$

$$\begin{bmatrix} s-3 & 1 & -1 \\ -2 & s & -1 \\ -1 & 1 & s-2 \end{bmatrix} \begin{bmatrix} y_1(s) \\ y_2(s) \\ y_3(s) \end{bmatrix} = \begin{bmatrix} \eta_1 \\ \eta_2 \\ \eta_3 \end{bmatrix}.$$

Expanding $\det(sI - A)$ by the first row, we have

$$\det(sE - A) = (s - 3)[s(s - 2) + 1] + [12(s - 2) + 1]$$
$$- [-2 + s] = s^3 - 5s^2 + 8s - 4 = (s - 1)(s - 2)^2.$$

By Cramer's rule, we have

$$Y_1(s) = \frac{\det \begin{bmatrix} \eta_1 & 1 & -1 \\ \eta_2 & s & -1 \\ \eta_3 & 1 & s-2 \end{bmatrix}}{(s-1)(s-2)^2}$$

$$= \frac{\eta_1[s(s-2)+1] - \eta_2(s-2+1) + \eta_3(-1+s)}{(s-1)(s-2)^2}$$

$$= \frac{\eta_1(s-1) - \eta_2 + \eta_3}{(s-2)^2},$$

$$Y_2(s) = \frac{\det \begin{bmatrix} s-3 & \eta_1 & -1 \\ -2 & \eta_2 & -1 \\ -1 & \eta_3 & s-2 \end{bmatrix}}{(s-1)(s-2)^2}$$

$$= \frac{\eta_1(2s-3) + \eta_2(s^2 - 5s + 5) + \eta_3(s-1)}{(s-1)(s-2)^2},$$

$$Y_3(s) = \frac{\det \begin{bmatrix} s-3 & 1 & \eta_1 \\ -2 & s & \eta_2 \\ -1 & 1 & \eta_3 \end{bmatrix}}{(s-1)(s-2)^2} = \frac{\eta_1(s-2) - \eta_2(s-2) + \eta_3(s^2 - 3s + 2)}{(s-1)(s-2)^2}$$

$$= \frac{\eta_1 - \eta_2}{(s-1)(s-2)} + \frac{\eta_3}{s-2}.$$

It is convenient to substitute specific values for η_1, η_2, η_3 at this point, rather than waiting until after taking the inverse transform as was suggested in the general procedure. When we take $\eta_1 = 1, \eta_2 = 0, \eta_3 = 0$, we obtain

$$Y_1(s) = \frac{s-1}{(s-2)^2} = \frac{A}{s-2} + \frac{B}{(s-2)^2}.$$

Then $(s-1) = A(s-2) + B$, which gives $A = 1, B = 1$. Thus

$$Y_1(s) = \frac{1}{s-2} + \frac{1}{(s-2)^2} \quad \text{and} \quad y_1(t) = e^{2t} + te^{2t} = (1+t)e^{2t}.$$

Next,

$$Y_2(s) = \frac{2s-3}{(s-1)(s-2)^2} = \frac{A}{s-1} + \frac{B}{s-2} + \frac{C}{(s-2)^2},$$

and $2s - 3 = A(s-2)^2 + B(s-2)(s-1) + C(s-1)$, which gives $A = -1, B = 1, C = 1$. Thus,

$$Y_2(s) = \frac{-1}{s-1} + \frac{1}{s-2} + \frac{1}{(s-2)^2},$$

and

$$y_2(t) = -e^t + e^{2t} + te^{2t} = (t+1)e^{2t} - e^t.$$

Also,

$$Y_3(s) = \frac{1}{(s-1)(s-2)} = \frac{1}{s-2} - \frac{1}{s-1}$$

and

$$y_3(t) = e^{2t} - e^t.$$

Thus,

$$\boldsymbol{\phi}_1(t) = \begin{bmatrix} (1+t)e^{2t} \\ (t+1)e^{2t} - e^t \\ e^{2t} - e^t \end{bmatrix}$$

is that solution of $\mathbf{y}' = A\mathbf{y}$ for which $\boldsymbol{\phi}_1(0) = \mathbf{e}_1 = [1, 0, 0]$.

8.5 Applications to Linear Systems

Now, consider the case $\eta_1 = 0$, $\eta_2 = 1$, $\eta_3 = 0$, then

$$Y_1(s) = -\frac{1}{(s-2)^2} \quad \text{and} \quad y_1(t) = -te^{2t},$$

$$Y_2(s) = \frac{s^2 - 5s + 5}{(s-1)(s-2)^2} = \frac{A}{s-1} + \frac{B}{s-2} + \frac{C}{(s-2)^2},$$

with

$$s^2 - 5s + 5 = A(s-2)^2 + B(s-1)(s-2) + C(s-1)$$

which implies $1 = A$, $A + B = 1$, so that $B = 0$, $-1 = C$.

$$Y_2(s) = \frac{1}{s-1} - \frac{1}{(s-2)^2}$$

and

$$y_2(t) = e^t - te^{2t}.$$

$$Y_3(s) = \frac{-1}{(s-1)(s-2)} = \frac{1}{s-1} - \frac{1}{s-2}$$

and

$$y_3(t) = e^t - e^{2t}.$$

Thus

$$\phi_2(t) = \begin{bmatrix} -te^{2t} \\ e^t - te^{2t} \\ e^t - e^{2t} \end{bmatrix}$$

$\phi_2(0) = e_2 = [0, 1, 0]$. Finally, taking $\eta_1 = \eta_2 = 0$, $\eta_3 = 1$, we have

$$Y_1(s) = \frac{1}{(s-2)^2} \quad \text{and} \quad y_1(t) = te^{2t},$$

$$Y_2(s) = \frac{1}{(s-2)^2} \quad \text{and} \quad y_2(t) = te^{2t},$$

$$Y_3(s) = \frac{1}{s-2} \quad \text{and} \quad y_3(t) = e^{2t}.$$

Thus,

$$\phi_3(t) = \begin{bmatrix} te^{2t} \\ te^{2t} \\ e^{2t} \end{bmatrix}$$

is that solution of $\mathbf{y}' = A\mathbf{y}$ for which $\boldsymbol{\phi}_3(0) = \mathbf{e}_3$. Therefore,

$$\Phi(t) = [\boldsymbol{\phi}_1(t), \boldsymbol{\phi}_2(t), \boldsymbol{\phi}_3(t)] = \begin{bmatrix} (1+t)e^{2t} & -te^{2t} & te^{2t} \\ (1+t)e^{2t} - e^t & e^t - te^{2t} & te^{2t} \\ e^{2t} - e^t & e^t - e^{2t} & e^{2t} \end{bmatrix}$$

is the fundamental matrix which is the identity at $t = 0$.

● **EXERCISES**

Find the solution $[\phi_1, \phi_2]$ of each of the following systems of equations which satisfies the given initial conditions

3. $y_1' + y_2' = 0$, $\phi_1(0) = 1$, $\phi_2(0) = 0$.
   $y_1' - y_2' = 1$,

4. $y_1' + 3y_2' = 0$, $\phi_1(0) = 1$, $\phi_2(0) = -1$.
   $y_1' - y_1 + 2y_2' = 0$,

5. $y_1'' + 3y_1' + 2y_1 + y_2' + y_2 = 0$, $y_1' + 2y_1 + y_2' - y_2 = 0$,
   $\phi_1(0) = 1$, $\phi_1'(0) = -1$, $\phi_2(0) = 0$.

6. $y_1'' - 3y_1' + 2y_1 + y_2'' - 2y_2' = 0$, $y_1' - y_1 + y_2' = 0$,
   $\phi_1(0) = 1$, $\phi_1'(0) = 1$, $\phi_2(0) = 1$, $\phi_2'(0) = 0$.

7. (a) Solve the system (1.18), (1.19) of Example 7, Chapter 1.
   (b) Solve the system (1.23) of Example 8, Chapter 1.

8. A mechanical system consisting of two coupled springs, assuming no friction and a periodic external force, is governed by a system of differential equations.

$$m_1 y_1'' + k_1 y_1 - k_2(y_2 - y_1) = 0$$
$$m_2 y_2'' + k_2(y_2 - y_1) = A \cos \omega t,$$

where m_1, m_2, k_1, k_2, and A are positive constants. Find the motion of the system corresponding to the solution (ϕ_1, ϕ_2), where $\phi_1(0) = 1$, $\phi_1'(0) = 0$, $\phi_2(0) = 0$, $\phi_2'(0) = 0$.

9. For a system as in Exercise 8 with $k_1 = k_2 = k$, $A = 0$, define the kinetic energy $T = \frac{1}{2} m_1 y_1'^2 + \frac{1}{2} m_2 y_2'^2$ and the potential energy $V = (k/2)(y_1^2 - 2y_1 y_2 + y_2^2))$. Show that the total energy $T + V$ is constant. [*Hint:* Show that $(d/dt\{T(t) + V(t)\}) = 0$, using the differential equations.]

10. Construct a fundamental matrix for the systems $\mathbf{y}' = A\mathbf{y}$, where A is each of the following matrices

(a) $A = \begin{bmatrix} 1 & 2 \\ 4 & 3 \end{bmatrix}$.

(b) $A = \begin{bmatrix} 2 & -3 \\ 3 & -4 \end{bmatrix}$.

(c) $A = \begin{bmatrix} 1 & 0 & 3 \\ 8 & 1 & -1 \\ 5 & 1 & -1 \end{bmatrix}$.

(d) $A = \begin{bmatrix} 3 & -1 & -4 & 2 \\ 2 & 3 & -2 & -4 \\ 2 & -1 & -3 & 2 \\ 1 & 2 & -1 & -3 \end{bmatrix}$.

8.6 A Table of Laplace Transforms

We present here a brief table of Laplace transforms, with references to this chapter where the transforms listed are derived. It is understood that $F(s)$ denotes the Laplace transform of $f(t)$, that $G(s)$ denotes the Laplace transform of $g(t)$, and that in transforms involving an arbitrary function f or g, the functions involved belong to a suitable class to guarantee the existence of the transform.

FUNCTION	TRANSFORM	REFERENCE
1	$\dfrac{1}{s}$	(8.2)
e^{zt}	$\dfrac{1}{s-z}$	(8.3)
$\cos \beta t$	$\dfrac{s}{s^2 + \beta^2}$	(8.7)
$\sin \beta t$	$\dfrac{\beta}{s^2 + \beta^2}$	(8.7)
$e^{\alpha t} \cos \beta t$	$\dfrac{s - \alpha}{(s - \alpha)^2 + \beta^2}$	(8.6)

(*table continued*)

FUNCTION	TRANSFORM	REFERENCE
$e^{\alpha t} \sin \beta t$	$\dfrac{\beta}{(s-\alpha)^2 + \beta^2}$	(8.6)
t^k	$\dfrac{k!}{s^{k+1}}$	(8.11)
$t^k e^{zt}$	$\dfrac{k!}{(s-z)^{k+1}}$	(8.12)
$t^k f(t)$	$(-1)^k F^{(k)}(s)$	(8.10)
$t^k \cos \beta t$	$\dfrac{k!\,(s+i\beta)^{k+1}}{[(s-\alpha)^2 + \beta^2]^{k+1}}$	(8.14)
$t^k \sin \beta t$	$\dfrac{k!\,(s+i\beta)^{k+1}}{[(s-\alpha)^2 + \beta^2]^{k+1}}$	(8.14)
$t^k e^{\alpha t} \cos \beta t$	$\dfrac{k!\,[(s-\alpha)+i\beta]^{k+1}}{[(s-\alpha)^2 + \beta^2]^{k+1}}$ *	(8.13)
$t^k e^{\alpha t} \sin \beta t$	$\dfrac{k!\,[(s-\alpha)+i\beta]^{k+1}}{[(s-\alpha)^2 + \beta^2]^{k+1}}$ *	(8.13)
$e^{at} f(t)$	$F(s-a)$	Theorem 3, Section 8.2
$f'(t)$	$sF(s) - f(0)$	(8.15)
$f^{(J)}(t)$	$s^J F(s) - s^{J-1} f(0) - \cdots - f^{(J-1)}(0)$	(8.16)
$\int_0^t f(t-u)g(u)\,du$	$F(s)G(s)$	Theorem 1, Section 8.3
$\int_0^t f(u)\,du$	$F(s)/s$	

* For the purpose of carrying out this calculation, assume that s is real, and then let s be complex again after obtaining the answer.

Formulas in the table involving $k!$ hold when k is not an integer if $k!$ is replaced by $\Gamma(k+1)$.

• MISCELLANEOUS EXERCISES

1. Find the Laplace transform of each of the following functions

(a) $f(t) = \begin{cases} \sin \omega t, & 0 \leq t \leq \dfrac{\pi}{\omega} \\ 0, & \dfrac{\pi}{\omega} < t \leq \dfrac{2\pi}{\omega}, \end{cases}$

and

$$f\left(t+\frac{2\pi}{\omega}\right)=f(t) \quad \text{for } 0 \leq t < \infty.$$

(b) $\int_0^t e^{-u} \sin(t-u)\, du$

(c) $t^{5/2}$.

(d) $t^2 \sin \omega t$.

2. Find the solution of each of the following initial value problems by means of Laplace transforms.

(a) $y'' + 4y' + 4y = 4 \cos 2t$, $y(0) = 2$, $y'(0) = 5$.

(b) $y'' + \omega^2 y = f(t)$, where f is the function in Exercise 1(a) above, $y(0) = 0$, $y'(0) = 0$.

(c) $y'' + y = H(t)$, where $H(t) = 3$ and $0 \leq t \leq 4$, $H(t) = 2t - 5$ and $t \geq 4$, $y(0) = 1$, $y'(0) = 0$.

(d) $y'' + 2y' + y = 2 + (t-3)U(t-3)$, where U is the unit step function

$$U(t) = \begin{cases} 0 & (t < 0) \\ 1 & (t \geq 0) \end{cases} \quad y(0) = 2,\ y'(0) = 1$$

3. (a) Show that if $f(t)$ belongs to the class Λ^1 and has Laplace transform $F(s)$, then $\lim_{s \to \infty} sF(s) = f(0)$. [*Hint:* Use Theorem 4, Section 8.2 and Theorem 3, Section 8.3.]

(b) Generalize this to a result for functions in the class Λ^k by using Theorem 5, Section 8.2 and Theorem 3, Section 8.3.

4. (a) Show that if $\phi(t)$ is a solution of the Bessel equation of index zero,

$$ty'' + y' + ty = 0$$

then the Laplace transform $Y(s)$ of $\phi(t)$ satisfies the first-order differential equation

$$(s+1)\frac{dY}{ds} + sY(s) = 0$$

regardless of the initial conditions prescribed.

(b) Solve the equation obtained in part (a) and use Exercise 3 above to show that $Y(s) = \phi(0)/(s^2+1)^{1/2}$. Explain why $\phi'(0)$ cannot be prescribed.

(c) By expanding $Y(s)$ in powers of $1/s$, show that

$$Y(s) = \phi(0) \sum_{k=0}^{\infty} (-1)^k \frac{1 \cdot 3 \cdot 5 \cdots (2k-1)}{2^k k!} \frac{1}{s^{2k+1}}.$$

(d) Show that

$$\phi(t) = \phi(0) \sum_{k=0}^{\infty} (-1)^k \frac{t^{2k}}{2^{2k}(k!)^2}.$$

Note that the step from (c) to (d) is purely formal. However, (d) is a "rigorous" solution of the Bessel equation, as can be justified by direct substitution and theorems on power series.

5. Solve Exercise 1, Miscellaneous Exercises for Chapter 7, by Laplace transforms.

6. Solve Exercise 3, Miscellaneous Exercises for Chapter 7, by Laplace transforms.

7. Solve Exercises 8 and 9, Miscellaneous Exercises for Chapter 7, by Laplace transforms.

8. Each of the following equations defines a function $\phi(t)$. Find $\phi(t)$ by using Laplace transforms. (Note that before we do this we should actually first prove that $\phi \in \Lambda$. This can, in fact, be done by estimating $|\phi(t)|$ and using Lemma 6, Appendix 2. Alternatively, we could just obtain the answer and then verify that if belongs to Λ.)

(a) $\phi(t) = 4t^2 - \int_0^t \phi(t-\tau)^{2-t} \, d\tau.$

(b) $\phi(t) = t^3 + \int_0^t \phi(\tau) \sin(t-\tau) \, d\tau.$

(c) $\phi(t) = 1 + 2\int_0^t \phi(t) - \tau) \cos \tau \, d\tau.$

(d) $\phi'(t) = \sin t + \int_0^t \phi(t-\tau) \cos \tau \, d\tau, \qquad \phi(0) = 0.$

(e) $\phi'(t) = t + \int_0^t \phi(t-\tau) \cos \tau \, d\tau, \qquad \phi(0) = 4.$

(f) $\int_0^t \phi(t-\tau)e^{-\tau} \, d\tau = t.$

Appendix 1 | THE EXPONENTIAL MATRIX

In order to justify the definition of exp M, where M is any matrix in $\mathscr{F}_{nn}$, we need the concept of convergence of a sequence of matrices. To measure distances between numbers, we use absolute values. To measure distances between matrices, we introduce the concept of **the norm of a matrix**. For the norm of a vector the reader should consult Appendix 2.

We define the **norm** (length) **of a matrix** A, denoted by $\|A\|$, by

$$\|A\| = \sum_{i,j=1}^{n} |a_{ij}|, \tag{1}$$

that is, as the sum of the absolute values of all the elements. We readily verify that the matrix norm satisfies the following properties:

(i) $\|A + B\| \le \|A\| + \|B\|$
(ii) $\|AB\| \le \|A\| \cdot \|B\|$

for matrices A, B of complex numbers. The above norm is convenient for our purposes; other matrix norms satisfying the properties (i) and (ii) are possible.

● **EXERCISE**

1. Prove properties (i) and (ii) of $\|A\|$.

We now use the matrix norm (1) to define the concept of convergence of a sequence of matrices.

Definition. *The sequence of $\{A^{(k)}\}$ converges to the matrix A, where $A^{(k)}$ and A belong to $\mathscr{F}_{nn}$ if and only if the sequence of real numbers $\{\|A - A^{(k)}\|\}$ has limit zero, and in this case we write*

$$\{A^{(k)}\} \to A \quad \text{or} \quad \lim_{k \to \infty} A^{(k)} = A.$$

Clearly, because of the definition of the norm, this means that $\{A^{(k)}\} \to A$ if and only if the sequence $\{a_{ij}^{(k)}\}$ of complex numbers, representing the element in the ith row and jth column in the matrices $\{A^{(k)}\}$, converges to the element a_{ij} of the matrix A as $k \to \infty$, for each of the n^2 elements $(i, j = 1, \ldots, n)$. To prove this, note that $|a_{ij}^{(k)} - a_{ij}| \le \|A^{(k)} - A\|$ for $i, j = 1, \ldots, n$.

A matrix function $A(t)$ is a correspondence that assigns to each point t of an interval I one and only one $n \times n$ matrix $A(t)$. Using the remark following the definition of convergence of a sequence of matrices, we see that it is consistent to say that a matrix function $A(t)$ is continuous, differentiable, or integrable on an interval I if and only if each of its n^2 elements $a_{ij}(t)$ is continuous, differentiable, or integrable, respectively, on I. This is precisely how these concepts were defined in Definitions 1 and 2 in Section 6.1. *We say that a series $\sum_{k=0}^{\infty} U_k$ of matrices converges if and only if the sequence $\{\sum_{k=0}^{n} U_k\}$ of partial sums converges.* The limit of this sequence of partial sums is called the **sum** of the series.

Combining the definition of convergence of a sequence of matrices with the Cauchy criterion for sequences of real or complex numbers, we can establish the following result:

Lemma 1. *A sequence $\{A_k\}$ of matrices converges if and only if given a number $\varepsilon > 0$, there exists an integer $N = N(\varepsilon) > 0$ such that $\|A_m - A_p\| < \varepsilon$ whenever $m, p > N$.*

We are now ready to prove that the definition

$$\exp M = I + M + \frac{M^2}{2!} + \cdots + \frac{M^j}{j!} + \cdots$$

makes sense for every matrix $M \in \mathscr{F}_{ni}$ as follows: We define the partial sums

$$S_k = I + \frac{M}{1!} + \frac{M^2}{2!} + \cdots + \frac{M^k}{k!}. \tag{2}$$

Using the matrix norm, we have for $m > p$

$$\|S_m - S_p\| = \left\|\sum_{k=p+1}^{m} \frac{M^k}{k!}\right\| \leq \sum_{k=p+1}^{m} \frac{\|M\|^k}{k!}. \tag{3}$$

- **EXERCISE**

    2. Use the properties of the matrix norm to justify the calculation (3).

[Note that the calculation is possible since the sums in (2) are finite.] Since, for any matrix M, $\|M\|$ is a real number, we can form

$$e^{\|M\|} = \sum_{k=0}^{\infty} (\|M\|^k/k!).$$

Hence, the sum

$$\sum_{k=0}^{m} (\|M\|^k/k!)$$

is a partial sum of a series of positive real numbers that is known to converge. By the Cauchy criterion for series of real numbers, we see that given $\varepsilon > 0$ we can choose an integer $N > 0$ such that the right-hand side of (3) is less than ε whenever $m, p > N$. By (3),

$$\|S_m - S_p\| < \varepsilon \quad \text{for } m, p > N.$$

By Lemma 1, the series converges, and thus $\exp M$ is well defined for every matrix M in $\mathscr{F}_{nn}$.

Appendix 2

THE EXISTENCE AND UNIQUENESS THEOREM FOR LINEAR SYSTEMS OF DIFFERENTIAL EQUATIONS

In order to prove Theorem 1, Section 6.2, we first need to familiarize ourselves with the properties of the **norm of a vector**, which will be needed to measure distances between vectors. In Appendix 1, we defined the norm of a matrix. For a vector in $\mathscr{F}_n$, the definition is similar: if $\mathbf{v} \in \mathscr{F}_n$, we define the norm of $\mathbf{v}$, denoted by $\|\mathbf{v}\|$, by

$$\|\mathbf{v}\| = \sum_{j=1}^{n} |v_j|,$$

that is the sum of the absolute values of all the components. If, for example,

$$\mathbf{v} = \begin{bmatrix} 1 \\ -2 \\ 1+i \end{bmatrix},$$

then $\|\mathbf{v}\| = |1| + |-2| + |1+i| = 1 + 2 + \sqrt{2} = 3 + \sqrt{2}$.

It is easy to prove that the vector norm satisfies the following properties: if $\mathbf{u}, \mathbf{v} \in \mathscr{F}_n$ and c is any complex scalar, then

(i) $\|\mathbf{v}\| \geq 0$; and $\|\mathbf{v}\| = 0$ if and only if $\mathbf{v} = \mathbf{0}$;

(ii) $\|c\mathbf{v}\| = |c| \|\mathbf{v}\|$;

(iii) $\|\mathbf{u} + \mathbf{v}\| \leq \|\mathbf{u}\| + \|\mathbf{v}\|$.

Property (iii) is called the **triangle inequality**. To prove, it we note that

$$\|\mathbf{u} + \mathbf{v}\| = \sum_{j=1}^{n} |u_j + v_j| \le \sum_{j=1}^{n} (|u_j| + |v_j|) = \sum_{j=1}^{n} |u_j| + \sum_{j=1}^{n} |v_j|$$
$$= \|\mathbf{u}\| + \|\mathbf{v}\|,$$

where the second step depends on the triangle inequality for complex numbers, $|u_j + v_j| \le |u_j| + |v_j|$.

The proofs of (i) and (ii) are equally simple. By mathematical induction, one easily proves the **extended triangle inequality**:

(iv) $\quad \left\| \sum_{j=1}^{n} \mathbf{u}_j \right\| \le \sum_{j=1}^{m} \|\mathbf{u}_j\|$

for any vectors $\mathbf{u}_1, \mathbf{u}_2, \ldots, \mathbf{u}_n \in \mathscr{F}_n$.

There is an inequality which connects the matrix norm and vector norm, namely

(v) $\quad \|A\mathbf{v}\| \le \|A\| \, \|\mathbf{v}\|$

for any $A \in \mathscr{F}_{nn}$ and $\mathbf{v} \in \mathscr{F}_n$.

● **EXERCISES**

    1. Prove the extended triangle inequality.
    2. Prove property (v). [*Hint:* To understand its significance, the reader might write out the case $A \in \mathscr{F}_{22}$, $\mathbf{v} \in \mathscr{F}_2$.]

The above procedure is not the only way to define the norm. For example the Euclidean norm of a vector $\mathbf{v}$ in $\mathscr{F}_n$ is defined to be

$$\|\mathbf{v}\|_E = [|v_1|^2 + |v_2|^2 + \cdots + |v_n|^2]^{1/2}$$

which reduces to the familiar Euclidean length in the plane $\mathscr{R}_2$ by virtue of the Pythagorean Theorem. However, the norm $\|\mathbf{v}\|$ defined previously is more convenient for our purposes.

The **distance $d(\mathbf{y}, \mathbf{z})$ between the two vectors $\mathbf{y}$ and $\mathbf{z}$** is defined by the relation

$$d(\mathbf{y}, \mathbf{z}) = \|\mathbf{y} - \mathbf{z}\|,$$

and we have immediately that the **distance function** d satisfies the following properties:

(i) $d(\mathbf{y}, \mathbf{z}) \geq 0$ and $d(\mathbf{y}, \mathbf{z}) = 0$ if and only if $\mathbf{y} = \mathbf{z}$;

(ii) $d(\mathbf{y}, \mathbf{z}) = d(\mathbf{z}, \mathbf{y})$;

(iii) $d(\mathbf{y}, \mathbf{z}) \leq d(\mathbf{y}, \mathbf{u}) + d(\mathbf{u}, \mathbf{z})$;

for any vectors $\mathbf{u}, \mathbf{y}, \mathbf{z} \in \mathscr{F}_n$.

● **EXERCISE**

3. Prove properties (i), (ii), and (iii) of the distance function d.

We shall need the concept of convergence of a sequence $\{\mathbf{v}_m\}_{m=0}^{\infty}$, where $\mathbf{v}_m \in \mathscr{F}_n$ for $m = 1, 2, \ldots$.

Definition *The sequence $\{\mathbf{v}_m\}$ of vectors in $\mathscr{F}_n$ is said to converge to a vector $\mathbf{v}$ in $\mathscr{F}_n$ if and only if*

$$\lim_{m \to \infty} d(\mathbf{v}_m, \mathbf{v}) = \lim_{m \to \infty} \|\mathbf{v}_m - \mathbf{v}\| = 0.$$

If

$$\mathbf{v}_m = \begin{bmatrix} v_1^m \\ v_2^m \\ \vdots \\ v_n^m \end{bmatrix} \quad \text{and} \quad \mathbf{v} = \begin{bmatrix} v_1 \\ v_2 \\ \vdots \\ v_n \end{bmatrix},$$

then $\|\mathbf{v}_m - \mathbf{v}\| = |v_1^m - v_1| + |v_2^m - v_2| + \cdots + |v_n^m - v_n|$. Thus, $\{\mathbf{v}_m\} \to \mathbf{v}$ if and only if every component v_j^m of the vector $\mathbf{v}_m$ approaches the corresponding component v_j of the vector $\mathbf{v}$ for $j = 1, 2, \ldots, n$. Since the components of $\mathbf{v}_m$ and $\mathbf{v}$ are complex numbers, it is clear that all properties of limits of sequences of complex numbers must hold for sequences of vectors. The same statements apply to sequences of **vector functions**. For example, if $\mathbf{g}(t)$ is a vector function defined on some interval $\mathscr{I}$, then first

$$\|\mathbf{g}(t)\| = |g_1(t)| + |g_2(t)| + \cdots + |g_n(t)|$$

$$= \sum_{j=1}^{n} |g_j(t)|,$$

where g_j for $j = 1, 2, \ldots, n$ are the n components of $\mathbf{g}$. A sequence $\{\mathbf{g}_m(t)\}$ of vector functions is said to *converge* to the vector function **g**(t) on the interval $\mathscr{I}$ if and only if

$$\lim_{m \to \infty} d(\mathbf{g}_m(t), \mathbf{g}(t)) = \lim_{m \to \infty} \|\mathbf{g}_m(t) - \mathbf{g}(t)\| = 0$$

for every t in $\mathscr{I}$. It is also clear that the sequence $\{\mathbf{g}_m(t)\}$ converges to the vector function $\mathbf{g}(t)$ if and only if every component of $\{\mathbf{g}_m(t)\}$ converges to the corresponding component of $\mathbf{g}(t)$. The careful reader will have noticed that the preceding definition can be used to give precise definitions of continuity, differentiability, and integrability of vector functions, rather than the informal ones given in Definitions 1, 2, and 3 in Section 6.1. These definitions are the same as those for scalar functions, with appropriate absolute values replaced by norms. For example, a vector function **g** defined on an interval $\mathscr{I}$ is continuous at a point t_0 in $\mathscr{I}$ if and only if for every number $\varepsilon > 0$ there exists a number $\delta > 0$ such that

$$d(\mathbf{g}(t), \mathbf{g}(t_0)) = \|\mathbf{g}(t) - \mathbf{g}(t_0)\| < \varepsilon$$

for all t in $\mathscr{I}$ with $|t - t_0| < \delta$. Similarly, such a vector function **g** is differentiable at t_0 if and only if

$$\lim_{h \to 0} \frac{\mathbf{g}(t_0 + h) - \mathbf{g}(t_0)}{h}$$

exists, and if it does, we denote this limit by $\mathbf{g}'(t_0)$.

We remark that if $\mathbf{g}(t)$ is continuous at a point, or on an interval, then so is $\|\mathbf{g}(t)\|$, because by definition every component of $\mathbf{g}(t)$ is continuous; hence the absolute value of every component is continuous; hence the sum of absolute values of the components is continuous.

We shall also need the easily proven inequality

$$\left\| \int_a^b \mathbf{f}(t)\, dt \right\| \leq \int_a^b \|\mathbf{f}(t)\|\, dt \qquad (1)$$

for every $b > a$ and every continuous vector function **f** on the interval $a \leq t \leq b$.

● **EXERCISE**

4. Prove the inequality (1). [*Hint:*

$$\left\| \int_a^b \mathbf{f}(t)\, dt \right\| = \left| \int_a^b f_1(t)\, dt \right| + \left| \int_a^b f_2(t)\, dt \right| + \cdots + \left| \int_a^b f_n(t)\, dt \right|.$$

Now, use the fact that

$$\left|\int_a^b f_j(t)\,dt\right| \leq \int_a^b |f_j(t)|\,dt, \qquad (j=1,\cdots,n),$$

and add up.]

We are now ready to prove the fundamental existence and uniqueness theorem for linear systems of differential equations.

Theorem 1. *Let $A(t)$ be a continuous $n \times n$ matrix, and let $\mathbf{g}(t)$ be a continuous vector with n components on some interval $\mathscr{I}$. Then for every t_0 in $\mathscr{I}$ and every constant vector $\mathbf{y}_0$, the initial value problem*

$$\mathbf{y}' = A(t)\mathbf{y} + \mathbf{g}(t), \qquad \mathbf{y}(t_0) = \mathbf{y}_0 \tag{2}$$

has a unique solution existing on the interval $\mathscr{I}$.

We divide the proof up into several steps. We first convert problem (2) to an equivalent problem for which we will prove the desired result.

Lemma 1. *The initial value problem (2) is equivalent to the problem of finding a continuous vector function $\boldsymbol{\phi}$ such that*

$$\boldsymbol{\phi}(t) = \mathbf{y}_0 + \int_{t_0}^t [A(s)\boldsymbol{\phi}(s) + \mathbf{g}(s)]\,ds$$

for every t in $\mathscr{I}$ (that is, problem (2) is equivalent to solving the equation

$$\mathbf{z}(t) = \mathbf{y}_0 + \int_{t_0}^t [A(s)\mathbf{z}(s) + \mathbf{g}(s)]\,ds, \tag{3}$$

which is called an **integral equation** because the unknown function appears under the integral sign).

Proof. Let $\boldsymbol{\phi}$ be a solution of the initial value problem (2). Then for all t for which $\boldsymbol{\phi}(t)$ exists, we have (by the definition of solution),

$$\boldsymbol{\phi}'(t) = A(t)\boldsymbol{\phi}(t) + \mathbf{g}(t), \qquad \boldsymbol{\phi}(t_0) = \mathbf{y}_0.$$

Integrating this equation from t_0 to t, we obtain (a continuous vector function can be integrated)

$$\boldsymbol{\phi}(t) - \boldsymbol{\phi}(t_0) = \int_{t_0}^t (A(s)\boldsymbol{\phi}(s) + \mathbf{g}(s))\,ds$$

or φ satisfies

$$\phi(t) = \mathbf{y}_0 + \int_{t_0}^{t} (A(s)\phi(s) + \mathbf{g}(s))\, ds; \qquad (4)$$

that is, φ is a solution of (3). Conversely, suppose φ is a continuous solution of (3). Then φ satisfies (4), and by the fundamental theorem of calculus

$$\phi'(t) = A(t)\phi(t) + \mathbf{g}(t);$$

moreover, from (4), $\phi(t_0) = \mathbf{y}_0$. Thus, φ is a solution of (2). ∎

We shall now solve the integral equation (3) by the so-called **method of successive approximations.** Let I be any closed finite subinterval of the given interval $\mathscr{I}$ (if $\mathscr{I}$ is closed, we may take $I = \mathscr{I}$), such that $t_0 \in I$. We define the vector functions $\phi_0, \phi_1, \phi_2, \ldots$ (successive approximations) as follows

$$\begin{aligned}
\phi_0(t) &= \mathbf{y}_0, \\
\phi_1(t) &= \mathbf{y}_0 + \int_{t_0}^{t} [A(s)\phi_0(s) + \mathbf{g}(s)]\, ds, \\
\phi_2(t) &= \mathbf{y}_0 + \int_{t_0}^{t} [A(s)\phi_1(s) + \mathbf{g}(s)]\, ds, \\
&\vdots \\
\phi_{m+1}(t) &= \mathbf{y}_0 + \int_{t_0}^{t} [A(s)\phi_m(s) + \mathbf{g}(s)]\, ds, \\
&\vdots
\end{aligned} \qquad (5)$$

We claim first that each vector function in the sequence $\{\phi_m\}$ is well-defined and continuous on the interval I. We proceed by induction. Clearly, ϕ_0 is well defined and continuous on I because it is a constant vector. Suppose that the vector function ϕ_m is well defined and continuous on I. Then $A(s)\phi_m(s) + \mathbf{g}(s)$ is a continuous vector on I, and its integral from t_0 to t is a continuous vector on I; hence, from (5), $\phi_{m+1}(t)$ is a continuous vector on I. By the principle of finite induction, each member of the sequence is well defined and continuous on I.

We next claim that the sequence of vector functions $\{\phi_m\}$, defined inductively by (5), converges to a vector function φ which is continuous on I and which satisfies the integral equation (3). Once we have done this, Lemma 1 shows that the function φ is also a solution of the given initial value problem (2).

Appendix 2

To turn to the proof of convergence of the sequence $\{\boldsymbol{\phi}_m\}$, we note first that for every j, the identity

$$\begin{aligned}\boldsymbol{\phi}_j(t) &= \boldsymbol{\phi}_0(t) + [\boldsymbol{\phi}_1(t) - \boldsymbol{\phi}_0(t)] + [\boldsymbol{\phi}_2(t) - \boldsymbol{\phi}_1(t)] \\ &\quad + \cdots + [\boldsymbol{\phi}_j(t) - \boldsymbol{\phi}_{j-1}(t)] \\ &= \boldsymbol{\phi}_0(t) + \sum_{m=0}^{j-1} [\boldsymbol{\phi}_{m+1}(t) - \boldsymbol{\phi}_m(t)]\end{aligned} \tag{6}$$

holds. From (6), it is obvious that the sequence $\{\boldsymbol{\phi}_j\}$ converges to a vector function Φ on I if and only if its sequence of partial sums (6) converges (in the sense of the norm, of course). We will prove the convergence of the sequence by obtaining a "good estimate" of the quantity $\|\boldsymbol{\phi}_{m+1}(t) - \boldsymbol{\phi}_m(t)\|$.

From the first two relations in (5), we have

$$\boldsymbol{\phi}_1(t) - \boldsymbol{\phi}_0(t) = \int_{t_0}^t [A(s)\boldsymbol{\phi}_0(s) + \mathbf{g}(s)]\, ds.$$

By properties of the norm for matrices (see Appendix 1) and vectors and using the inequality (see Exercise 4)

$$\left\| \int_{t_0}^t \mathbf{f}(s)\, ds \right\| \leq \int_{t_0}^t \|\mathbf{f}(s)\|\, ds$$

for any continuous vector $\mathbf{f}$ we have for $t > t_0$

$$\begin{aligned}\|\boldsymbol{\phi}_1(t) - \boldsymbol{\phi}_0(t)\| &\leq \left\| \int_{t_0}^t [A(s)\boldsymbol{\phi}_0(s) + \mathbf{g}(s)]\, ds \right\| \\ &\leq \int_{t_0}^t \|A(s)\boldsymbol{\phi}_0(s) + \mathbf{g}(s)\|\, ds \\ &\leq \int_{t_0}^t \|A(s)\|\, \|\boldsymbol{\phi}_0(s)\|\, ds + \int_{t_0}^t \|\mathbf{g}(s)\|\, ds.\end{aligned}$$

Let $\alpha = \max_{s \in I} \|A(s)\|$, $\beta = \max_{s \in I} \|\mathbf{g}(s)\|$. Since A, $\mathbf{g}$ are continuous, so are $\|A(s)\|$ and $\|\mathbf{g}(s)\|$, and α and β exist because the interval I is closed. Thus,

$$\begin{aligned}\|\boldsymbol{\phi}_1(t) - \boldsymbol{\phi}_0(t)\| &\leq \int_{t_0}^t \alpha \|\boldsymbol{\phi}_0(s)\|\, ds + \int_{t_0}^t \beta\, ds \\ &= (\alpha \|\mathbf{y}_0\| + \beta)(t - t_0) \qquad \text{for } t > t_0.\end{aligned} \tag{7}$$

The Existence and Uniqueness Theorem

Let $\gamma = \alpha \|y_0\| + \beta$ and notice that the constant $\gamma > 0$ is completely determined by the given matrix A, the given vector $\mathbf{g}$ and the given constant vector $\mathbf{y}_0$. From (5) again we have

$$\boldsymbol{\phi}_2(t) - \boldsymbol{\phi}_1(t) = \int_{t_0}^{t} A(s)[\boldsymbol{\phi}_1(s) - \boldsymbol{\phi}_0(s)] \, ds.$$

Hence, taking norms and using (7), we have

$$\|\boldsymbol{\phi}_2(t) - \boldsymbol{\phi}_1(t)\| = \left\| \int_{t_0}^{t} A(s)[\boldsymbol{\phi}_1(s) - \boldsymbol{\phi}_0(s)] \, ds \right\|$$

$$\leq \int_{t_0}^{t} \|A(s)\| \, \|\boldsymbol{\phi}_1(s) - \boldsymbol{\phi}_0(s)\| \, ds$$

$$\leq \int_{t_0}^{t} \alpha(\gamma)(s - t_0) \, ds \qquad (8)$$

$$= \frac{\alpha \gamma}{2}(t - t_0)^2 \qquad \text{for } t > t_0.$$

Doing this, once more we find, using (8)

$$\|\boldsymbol{\phi}_3(t) - \boldsymbol{\phi}_2(t)\| = \left\| \int_{t_0}^{t} A(s)[\boldsymbol{\phi}_2(s) - \boldsymbol{\phi}_1(s)] \, ds \right\|$$

$$\leq \int_{t_0}^{t} \|A(s)\| \, \|\boldsymbol{\phi}_2(s) - \boldsymbol{\phi}_1(s)\| \, ds \leq \int_{t_0}^{t} \alpha \cdot \frac{\alpha \gamma}{2}(s - t_0)^2 \, ds$$

$$= \gamma \frac{\alpha^2}{3!}(t - t_0)^3 \qquad \text{for } t > t_0.$$

These calculations suggest the following result:

Lemma 2. *The successive approximations $\{\boldsymbol{\phi}_m\}$ satisfy the estimate*

$$\|\boldsymbol{\phi}_{m+1}(t) - \boldsymbol{\phi}_m(t)\| \leq \left(\frac{\gamma}{\alpha}\right) \frac{\alpha^{m+1}}{(m+1)!}(t - t_0)^{m+1} \qquad \text{for } t > t_0, \qquad (9)$$

and for $m = 0, 1, 2, \ldots$.

Proof. We verify our guess by induction as follows: (9) is certainly true for $m = 0$ (proved in (7); in fact, it holds for $m = 1, 2$, as well). Now, suppose that (9) has been proved for any value of the index $m = j - 1$ for any integer $j \geq 1$. Then for the value $m = j$ we have, from (5) and the induction assumption (that is, (9) with $m = j - 1$)

$$\|\phi_{j+1}(t) - \phi_j(t)\| = \left\| \int_{t_0}^{t} A(s)[\phi_j(s) - \phi_{j-1}(s)] \, ds \right\|$$

$$\leq \int_{t_0}^{t} \|A(s)\| \, \|\phi_j(s) - \phi_{j-1}(s)\| \, ds$$

$$\leq \int_{t_0}^{t} \alpha \left(\frac{\gamma}{\alpha}\right) \frac{\alpha^j}{j!} (s - t_0)^j \, ds \qquad (10)$$

$$= \left(\frac{\gamma}{\alpha}\right) \frac{\alpha^{j+1}}{j!} \frac{1}{j+1} (t - t_0)^{j+1}$$

$$= \left(\frac{\gamma}{\alpha}\right) \frac{\alpha^{j+1}}{(j+1)!} (t - t_0)^{j+1} \qquad \text{for } t > t_0.$$

But this is precisely (9) with m replaced by j. Since (10) holds for any integer $j \geq 1$, the principle of finite induction tells us that the desired estimate (9) holds for $m = 0, 1, \ldots$, as asserted. ∎

● **EXERCISE**

5. Show that

$$\|\phi_{m+1}(t) - \phi_m(t)\| \leq \frac{\gamma}{\alpha} \frac{\alpha^{m+1}}{(m+1)!} (t_0 - t)^{m+1}$$

for $t < t_0$ and $m = 0, 1, \ldots$.

Combining Lemma 2 and Exercise 5, we have the estimate

$$\|\phi_{m+1}(t) - \phi_m(t)\| \leq \frac{\gamma}{\alpha} \frac{\alpha^{m+1}}{(m+1)!} |t - t_0|^{m+1} \qquad (11)$$

for $m = 0, 1, \ldots$ and every t in I.

We now study the infinite series

$$\phi_0(t) + \sum_{m=0}^{\infty} (\phi_{m+1}(t) - \phi_m(t)) \qquad (12)$$

defined in the discussion following (6), and we begin by looking at the **series of real positive numbers**:

$$\|\phi_0(t)\| + \sum_{m=0}^{\infty} \|\phi_{m+1}(t) - \phi_m(t)\|. \tag{13}$$

By the definition of $\phi_0(t)$ in (5) and by the estimate (11), we see that every term in the series (13) is dominated by the corresponding term of the series

$$\|y_0\| + \frac{\gamma}{\alpha} \sum_{m=0}^{\infty} \frac{\alpha^{m+1}}{(m+1)!} |t - t_0|^{m+1}. \tag{14}$$

But (14) is a convergent series (of positive real numbers) for every $t \in I$, because, by **the ratio test,** the infinite series in (14) converges (the ratio of the $(m+1)$st to the mth terms is

$$\frac{\alpha}{m+1} |t - t_0|$$

which tends to 0 as $m \to +\infty$). Hence, by the ordinary comparison theorem, the series (13) of continuous function converges for every t on I. By definition of convergence in terms of the norm, the series (12) of vectors also converges for every t on I. Let $\phi(t)$ be its sum, that is

$$\phi(t) = \phi_0(t) + \sum_{m=0}^{\infty} (\phi_{m+1}(t) - \phi_m(t)). \tag{15}$$

Lemma 3. *The function ϕ defined by* (15) *is continuous at each point of the finite* (closed) *interval I.*

Proof. We first obtain another estimate which describes an upper bound for the error if the successive approximations are stopped after j steps. Thus, we compute, using (6), (15), the triangle inequality, and the estimate (11).

$$\|\phi(t) - \phi_j(t)\| = \left\| \sum_{m=j}^{\infty} \phi_{m+1}(t) - \phi_m(t) \right\|$$

$$\leq \sum_{m=j}^{\infty} \|\phi_{m+1}(t) - \phi_m(t)\|$$

$$\leq \sum_{m=j}^{\infty} \frac{\gamma}{\alpha} \frac{\alpha^{m+1}}{(m+1)!} |t - t_0|^{m+1}$$

$$\leq \frac{\gamma}{\alpha} \frac{\alpha^{j+1}}{(j+1)!} |t - t_0|^{j+1} \sum_{m=0}^{\infty} \frac{\alpha |t - t_0|^m}{m!}$$

318 *Appendix 2*

or, finally,

$$\|\phi(t) - \phi_j(t)\| \leq \frac{\gamma}{\alpha} \frac{\alpha^{j+1}}{(j+1)!} |t - t_0|^{j+1} e^{\alpha|t-t_0|}. \tag{16}$$

By elementary calculus,

$$\lim_{j \to \infty} \frac{[\alpha|t - t_0|]^{j+1}}{(j+1)!} = 0,$$

because the quantity whose limit is desired is the general term of a series which is known to converge by the ratio test. Thus, for any fixed t in the interval I the error $\|\phi(t) - \phi_j(t)\|$ can be made as small as desired by making j sufficiently large, that is, by taking sufficiently many successive approximations $\phi_j(t)$. But more is true. Since I is a finite interval, say of length d, the use of $|t - t_0| \leq d$ in (16) gives

$$\|\phi(t) - \phi_j(t)\| \leq \frac{\gamma}{\alpha} \frac{\alpha^{j+1}}{(j+1)!} d^{j+1} e^{\alpha d}, \tag{17}$$

and this tends to zero as $j \to +\infty$ independent of the choice of t in I (this is called **uniform convergence**).

To prove the continuity of the vector ϕ on I, we let t be a point of I, and then the triangle inequality gives

$$\|\phi(t + h) - \phi(t)\|$$
$$= \|\phi(t + h) - \phi_j(t + h) + \phi_j(t + h) - \phi_j(t) + \phi_j(t) - \phi(t)\|$$
$$\leq \|\phi(t + h) - \phi_j(t + h)\| + \|\phi_j(t + h) - \phi_j(t)\| + \|\phi_j(t) - \phi(t)\|.$$

By the continuity of ϕ_j, already proved in the first step of the proof, the middle term $\|\phi_j(t + h) - \phi_j(t)\| \to 0$ as $h \to 0$. By the estimate (17) the first and last term can be made as small as desired (irrespective of the position of $t \in I$), by taking j sufficiently large. Hence, for every $t \in I$,

$$\lim_{h \to 0} \|\phi(t + h) - \phi(t)\| = 0,$$

and ϕ is continuous in I.

We next come to the crucial step that ϕ is the solution we seek.

Lemma 4. *The vector function* ϕ *(which can be thought of either as the sum of the series (15) or as the limit of the sequence of successive approximations (5)) satisfies the Volterra integral equation (3).*

The Existence and Uniqueness Theorem

Proof. Notice that from (4)

$$\phi_{j+1}(t) = y_0 + \int_{t_0}^{t} A(s)\phi_j(s)\, ds + \int_{0}^{t} g(s)\, ds. \tag{18}$$

Now, if we let $j \to +\infty$ and if we assume blindly that $\lim_{j\to+\infty} \phi_{j+1}(t) = \phi(t)$ may be used to replace ϕ_{j+1} in (18), then this is exactly what is needed. But where is the justification? Suppose the result is, in fact, true. Then we would also have

$$\phi(t) = y_0 + \int_{t_0}^{t} A(s)\phi(s)\, ds + \int_{0}^{t} g(s)\, ds. \tag{19}$$

All that we need to do to justify the step from (18) to (19) is to prove that

$$\lim_{j\to+\infty} \int_{t_0}^{t} A(s)\phi_j(s)\, ds = \int_{t_0}^{t} A(s)\phi(s)\, ds$$

for t, t_0 in I. But for $t > t_0$ we have, (by (17), and the property (1))

$$\left\| \int_{t_0}^{t} [A(s)\phi_j(s) - A(s)\phi(s)]\, ds \right\|$$

$$\leq \int_{t_0}^{t} \| A(s)[\phi_j(s) - \phi(s)] \|\, ds$$

$$\leq \int_{t_0}^{t} \| A(s) \| \cdot \| \phi_j(s) - \phi(s) \|\, ds$$

$$\leq \int_{t_0}^{t} d\, \frac{\gamma}{\alpha} \frac{\alpha^{j+1}}{(j+1)!} d^{j+1} e^{\alpha d}\, ds = \frac{\gamma}{\alpha} \frac{\alpha^{j+1}}{(j+1)!} d^{j+2} e^{\alpha d},$$

which tends to zero as $j \to +\infty$. The same result holds for $t < t_0$. This justifies passing to the limit under the integral sign in (18) to obtain (19). ∎

Since I is an arbitrary finite closed subinterval of $\mathscr{I}$, Lemmas 1, 2, 3, 4 establish that a solution ϕ of the initial value problem (1) exists on any interval $\mathscr{I}$ under the given conditions that $A(t)$, $g(t)$ are both continuous on $\mathscr{I}$. We now ask whether this solution ϕ is unique.

Lemma 5. *The initial value problem* (1) *has at most one solution ϕ on the interval $\mathscr{I}$.*

Proof. Suppose ψ is another solution of the **same** initial value problem (2) on the interval $\mathscr{I}$. We wish to prove that ϕ and ψ are identical. Then by

Lemma 1, ϕ and ψ both satisfy the equivalent integral equation (3). Writing down this fact and subtracting we obtain

$$\phi(t) - \psi(t) = \int_{t_0}^{t} A(s)(\phi(s) - \psi(s))\, ds.$$

Thus,

$$\|\phi(t) - \psi(t)\| \leq \int_{t_0}^{t} \|A(s)\|\, \|\phi(s) - \psi(s)\|\, ds$$

$$\leq \alpha \int_{t_0}^{t} \|\phi(s) - \psi(s)\|\, ds, \tag{20}$$

for $t \geq t_0$. Let $r(t) = \|\phi(t) - \psi(t)\|$. Then the inequality (20) states that

$$r(t) \leq \alpha \int_{t_0}^{t} r(s)\, ds, \qquad t \geq t_0. \tag{21}$$

A similar inequality holds for $t < t_0$. To resolve (21) observe that by definition $0 \leq r(t)$ and hence from (21) $r(t_0) = 0$. Let $u(t) = \int_{t_0}^{t} r(s)\, ds$. Then $u(t) \geq 0$, $u'(t) = r(t)$ for all $t > t_0$ on $\mathscr{I}$. Clearly, $u(t_0) = 0$. Thus, $u'(t) - \alpha u(t) \leq 0$, $u(t_0) = 0$. Multiplication by the nonnegative function $e^{-\alpha t}$ gives

$$\frac{d}{dt}[u(t)e^{-\alpha t}] \leq 0,$$

and thus

$$u(t)e^{-\alpha t} - u(t_0)\exp(-\alpha t_0) \leq 0.$$

But $u(t_0) = 0$; thus

$$0 \leq u(t) \leq 0 \quad \text{and} \quad u(t) \equiv 0,$$

which implies, from (21),

$$0 \leq r(t) = \|\phi(t) - \psi(t)\| \leq \alpha u(t) \leq 0 \qquad \text{for } t > t_0.$$

Thus, $r(t) \equiv 0$ and $\phi(t) \equiv \psi(t)$, for $t \geq t_0$. ∎

● **EXERCISE**

6. Prove that $\phi(t) \equiv \psi(t)$ for $t \leq t_0$.

Combining Lemmas 1, 2, 3, 4, and 5, we now obtain Theorem 1.

The method of dealing with the inequality (21) which was used in Lemma 5 to prove the uniqueness of solutions has many other applications. For example, it may be used to justify the use of Laplace transforms in Chapter 8 to solve initial value problems for linear systems of differential equations with constant coefficients, provided the nonhomogeneous terms satisfy a certain growth condition. In order to do this, we must first generalize the result slightly. This result is usually called Gronwall's lemma.

Lemma 6. *Let $K > 0$ and $a \geq 0$ be constants. Suppose that $r(t)$ is a continuous nonnegative function for $t \geq t_0$ which satisfies the inequality*

$$r(t) \leq a + K \int_{t_0}^{t} r(\tau) \, d\tau \tag{22}$$

on some interval I. Then

$$r(t) \leq a[\exp(K(t - t_0))] \tag{23}$$

for $t \geq t_0$, and t in the interval I.

Proof. We let

$$u(t) = a + K \int_{t_0}^{t} r(\tau) \, d\tau.$$

Then $r(t) \leq u(t)$ by (22) and $u'(t) = Kr(t)$ by the fundamental theorem of calculus. Thus, (22) becomes

$$u'(t) - Ku(t) \leq 0.$$

Multiplying by e^{-Kt} and integrating from t_0 to t, and using the fact that

$$(u(t)e^{-Kt})' = [u'(t) - Ku(t)]e^{-Kt},$$

and $u(t_0) = a$, we obtain

$$u(t)e^{-Kt} - u(t_0) \exp(-Kt_0) \leq 0,$$

on

$$u(t) \leq a \exp(K(t - t_0)).$$

Since $r(t) \leq u(t)$, this establishes (23). ∎

We are now ready to prove the result needed in Chapter 8. The reader will recall that in Chapter 8 we considered **functions with exponential growth at infinity**, that is, functions $\mathbf{f}(t)$ defined for $t \geq 0$ such that there exist constants $M > 0$ and c such that

$$\|f(t)\| \leq Me^{ct}$$

for all sufficiently large t.

Theorem 2. *Consider the initial value problem*

$$\mathbf{y}' = A\mathbf{y} + \mathbf{g}(t), \qquad \mathbf{y}(0) = \mathbf{y}_0, \tag{24}$$

where A is a constant matrix in $\mathscr{F}_{nn}$, $\mathbf{g}(t)$ is continuous and of exponential growth at infinity. Then the solution $\boldsymbol{\phi}(t)$ of (24) (which exists and is unique for $0 \leq t < \infty$ by Theorem 1) and its derivative $\boldsymbol{\phi}'(t)$ are of exponential growth at infinity.

Proof. By Lemma 1, the solution $\boldsymbol{\phi}(t)$ of the given initial value problem satisfies

$$\boldsymbol{\phi}(t) = \mathbf{y}_0 + \int_0^t A\boldsymbol{\phi}(s)\, ds + \int_{t_0}^t \mathbf{g}(s)\, ds. \tag{25}$$

By hypothesis, there exist constants $M > 0$ and c and a time T such that

$$\|\mathbf{g}(t)\| \leq Me^{ct} \tag{26}$$

for $t \geq T$. We may assume $c > 0$, since increasing c increases the right-hand side of (26) and does not affect the truth of the inequality. We may rewrite (25) as

$$\boldsymbol{\phi}(t) = \mathbf{y}_0 + \int_0^T A\boldsymbol{\phi}(s)\, ds + \int_0^T \mathbf{g}(s)\, ds + \int_T^t A\boldsymbol{\phi}(s)\, ds + \int_T^t \mathbf{g}(s)\, ds. \tag{27}$$

We have shown in Theorem 1 that $\boldsymbol{\phi}(t)$ is bounded on the bounded interval $0 \leq t \leq T$. In view of this fact, and the continuity of $\mathbf{g}(t)$, there exists a constant K such that

$$\left\| \mathbf{y}_0 + \int_0^T A\boldsymbol{\phi}(s)\, ds + \int_0^T \mathbf{g}(s)\, ds \right\| \leq K. \tag{28}$$

Now, taking norms in (27) and using (26), (28), and the properties (v), (1), and the triangle inequality for vector norms, we obtain

$$\|\phi(t)\| \leq K + \int_T^t \|A\| \cdot \|\phi(s)\| \, ds + \int_T^t M e^{cs} \, ds$$

$$\leq K + \int_T^t \|A\| \cdot \|\phi(s)\| \, ds + \frac{M}{c}(e^{ct} - e^{cT}) \qquad (29)$$

$$\leq K + \frac{M}{c} e^{ct} + \int_T^t \|A\| \cdot \|\phi(s)\| \, ds.$$

Multiplying (29) by e^{-ct}, we obtain

$$\|\phi(t)\| e^{-ct} \leq \left(K e^{-ct} + \frac{M}{c}\right) + \int_T^t \|A\| \cdot \|\phi(s)\| e^{-ct} \, ds$$

$$\leq \left(K e^{-ct} + \frac{M}{c}\right) + \int_T^t \|A\| \cdot \|\phi(s)\| e^{-cs} \, ds, \qquad (30)$$

since $e^{-ct} \leq e^{-cs}$ for $t \geq s$. Since $c > 0$, there exists a constant L such that

$$K e^{-ct} + \frac{M}{c} \leq L \quad \text{for } t \geq T.$$

Thus, (30) becomes

$$\|\phi(t)\| e^{-ct} \leq L + \|A\| \int_T^t \|\phi(s)\| e^{-cs} \, ds.$$

Application of Lemma 6 with $r(t) = \|\phi(t)\| e^{-ct}$ gives

$$\|\phi(t)\| e^{-ct} \leq L e^{\|A\|(t-T)}$$

or

$$\|\phi(t)\| \leq L e^{-\|A\|T} e^{(\|A\|+c)t},$$

for $t \geq T$. Thus, $\|\phi(t)\|$ has exponential growth at infinity. Since $\phi'(t) = A\phi(t) + g(t)$,

$$\|\phi'(t)\| \leq \|A\| \cdot \|\phi(t)\| + \|g(t)\| \leq \|A\| L e^{-\|A\|T} e^{(\|A\|-c)t} + M e^{ct}$$

$$\leq (\|A\| L e^{-\|A\|T} + M) e^{(\|A\|+c)t},$$

and thus $\|\phi'(t)\|$ also has exponential growth at infinity. ∎

Appendix 3

GENERALIZED EIGENVECTORS, INVARIANT SUBSPACES, AND CANONICAL FORMS OF MATRICES

In our study of a linear system of differential equations with constant coefficients, $\mathbf{y}' = A\mathbf{y}$, when we treated the general case where A has fewer than n linearly independent eigenvectors in Section 7.5, we stated a theorem of linear algebra without proof. This appendix is devoted to the proof of that theorem, and some interesting algebraic consequences.

In order to prove the desired theorem, we must introduce the concept of a **direct sum of subspaces of a vector space.** Let V be a vector space and let X and Y be subspaces of V. We make the following definitions:

Definition 1. *The vector sum of the subspaces X and Y, denoted by $X + Y$, is the set of all vectors $\mathbf{v} \in V$ of the form $\mathbf{v} = \mathbf{x} + \mathbf{y}$ for some $\mathbf{x} \in X$, $\mathbf{y} \in Y$.*

Definition 2. *The intersection of the subspaces X and Y, denoted by $X \cap Y$, is the set of all vectors $\mathbf{v} \in V$ which are in both subspaces X and Y.*

It is easy to verify that $X + Y$ and $X \cap Y$ are subspaces of V.

● **EXERCISE**

    1. Let X and Y be subspaces of a vector space V. Show that $X + Y$ and $X \cap Y$ are also subspaces of V.

For any subspaces X, Y of V, the vector $\mathbf{0}$ belongs to $X \cap Y$.

Generalized Eigenvectors, Invariant Subspaces, and Forms of Matrices

Definition 3. *The sum $X + Y$ is said to be a* **direct sum***, written $X \oplus Y$, if $X \cap Y = \{\mathbf{0}\}$.*

● **EXERCISE**

2. (a) If X and Y are finite-dimensional subspaces of a vector space V, show that

$$\dim(X + Y) + \dim(X \cap Y) = \dim X + \dim Y.$$

   [*Hint:* Begin with a basis for $X \cap Y$ and extend it to a basis for X and a basis for Y.]
   (b) Deduce that the sum $X + Y$ is direct if and only if

$$\dim(X + Y) = \dim X + \dim Y.$$

We now extend the definitions of vector sums, intersections, and direct sums to any finite number of subspaces. Let $X_1, X_2, \ldots, X_k$ be subspaces of a vector space V.

Definition 4. *The vector sum*

$$\sum_{i=1}^{k} X_i$$

of the subspaces $X_1, X_2, \ldots, X_k$ is the set of all vectors $\mathbf{v} \in V$ of the form

$$\mathbf{v} = \sum_{i=1}^{k} \mathbf{x}_i$$

for some $\mathbf{x}_i \in X_i$, where $i = 1, 2, \ldots, k$.

Definition 5. *The intersection*

$$\bigcap_{i=1}^{k} X_i$$

of the subspaces $X_1, X_2, \ldots, X_k$ is the set of all vectors $\mathbf{v} \in V$ which is in each of the subspaces $X_1, X_2, \ldots, X_k$.

Definition 6. *The sum*

$$\sum_{i=1}^{k} X_i$$

is a direct sum, written

$$\bigoplus_{i=1}^{k} X_i$$

if and only if

$$X_j \cap \sum_{i \neq j} X_i = 0$$

for every j, where $j = 1, 2, \ldots, k$.

● **EXERCISES**

3. Let $X_1, X_2, \ldots, X_k$ be subspaces of a vector space V. Show that if

$$W = \sum_{i=2}^{k} X_i$$

is a direct sum and if $X_1 + W$ is a direct sum, then

$$\sum_{i=1}^{k} X_i$$

is a direct sum.

4. Show that a vector space V is a direct sum of subspaces $X_1, X_2, \ldots, X_k$,

$$V = \bigoplus_{i=1}^{k} X_i,$$

if and only if every vector $\mathbf{v} \in V$ has a unique representation

$$\mathbf{v} = \sum_{i=1}^{k} \mathbf{x}_i,$$

where $\mathbf{x}_i \in X_i$ for $i = 1, 2, \ldots, k$.

5. Let $\mathbf{x}_{i,1}, \ldots, \mathbf{x}_{i,n_i}$ be a basis for the subspace X_i, where $i = 1, \ldots, k$. If

$$V = \bigoplus_{i=1}^{k} X_i,$$

show that $\{\mathbf{x}_{11}, \ldots, \mathbf{x}_{1n_1}, \mathbf{x}_{21_2}, \ldots, \mathbf{x}_{2n_2}, \ldots, \mathbf{x}_{k1}, \ldots, \mathbf{x}_{kn_k}\}$ is a basis for V.

We recall that if T is a linear transformation of $\mathscr{C}_n$ into itself, then corresponding to any basis $\{\mathbf{v}_1, \mathbf{v}_2, \ldots, \mathbf{v}_n\}$ of $\mathscr{C}_n$ there is an $n \times n$ matrix that represents the linear transformation T with respect to this basis. The elements a_{ij} with $i, j = 1, \ldots, n$, of this matrix A are defined by

$$T\mathbf{v}_i = \sum_{j=1}^{n} a_{ji} \mathbf{v}_j \quad (i = 1, \ldots, n). \tag{1}$$

Corresponding to a different basis $\{\mathbf{w}_1, \mathbf{w}_2, \ldots, \mathbf{w}_n\}$ of $\mathscr{C}_n$ there is (perhaps) a different matrix B that represents T. We recall that the matrices A and B are similar; that is, there exists a nonsingular matrix P such that

$$B = P^{-1}AP \quad \text{(Theorem 2, Section 5.7).}$$

The study of canonical forms of matrices involves the choice of a basis of $\mathscr{C}_n$, relative to which the matrix of a given linear transformation takes a particularly simple form. Equivalently, this study involves the determination of a matrix of a particularly simple form that is similar to a given matrix.

To discuss the situation when there are fewer than n linearly independent eigenvectors, we introduce the concept of **generalized eigenvector.** If for some value λ and some $p \geq 1$, there is a vector $\mathbf{v}$ such that

$$(A - \lambda I)^p \mathbf{v} = 0 \quad \text{but} \quad (A - \lambda I)^{p-1} \mathbf{v} \neq 0, \tag{2}$$

then $\mathbf{v}$ is said to be a generalized eigenvector of index p corresponding to the generalized eigenvalue λ. When $p = 1$, λ is an eigenvalue and $\mathbf{v}$ a corresponding eigenvector. We note that since $\mathbf{u} = (A - \lambda I)^{p-1} \mathbf{v} \neq 0$, and since because of (2), $(A - \lambda I)\mathbf{u} = (A - \lambda I)^p \mathbf{v} = 0$, the "generalized eigenvalue" λ must be an eigenvalue of A with a corresponding eigenvector $\mathbf{u}$.

Lemma 1. *If $\mathbf{v}$ is a generalized eigenvector of index p, then the vectors $\mathbf{v}$, $(A - \lambda I)\mathbf{v}, \ldots, (A - \lambda I)^{p-1}\mathbf{v}$ are linearly independent.*

Proof. If the given vectors are linearly dependent, then one of them, say $(A - \lambda I)^k \mathbf{v}$ for some $k (0 \leq k < p - 1)$, can be written as a linear combination of the later ones:

$$(A - \lambda I)^k \mathbf{v} = \sum_{j=k+1}^{p-1} c_j (A - \lambda I)^j \mathbf{v}.$$

Since $(A - \lambda I)^q \mathbf{v} = 0$ for $q \geq p$, application of $(A - \lambda I)^{p-1-k}$ to both sides of this equation gives $(A - \lambda I)^{p-1}\mathbf{v} = \mathbf{0}$, which is a contradiction, proving the linear independence of the given vectors. ∎

Given an eigenvalue λ of A, we consider the subset X of $\mathscr{C}_n$ consisting of all generalized eigenvectors corresponding to λ together with the zero vector. It is easily shown that X is a subspace of $\mathscr{C}_n$. Let r be the largest index of any such generalized eigenvector, so that $(A - \lambda I)^r \mathbf{x} = \mathbf{0}$ for every $\mathbf{x} \in X$, but $(A - \lambda I)^{r-1}\mathbf{y} \neq \mathbf{0}$ for some $\mathbf{y} \in X$. Since, by Lemma 1, X contains at least r linearly independent vectors, r is finite (in fact, $r \leq n$). The integer r is called the index of X, and dim $X \geq r$.

The subspace X may also be described as the null space of the linear transformation $(T - \lambda I)^r$ (or the null space of the matrix $(A - \lambda I)^r$). Let Y be the range of the linear transformation $(T - \lambda I)^r$. We recall that the range of the linear transformation $(T - \lambda I)^r$ is the set of vectors $\mathbf{y} \in \mathscr{C}_n$ such that $(T - \lambda I)^r \mathbf{z} = \mathbf{y}$ for some $\mathbf{z} \in \mathscr{C}_n$. By the theory of linear algebraic systems (namely Theorems 2 and 3, Section 5.6), dim X + dim $Y = n$. Next, we observe that X and Y are disjoint subspaces; for if $\mathbf{v}$ is in both X and Y, then $(T - \lambda I)^r \mathbf{v} = \mathbf{0}$ and $\mathbf{v} = (T - \lambda I)^r \mathbf{u}$ for some $\mathbf{u} \in \mathscr{C}_n$. Thus, $(T - \lambda I)^{2r}\mathbf{u} = (T - \lambda I)^r \mathbf{v} = \mathbf{0}$, and thus $\mathbf{u}$ is a generalized eigenvector; that is, $\mathbf{u} \in X$. But since r is the largest index of any generalized eigenvector corresponding to λ, there exists an integer l, $0 \leq l \leq r$ such that $(T - \lambda I)^l \mathbf{u} = \mathbf{0}$; hence also $\mathbf{v} = (T - \lambda I)^r \mathbf{u} = \mathbf{0}$. This shows that X and Y are disjoint subspaces of $\mathscr{C}_n$ whose dimensions add up to n. It follows that $\mathscr{C}_n$ is the direct sum of X and Y (written $\mathscr{C}_n = X \oplus Y$), that is, that every vector $\mathbf{v} \in \mathscr{C}_n$ has a unique decomposition $\mathbf{v} = \mathbf{x} + \mathbf{y}$, with $\mathbf{x} \in X$, $\mathbf{y} \in Y$.

The subspaces X and Y are **invariant** under T, that is, $T\mathbf{x} \in X$ for every $\mathbf{x} \in X$ and $T\mathbf{y} \in Y$ for every $\mathbf{y} \in Y$. To see this we observe that X and Y are invariant under $(T - \lambda I)$, and it follows easily that they are invariant under T.

Next, suppose that the subspace X has dimension k, and let $\mathbf{v}_1, \ldots, \mathbf{v}_k$ be a basis of X. Then Y has dimension $(n - k)$, and if $\mathbf{v}_{k+1}, \ldots, \mathbf{v}_n$ is a basis of Y, the fact that $\mathscr{C}_n = X \oplus Y$ implies that $\mathbf{v}_1, \ldots, \mathbf{v}_k, \mathbf{v}_{k+1}, \ldots, \mathbf{v}_n$ is a basis of $\mathscr{C}_n$. Since X is invariant under T, $T\mathbf{v}_i \in X$, where $(i = 1, \ldots, k)$, and since $\mathbf{v}_1, \ldots, \mathbf{v}_k$ is a basis of X,

$$T\mathbf{v}_i = \sum_{j=1}^{k} a_{ji} \mathbf{v}_j \qquad (i = 1, \ldots, k).$$

Note that the sum is from $j = 1$ to $j = k$, not to $j = n$. Similarly,

$$T\mathbf{v}_i = \sum_{j=k+1}^{n} a_{ji} \mathbf{v}_j \qquad (i = k+1, \ldots, n).$$

Comparing these formulas to (1), we see that the matrix A of T with respect to the basis $v_1, \ldots, v_n$ of $\mathscr{C}_n$ is a "block diagonal" matrix,

$$A = \begin{bmatrix} A_1 & 0 \\ 0 & A_2 \end{bmatrix}.$$

Here, the $k \times k$ matrix A_1 represents the restriction T_1 of the transformation T to the subspace X and the $(n-k) \times (n-k)$ matrix A_2 represents the restriction T_2 of T to the subspace Y.

Since for the block diagonal matrix A,

$$\det(A - \lambda I) = \det(A_1 - \lambda I_1) \det(A_2 - \lambda I_2),$$

where I_1 is the $k \times k$ identity matrix and I_2 is the $(n-k) \times (n-k)$ identity matrix, the characteristic polynomial of A (or of T) is the product of the characteristic polynomials of A_1 and A_2 (or of T_1 and of T_2).

On X, the transformation T has only the eigenvalue λ. For, if μ is an eigenvalue and v a corresponding eigenvector in X, and if r is the index of X, then

$$0 = (A - \lambda I)^r v = (A - \lambda I)^{r-1}(A - \lambda I)v = (A - \lambda I)^{r-1}(Av - \lambda v)$$
$$= (A - \lambda I)^{r-1}(\mu - \lambda)v = (\mu - \lambda)(A - \lambda I)^{r-1}v = \cdots = (\mu - \lambda)^r v,$$

and $\lambda = \mu$. Thus, X contains all the eigenvectors of A corresponding to the eigenvalue λ, but no other eigenvector. In particular, if the linear transformation T has only one distinct eigenvalue λ, then the subspace Y (the range of $(T - \lambda I)^r$) contains no eigenvectors of T. Thus, the restriction T_2 of T to Y has no eigenvalues or eigenvectors. Therefore, Y must consist of only the zero vector, and $X = \mathscr{C}_n$. Observe that in this case, the linear transformation $T - \lambda I$ (or the matrix $A - \lambda I$) is nilpotent: For every vector $x \in \mathscr{C}_n$, $(A - \lambda I)^r x = 0$.

We have now developed the algebraic machinery needed to prove the following basic theorem, which is used in Section 7.5.

Theorem 1. Let $\lambda_1, \lambda_2, \ldots, \lambda_k$ be the distinct eigenvalues of a matrix A, with multiplicities $n_1, n_2, \ldots, n_k$, respectively. Then $\mathscr{C}_n$ is the direct sum of the subspaces $X_1, X_2, \ldots, X_k$ of generalized eigenvectors corresponding to the eigenvalues $\lambda_1, \lambda_2, \ldots, \lambda_k$, respectively. The subspace X_j is invariant under A, has dimension n_j, and $(A - \lambda_j I)^{n_j} x = 0$ for every $x \in X_j$, where $j = 1, 2, \ldots, n$.

Proof. We prove the result by induction on k, the number of distinct eigenvalues. The case $k = 1$ has been covered in the remarks preceding the statement of the theorem. Now, suppose that the result has been established for every matrix with fewer than k distinct eigenvalues, and let A have k distinct eigenvalues $\lambda_1, \ldots, \lambda_k$. On X_1, the transformation T corresponding to the matrix A has only the eigenvalue λ_1. Let Y_1 be the range of $(T - \lambda_1 I)^{r_1}$, where r_1 is the largest index of the generalized eigenvectors corresponding to λ_1. The characteristic polynomial of A is

$$p(\lambda) = \prod_{i=1}^{k} (\lambda - \lambda_i)^{n_i}.$$

As we have seen in the remarks preceding the statement of Theorem 1, the characteristic polynomial of A is the product of the characteristic polynomial of A on X_1 and the characteristic polynomial of A on Y_1. The characteristic polynomial of A on X_1 contains all the factors $(\lambda - \lambda_1)$ in $p(\lambda)$ because X_1 contains all the eigenvectors of A corresponding to λ_1, and it contains no factor $(\lambda - \lambda_j)$ with $j \neq 1$ because X_1 contains no eigenvector of A corresponding to an eigenvalue $\lambda_j \neq \lambda_1$. Therefore, the characteristic polynomial of A on X_1 is exactly $\pm(\lambda - \lambda_1)^{n_1}$. Since the degree of the characteristic polynomial of a linear transformation is equal to the dimension of the space, this shows that X_1 has dimension n_1.

We have seen that $\mathscr{C}_n$ is the direct sum of X_1 and Y_1. By the induction assumption applied to the restriction T_2 of T to the subspace Y_1, which has only $(k - 1)$ distinct eigenvalues $\lambda_2, \lambda_3, \ldots, \lambda_k$, Y_1 is the direct sum of $X_2, X_3, \ldots, X_k$. It now follows from Exercise 3 that $\mathscr{C}_n$ is the direct sum of $X_1, X_2, \ldots, X_k$, and the theorem is proved by induction.

By the argument preceding the statement of Theorem 1, we now obtain the following interpretation of Theorem 1.

Corollary 1 to Theorem 1. *The matrix A of the linear transformation T relative to a basis of $\mathscr{C}_n$ made up of bases of the subspaces $X_1, \ldots, X_k$ is a block diagonal matrix*

$$A = \begin{bmatrix} A_1 & & & \\ & A_2 & & \\ & & \ddots & \\ & & & A_k \end{bmatrix},$$

where A_j is an $n_j \times n_j$ matrix that represents T on X_j for $j = 1, \ldots, k$.

Generalized Eigenvectors, Invariant Subspaces, and Forms of Matrices 331

Corollary 2 to Theorem 1. *If $A \in \mathscr{C}_{nn}$, then there exists a nonsingular matrix $P \in \mathscr{C}_{nn}$ such that $P^{-1}AP$ has the block diagonal form given in Corollary 1.*

The **Jordan canonical form** of a matrix is obtained from the above representation by choosing bases of the subspaces $X_1, \ldots, X_k$ in a suitable manner. This requires a careful study of nilpotent transformations. A linear transformation L such that $L^r = 0$ but $L^{r-1} \neq 0$ is said to be **nilpotent** of index r. We recall that the subspace X_j is the null space of the transformation represented by the matrix $(A - \lambda_j I)^{r_j}$, where r_j is the largest index of the generalized eigenvectors corresponding to λ_j. Since X_j is invariant under $(A - \lambda_j I)$, we may regard $(A - \lambda_j I)$ as a linear transformation on X_j that is nilpotent of index r_j.

Let L be a nilpotent linear transformation of index r on a vector space X of dimension n. Then there is a vector $\mathbf{u}$ such that $L^r\mathbf{u} = \mathbf{0}$ but $L^{r-1}\mathbf{u} \neq \mathbf{0}$. By Lemma 1, the chain of vectors $\mathbf{u}, L\mathbf{u}, \ldots, L^{r-1}\mathbf{u}$ is linearly independent. We will form a basis for X consisting of several chains of this type. If $r = n$, then we have a basis of X consisting of the vectors $\mathbf{u}, L\mathbf{u}, \ldots, L^{r-1}\mathbf{u}$. If $r < n$, let U_1 be the subspace of X spanned by these vectors. For every $\mathbf{v} \notin U_1$, consider the chain $\mathbf{v}, L\mathbf{v}, \ldots, L^{s-1}\mathbf{v}$, where each vector $L^p\mathbf{v}$, $0 \leq p \leq s - 1$, lies outside U_1 but $L^s\mathbf{v} \in U_1$. We choose a $\mathbf{v}$ that maximizes the length s of this chain, and we let $r_2 \leq r$ be the length of this maximal chain. Then $\mathbf{v}, L\mathbf{v}, \ldots, L^{r_2-1}\mathbf{v}$ are outside U_1 but $L^{r_2}\mathbf{v} \in U_1$. Since $\mathbf{u}, L\mathbf{u}, \ldots, L^{r-1}\mathbf{u}$ is a basis of U_1, we may write

$$L^{r_2}\mathbf{v} = \sum_{j=0}^{r-1} c_j L^j \mathbf{u}. \tag{3}$$

We apply L^{r-r_2} to both sides of this equation and use $L^r\mathbf{u} = L^r\mathbf{v} = \mathbf{0}$ to see that

$$0 = L^r\mathbf{v} = \sum_{j=0}^{r-1} c_j L^{j+r-r_2}\mathbf{u} = \sum_{j=0}^{r_2-1} c_j L^{j+r-r_2}\mathbf{u}.$$

Since $L^{r-r_2}\mathbf{u}, L^{r-r_2+1}\mathbf{u}, \ldots, L^{r-1}\mathbf{u}$ are linearly independent, $c_j = 0$ for $j = 0, 1, \ldots, (r_2 - 1)$. Thus, (3) becomes

$$L^{r_2}\mathbf{v} = \sum_{j=r_2}^{r-1} c_j L^j \mathbf{u}. \tag{4}$$

We define

$$\mathbf{u}_2 = \mathbf{v} - \sum_{j=r_2}^{r-1} c_j L^{j-r_2}\mathbf{u}.$$

Appendix 3

Then it is clear from (4) that $L^{r_2}u_2 = 0$. On the other hand,

$$L^k u_2 = L^k v - \sum_{j=r_2}^{r-1} c_j L^{k+j-r_2} u \qquad (k = 0, 1, \ldots, (r_2 - 1)).$$

Since $L^{k+j-r_2}u$ is in U_1 but $L^k v$ is outside U_1, $L^k u_2$ is outside U_1 for $k = 0, 1, \ldots, (r_2 - 1)$. Thus, every nonzero linear combination of u_2, Lu_2, $\ldots$, $L^{r_2-1}u_2$ is outside U_1. Let U_2 be the subspace of X spanned by u_2, Lu_2, $\ldots$, $L^{r_2-1}u_2$; then U_1 and U_2 are disjoint. The direct sum $U_1 \oplus U_2$ is invariant under L.

● **EXERCISE**

6. Prove the last statement.

If this direct sum is not all of X, we construct a maximal chain outside $U_1 \oplus U_2$ by the same method. Continuing in this manner, we can write X as a direct sum of a finite number of subspaces $U_1, U_2, \ldots, U_r$, each of which is spanned by a chain of the type given above. Thus, we have proved the following result.

Theorem 2. *Let L be a nilpotent linear transformation of index r_1 on a vector space X of dimension n. Then X has a basis of the form*

$$L^{r_1-1}u, L^{r_1-2}u, \ldots, Lu, u, L^{r_2-1}u_2, L^{r_2-2}u_2, \ldots, u_2, \ldots, L^{r_t-1}u_t, \ldots, u_t$$

with

$$r_1 \geq r_2 \geq r_3 \geq \cdots \geq r_t \geq 1 \quad \text{and} \quad L^{r_k}u_t = 0 \qquad (k = 1, 2, \ldots, t).$$

To construct the matrix B that represents L with respect to the basis given by Theorem 2, we denote the basis elements respectively by $v_1, v_2, \ldots, v_n$. Then we have

$$Lv_1 = 0, Lv_2 = v_1, \ldots, Lv_{r_1} = v_{r_1-1},$$
$$Lv_{r_1+1} = 0, Lv_{r_1+2} = v_{r_1+1}, \ldots, Lv_n = v_{n-1}.$$

Generalized Eigenvectors, Invariant Subspaces, and Forms of Matrices 333

From the definition (1) of the matrix of a linear transformation with respect to a given basis, we see that B has the form

$$B = \begin{bmatrix} B_1 & & & \\ & B_2 & & \\ & & \ddots & \\ & & & B_t \end{bmatrix}, \tag{5}$$

where B_k is the $r_k \times r_k$ matrix given by

$$B_k = \begin{bmatrix} 0 & 1 & 0 & \cdots & 0 \\ 0 & & \ddots & \ddots & \vdots \\ \vdots & & \ddots & \ddots & 0 \\ \vdots & & & \ddots & 1 \\ 0 & & & 0 & 0 \end{bmatrix}.$$

The **Jordan canonical form** of a linear transformation T is now obtained by combining Theorems 1 and 2. According to Theorem 1, we can decompose $\mathscr{C}_n$ into a direct sum of subspaces $X_1, \ldots, X_k$. On the subspace X_j, the transformation $T - \lambda_j I$ is nilpotent of index r_j. We now use Theorem 2 with $L = T - \lambda_j I$ to construct a basis for the subspace X_j. The matrix of the transformation $T - \lambda_j I$ restricted to X_j has the form (5). Thus, with respect to this basis, the matrix of the restriction of T to X_j has the form

$$C = \begin{bmatrix} C_1 & & & \\ & C_2 & & \\ & & \ddots & \\ & & & C_t \end{bmatrix},$$

where C_l is the $r_l \times r_l$ matrix

$$C_l = \begin{bmatrix} \lambda_j & 1 & 0 & \cdots & 0 \\ 0 & & \ddots & \ddots & \vdots \\ \vdots & & \ddots & \ddots & 0 \\ \vdots & & & \ddots & 1 \\ 0 & \cdots & & 0 & \lambda_j \end{bmatrix}.$$

This gives the following important result.

Theorem 3. (Jordan Canonical Form) *Let T be a linear transformation of $\mathscr{C}_n$ with eigenvalues $\lambda_1, \ldots, \lambda_k$ of multiplicities $n_1, \ldots, n_k$, respectively. Then there exists a basis of $\mathscr{C}_n$ relative to which T is represented by a Jordan canonical matrix*

$$A = \begin{bmatrix} A_1 & & & \\ & A_2 & & \\ & & \ddots & \\ & & & A_k \end{bmatrix}.$$

Here A_j is an $n_j \times n_j$ matrix that has all diagonal elements equal to λ_j, and that has chains of 1's separated by single 0's immediately above the main diagonal, and all other elements zero.

Corollary. *Every matrix is similar to a Jordan canonical matrix.*

We remark that the length of the chains of 1's in A_j depends on the integers $r_1, \ldots, r_k$ in Theorem 2. It can also be shown that, except for the order of the blocks A_j, the Jordan canonical form is unique.

SOLUTIONS TO SELECTED EXERCISES

§2.1

1. No. **2.** No.

§2.2

1. $A + B = \begin{bmatrix} 3 & 2 & 4 \\ 11 & 6 & 8 \\ -2 & 1 & -1 \end{bmatrix}, C + D = \begin{bmatrix} 3 & 9 & 0 & 1 \\ 1 & 7 & 1 & 2 \\ 6 & 3 & 3 & -6 \end{bmatrix}$

$C + F = \begin{bmatrix} 1 & 0 & 0 & 1 \\ 1 & 1 & 0 & 1 \\ 0 & -1 & 1 & 1 \end{bmatrix}.$

§2.3

1. (a) $\begin{bmatrix} 5 & 10 \\ 15 & 20 \end{bmatrix}.$ (b) $\begin{bmatrix} 5 & 10 \\ 15 & 20 \end{bmatrix}.$ (c) $[32]$ (d) $\begin{bmatrix} 4 & 8 & 12 \\ 5 & 10 & 15 \\ 6 & 12 & 18 \end{bmatrix}.$

(g) $\begin{bmatrix} 1 & 0 \\ 0 & 1 \end{bmatrix}.$ (k) $\begin{bmatrix} 1 & 0 \\ 0 & 1 \end{bmatrix}.$

2. Let 0_{qm} be the zero matrix in $\mathscr{F}_{qm}$. Then $0_{qm} A = 0_{qn}$, the zero matrix in $\mathscr{F}_{qn}$.

3. No. $\begin{bmatrix} 1 & 0 \\ 0 & 0 \end{bmatrix} \begin{bmatrix} 1 & 1 \\ 1 & 1 \end{bmatrix} \neq \begin{bmatrix} 1 & 1 \\ 1 & 1 \end{bmatrix} \begin{bmatrix} 1 & 0 \\ 0 & 0 \end{bmatrix}.$ **4.** $\sum_{k=1}^{n} \sum_{l=1}^{n} a_{ik}(b_{kl} c_{lj}).$

§2.4

6. (a) $\begin{bmatrix} 1 & 0 \\ 0 & \frac{1}{2} \end{bmatrix}$. (b) $\begin{bmatrix} 1 & 0 & 0 \\ 0 & \frac{1}{2} & 0 \\ 0 & 0 & \frac{1}{3} \end{bmatrix}$. (c) $\begin{bmatrix} 1 & -\frac{1}{2} & -\frac{1}{12} \\ 0 & \frac{1}{4} & -\frac{5}{24} \\ 0 & 0 & \frac{1}{6} \end{bmatrix}$. (d) $\begin{bmatrix} \frac{5}{17} & -\frac{3}{17} \\ -\frac{1}{17} & \frac{4}{17} \end{bmatrix}$.

16. Let C, D be any two matrices in $\mathscr{F}_{n,n}$ such that $CD \neq DC$; then $2CD \neq CD + DC$ and hence $(C+D)^2 \neq C^2 + 2CD + D^2$.

17. Any matrix $\begin{bmatrix} x_{11} & x_{12} \\ x_{21} & x_{22} \end{bmatrix}$ such that $x_{11} = -x_{22}$ and $x_{12}x_{21} = 1 - x_{11}^2$, and also any matrix $\begin{bmatrix} y_{11} & y_{12} \\ y_{21} & y_{22} \end{bmatrix}$ such that $y_{21} = y_{12} = 0$, $y_{11}^2 = y_{22}^2 = 1$.

§3.4

3. (a) $\begin{bmatrix} 1 & 0 & 0 \\ -2 & 1 & 0 \\ -1 & 0 & 1 \end{bmatrix}$. (b) $\begin{bmatrix} 1 & 0 & 0 \\ 0 & 2 & 0 \\ 1 & 1 & 1 \end{bmatrix}$. (c) $\begin{bmatrix} 1 & -1 & 0 \\ 0 & 1 & 0 \\ 0 & -1 & 1 \end{bmatrix}$.

§3.5

1. (a) $\begin{bmatrix} 1 & 0 & 0 & 1 \\ 0 & 1 & 0 & -\frac{1}{3} \\ 0 & 0 & 1 & -1 \end{bmatrix}$. (b) $\begin{bmatrix} 1 & 0 & -1 & -2 \\ 0 & 1 & 2 & 3 \\ 0 & 0 & 0 & 0 \end{bmatrix}$. (c) $\begin{bmatrix} 1 & 0 \\ 0 & 1 \end{bmatrix}$.

(d) $\begin{bmatrix} 1 & -1 & 0 \\ 0 & 0 & 1 \\ 0 & 0 & 0 \end{bmatrix}$. (e) $\begin{bmatrix} 1 & 2 & 3 \\ 0 & 0 & 0 \\ 0 & 0 & 0 \end{bmatrix}$. (f) $\begin{bmatrix} 1 & 0 & 0 \\ 0 & 1 & 0 \\ 0 & 0 & 1 \end{bmatrix}$.

(g) $\begin{bmatrix} 1 & 0 & 0 & 4/17 \\ 0 & 1 & 0 & 19/17 \\ 0 & 0 & 1 & 11/17 \end{bmatrix}$.

3. (a) $\begin{bmatrix} 1 & 0 & 0 & 1 & -2 & -11 \\ 0 & 1 & 0 & 0 & 1 & 4 \\ 0 & 0 & 1 & 0 & 0 & 1 \end{bmatrix}$.

§3.7

2. (a) $\begin{bmatrix} 23/17 \\ 3/17 \\ -9/17 \end{bmatrix}$. (b) Inconsistent. (c) $\begin{bmatrix} 1/2 \\ -4/5 \\ -17/10 \end{bmatrix}$. (d) Inconsistent.

(e) $\begin{bmatrix} 15/101 \\ -153/101 \\ -34/101 \\ 0 \end{bmatrix}$. (f) Inconsistent. (g) Inconsistent. (h) Inconsistent.

(i) $\begin{bmatrix} 2 \\ 0 \\ 1 \end{bmatrix}$.

§3.8

7. (a) $\begin{bmatrix} -3/8 & 7/8 & -1 \\ 3/4 & -3/4 & 1 \\ -5/8 & 9/8 & -1 \end{bmatrix}$. (b) $\begin{bmatrix} 2/3 & -1/3 \\ -1/3 & 2/3 \end{bmatrix}$. (c) $\begin{bmatrix} 2/3 & -1/3 & -1/3 \\ 1/3 & 1/3 & 0 \\ 0 & 0 & 1/3 \end{bmatrix}$.

(d) $\begin{bmatrix} -2 & 1 \\ 3/2 & -1/2 \end{bmatrix}$. (e) $\begin{bmatrix} -7 & 0 & 5 & 0 \\ 0 & 1/7 & 0 & -1/21 \\ 3 & 0 & -2 & 0 \\ 0 & 0 & 0 & 1/6 \end{bmatrix}$.

(f) $\begin{bmatrix} -1 & 4/3 & 0 & 0 \\ 1 & -1 & 0 & 0 \\ 0 & 0 & 1/7 & 0 \\ 3 & 0 & 0 & 1/6 \end{bmatrix}$. (g) $\begin{bmatrix} 2 & -4 & 5 & -3 & -1/2 \\ -1 & 3 & -3 & 1 & 1/2 \\ -2 & 2 & -3 & 4 & 0 \\ 0 & 1 & -1 & 0 & 1/2 \\ 1 & -2 & 2 & -1 & -1/2 \end{bmatrix}$.

(h) No inverse. (i) $\begin{bmatrix} 1 & 0 & 1 \\ -3/4 & 1/4 & -1/4 \\ -3/8 & 1/8 & -5/8 \end{bmatrix}$.

Miscellaneous Exercises on Chapter 3 (p. 69)

2. (a) $\alpha \begin{pmatrix} 11/19 \\ -18/19 \\ 1 \end{pmatrix}$, $\alpha \in \mathscr{F}$. (b) $\begin{pmatrix} 11/7 \\ -13/7 \\ 0 \end{pmatrix} + \alpha \begin{pmatrix} -4/7 \\ 13/7 \\ 1 \end{pmatrix}$, $\alpha \in \mathscr{F}$. (c) $\begin{pmatrix} 8/5 \\ -11/5 \end{pmatrix}$.

(d) $\begin{pmatrix} -1 \\ 0 \end{pmatrix} + \alpha \begin{pmatrix} -2 \\ 1 \end{pmatrix}$, $\alpha \in \mathscr{F}$.

5. (a) For $A\mathbf{x} = \mathbf{b}_1$; if $\lambda \neq 4$, only the trivial solution;
   if $\lambda = 4$, more than one solution.
   For $A\mathbf{x} = \mathbf{b}_2$; unique solution if $\lambda \neq 4$,
   more than one solution if $\lambda = 4$.
   For $A\mathbf{x} = \mathbf{b}_3$; unique solution if $\lambda \neq 4$,
   no solution if $\lambda = 4$.

(b) For $A\mathbf{x} = \mathbf{b}_1$; more than one solution if $\lambda = 1$, only trivial solution if $\lambda \neq 1$. For $A\mathbf{x} = \mathbf{b}_2$ or $A\mathbf{x} = \mathbf{b}_3$; no solution if $\lambda = 1$, unique solution if $\lambda \neq 1$.

§4.1

5. Let $A = \begin{bmatrix} 1 & 1 \\ 0 & 2 \end{bmatrix}$, $B = \begin{bmatrix} 1 & 2 \\ 3 & 4 \end{bmatrix}$.

7. (a) 1. (b) 1. (c) 6. (d) -88. (e) -48. (f) 1.0.

§4.3

6. (a) 20. (c) 27. (e) 9. (g) $(c - a)(c - b)(b - a)$.

§5.3

4. (a) Not a subspace. (b) Subspace. (c) Not a subspace. (d) Not a subspace.
   (e) Not a subspace.
5. (a) Not a subspace. (b) Subspace. (c) Not a subspace. (d) Subspace.

§5.4

3. (a) Independent. (b) Dependent. (c) Independent.
4. (a) Independent. (b) Independent. (c) Independent if $\lambda \neq \mu$, dependent if $\lambda = \mu$.

§5.5

11. 4. 12. 2. 13. (d) S^* is infinite dimensional. (e) S_{50} is infinite dimensional.
14. W is infinite dimensional.

§5.7

5. (a) Yes. (b) Yes. (c) No. (d) Yes. (e) Yes. (f) No. (g) Yes.

9. Let $P = \begin{pmatrix} 1 & 1 \\ 2 & 0 \end{pmatrix}$ and calculate $P^{-1}CP$.

10. Let $P = \begin{pmatrix} 1 & 1 \\ 0 & 1 \end{pmatrix}$ and calculate $P^{-1}AP$.

§6.1

3. (a) $\begin{bmatrix} \frac{3}{2} & \frac{1}{e} - \frac{1}{e^2} & 7 \\ \cos(1) - \cos(2) & 0 & \sin(2) - \sin(1) \\ \frac{7}{3} & \frac{3}{2} & 1 \end{bmatrix}$.

(c) $e^2 \begin{bmatrix} \frac{1}{2}(e^2 - 1) & \frac{1}{4}(3e^2 - 1) \\ e^2 - 1 & 2e^2 - 1 \end{bmatrix}$, (d) $\begin{bmatrix} \ln 4 - 1 \\ \ln 4 - \frac{3}{4} \\ (8/3) \ln 2 - 7/9 \end{bmatrix}$. (e) $\begin{bmatrix} 1 \\ 2 \\ 3 \\ 4 \end{bmatrix}$.

5. $\dfrac{1}{t} - \dfrac{1}{t^2}$ on $0 < t < \infty$.

6. $\tfrac{1}{3}e^t + \tfrac{2}{3}e^{-2t}$ on $-\infty < t < \infty$.
10. All solutions tend toward zero if $\lambda < 0$; constant if $\lambda = 0$; all solutions grow exponentially if $\lambda > 0$.
15. $\varphi(t) = (2e^{-t}, 2e^t - e^{-t})$ on $-\infty < t < \infty$.

18. $\varphi(t) = \left(\int_0^t (g(u) + f(u))\, du,\ \int_0^t g(u)\, du\right)$, where $g(u) = e^{2u} \int_0^u e^{-2s} f(s)\, ds$.

§6.2

1. (a) There is a unique solution valid on $0 < t < \infty$.
   (d) There is a unique solution valid on $0 < t < 3$.
4. (a) There is a unique solution valid on $-\pi/2 < t < \pi/2$.

§6.4

2. (a) $\tfrac{1}{2}(e^t - e^{-t})$. (b) $e^{3t} - e^{2t}$. (c) $\tfrac{1}{2}e^t - e^{2t} + \tfrac{1}{2}e^{3t}$.
3. The condition for distinct real roots is $p^2 > 4q$. All solutions tend to zero.
8. (a) $\sin t + \cos t$. (b) $e^{2t}(\cos 3t - \tfrac{1}{3}\sin 3t)$. (c) $\cos 2t + \tfrac{1}{2}\sin 2t$.
   (d) $e^{-t}(\cos t + 2 \sin t)$.
9. $p^2 < 4q$; all solutions tend toward zero.
10. (a) $c_1 \sin 3t + c_2 \cos 3t$. (b) $c_1 e^{2t} + c_2 e^{3t}$.
    (d) $c_1 \exp(\sqrt{2} - 1)it + c_2 \exp -(\sqrt{2} + 1)it$.
    (f) $e^{-5t/2}\left(c_1 \cos \dfrac{\sqrt{15}}{2}t + c_2 \sin \dfrac{\sqrt{15}}{2}t\right)$.
    (g) $c_1 \exp[(-1 + \sqrt{1-\varepsilon})/\varepsilon]t + c_2 \exp[(-1 - \sqrt{1-\varepsilon})/\varepsilon]t$,
    (h) $e^{-t/2}(c_1 + c_2 t)$.
12. (a) Parts a, c, d, f, g, h. (b) Parts a, d.
13. (b) $c_1 e^{2t} + c_2 e^{-2t} + c_3 \cos 2t + c_4 \sin 2t$.
    (d) $c_1 \sin t + c_2 \cos t + c_3 \sin 2t + c_4 \cos 2t$.
15. (a) $c_1 + c_2 t + c_3 t^2 + c_4 t^3$.

(b) $\left[c_1 \sin \dfrac{\sqrt[4]{\lambda}}{\sqrt{2}} t + c_2 \cos \dfrac{\sqrt[4]{\lambda}}{\sqrt{2}} t\right] \exp(\sqrt[4]{\lambda}/\sqrt{2})t$

$+ \left[c_3 \sin \dfrac{\sqrt[4]{\lambda}}{\sqrt{2}} t + c_2 \cos \dfrac{\sqrt[4]{\lambda}}{\sqrt{2}} t\right] \exp -(\sqrt[4]{\lambda}/\sqrt{2})t$.

(c) $c_1 \exp \sqrt[4]{-\lambda}\, t + c_2 \exp -\sqrt[4]{-\lambda}\, t + c_3 \cos \sqrt[4]{-\lambda}\, t + c_4 \sin \sqrt[4]{-\lambda}\, t$.

§6.5

2. $\varphi(t) = \left(-\dfrac{2}{25}\cos t - \dfrac{14}{25}\sin t + \dfrac{27}{25}e^{2t} - \dfrac{3}{5}te^{2t}, \quad -\dfrac{2}{5}\cos t + \dfrac{1}{5}\sin t - \dfrac{3}{5}e^{2t}\right).$

6. (a) $(\cos t)(\log \cos |t|) + t \sin t + c_1 \cos t + c_2 \sin t.$

(c) $c_1 \cos 2t + c_2 \sin 2t - \dfrac{1}{2}\cos 2t \int_{t_0}^{t} f(s) \sin 2s \, ds + \dfrac{1}{2}\sin 2t \int_{t_0}^{t} \cos 2s f(s) \, ds.$

(e) $C_1|t|^i + C_2|t|^{-i} + \dfrac{1}{10}t^3.$

10. (a) $e^{-t}(C_1 \cos \sqrt{3}t + C_2 \sin \sqrt{3}t) + C_3 e^{2t} + \dfrac{1}{12}te^{2t}.$

§7.1

5. $\varphi(t) = e^{-2t}\begin{pmatrix} 1 & t & t^2/2 \\ 0 & 1 & t \\ 0 & 0 & 1 \end{pmatrix}$

6. $\varphi(t) = \begin{pmatrix} 1 & t & \dfrac{t^2}{2} & \cdots & \dfrac{t^{n-1}}{(n-1)!} \\ 0 & 1 & t & \cdots & \\ 0 & 0 & 1 & \cdots & \\ \vdots & \vdots & \vdots & & \vdots \\ 0 & 0 & 0 & \cdots & 1 \end{pmatrix}$

§7.3

1. (c) A fundamental matrix:

$$\begin{pmatrix} e^{-3t} & \exp(2+\sqrt{7})t & \exp(2-\sqrt{7})t \\ -\dfrac{7}{3}e^{-3t} & \dfrac{-5+4\sqrt{7}}{3}\exp(2+\sqrt{7})t & \dfrac{-5-4\sqrt{7}}{3}\exp(2-\sqrt{7})t \\ -\dfrac{4}{3}e^{-3t} & \dfrac{1+\sqrt{7}}{3}\exp(2+\sqrt{7})t & \dfrac{1-\sqrt{7}}{3}\exp(2-\sqrt{7})t \end{pmatrix}.$$

§7.4

3.

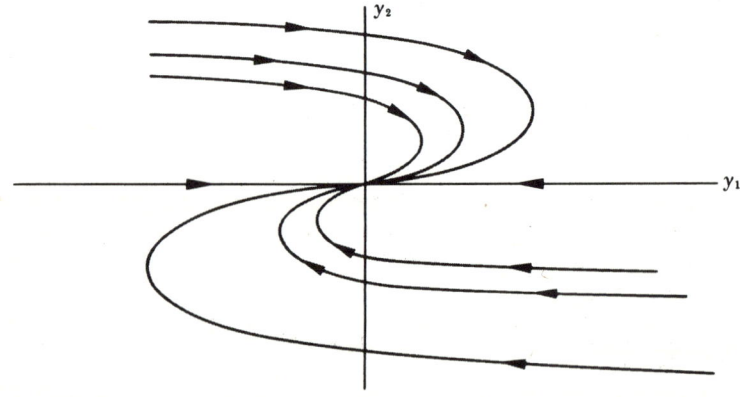

13. Node, attractor. **15.** Saddle point, not an attractor.
17. Node, attractor. **19.** Saddle point, not an attractor.

Miscellaneous Exercises on Chapter 7 (p. 261)

2. (a) $\begin{pmatrix} e^t & e^{5t} \\ -e^t & 3e^{5t} \end{pmatrix}$. (c) $\begin{pmatrix} e^{2t}\cos t & e^{2t}\sin t \\ e^{2t}(\cos t - \sin t) & e^{2t}(\cos t + \sin t) \end{pmatrix}$.

(g) $\begin{pmatrix} e^t & 0 & e^{-t} \\ e^t & e^{2t} & 0 \\ 0 & 2e^{2t} & -e^{-t} \end{pmatrix}$. (k) $\begin{pmatrix} 1 & t & 4e^{3t} \\ -2 & 1-2t & 4e^{3t} \\ 1 & -1+t & e^{3t} \end{pmatrix}$.

(m) $e^{2t}\begin{pmatrix} 1 & t & t^2 \\ 2 & -1+2t & -2t+2t^2 \\ 1 & -1+t & 2-2t+t^2 \end{pmatrix}$.

4. (a) $\begin{pmatrix} e^t & e^{-t} \\ e^t & -e^{-t} \end{pmatrix} + \mathbf{c}\begin{pmatrix} te^t - t^2 - 2 \\ (t-1)e^t - 2t \end{pmatrix}$,

(c) $\begin{pmatrix} e^t & 3e^{2t} \\ e^t & 2e^{2t} \end{pmatrix} + \mathbf{c}\begin{pmatrix} \cos t - 2\sin t \\ 2(\cos t - \sin t) \end{pmatrix}$.

8. $\mathbf{c}t^A + t^A \mathbf{c}\int_{t_0}^{t} s^{-A}\mathbf{b}(s)\,ds$ where $t \cdot t_0 > 0$.

§8.2

3. (a) $\dfrac{s_1 + 2\beta^2}{s(s^2 + 4\beta)}$, $(Rs > 0)$. (b) $\dfrac{\beta}{s^2 + 4\beta^2}$, $(Rs > 0)$.

4. $\left(\dfrac{1}{s} + \dfrac{1}{s^2}\right)e^{-s} - \left(\dfrac{2}{s} + \dfrac{1}{s^2}\right)e^{-2s}$, $(Rs > 0)$.

5. $\dfrac{2(1 - e^{-\pi s})}{s^2 + 4}$, $(Rs > 0)$.

9. $\dfrac{\int_0^a \phi(\sigma)\exp(-s\sigma)\,d\sigma}{1 - \exp(-as)}$, $(Rs > 0)$.

10. $\dfrac{1 - 2\exp(-as/2) + \exp(-as)}{s(1 - \exp(-as))}$, $(Rs > 0)$.

11. $\dfrac{1 - \exp(-as/2)}{s^2(1 + \exp(-as^2/))}$, $(Rs > 0)$.

§8.3

7. (a) $-2 + \dfrac{5}{2}e^{-t} + \dfrac{1}{2}e^{-3t}$. (c) $1 - \cos 2t - t\sin 2t$. (e) $\int_0^t \sin(t-u)f(u)\,du$

8. $\cos 2t$.

§8.4

3. $\dfrac{1}{\sigma(z_2 - z_1)}(e^{-z_1(t-\sigma)} - e^{-z_2(t-\sigma)} + e^{-z_2 t} - e^{-z_1 t})$. **5.** $e^{3t} - e^{2t}$.

6. $\dfrac{1}{2}e^t - e^{2t} + \dfrac{1}{2}e^{3t}$. **7.** $\dfrac{1}{24}e^{2t} - \dfrac{1}{24}e^{-t}(\cos\sqrt{3}\,t + \dfrac{5\sqrt{3}}{3}\sin\sqrt{3}\,t) + \dfrac{t}{12}e^{2t}$.

9. $\dfrac{1}{48}(25e^{3t} + 23e^{-3t}) - \dfrac{1}{8}\sin t$. **11.** $\dfrac{1}{2}(e^{t-1} - e^{-(t-1)})$.

13. $\cos 2t + \dfrac{1}{2}\sin 2t + \dfrac{1}{2}\int_0^t \sin 2(t-x) f(x)\, dx$. **15.** $\dfrac{1}{13}3e^{2t} + \alpha \cos + 3t + \beta \sin 3t$.

§8.5

3. $\phi_1(t) = \dfrac{1}{2}t + 1,\ \phi_2(t) = -\dfrac{1}{2}t$.

5. $\phi_1(t) = \dfrac{1}{4}e^{-2t} + \dfrac{2}{3}e^{-t} + \dfrac{1}{12}e^{2t},\ \phi_2(t) = \dfrac{1}{3}(e^{-t} - e^{2t})$.

6. $\phi_2(t)$ is any function for which $\phi_2(0) = 1$ and $\phi_2'(0) = 0$,

$$\phi_1(t) = e^t - \int_0^t e^{(t-u)}\phi_2'(u)\, du.$$

Miscellaneous Exercises on Chapter 8 (p. 302)

1. (a) $\dfrac{\omega}{(s^2 + \omega^2)(1 - e^{-\pi s/\omega})}$. (b) $\dfrac{1}{(s^2 + 1)(s + 1)}$. (c) $\dfrac{15}{8s^3}\left(\dfrac{\pi}{s}\right)^{1/2}$, $(Rs > 0)$.

(d) $\dfrac{2\omega(3s^2 - \omega^2)}{(s^2 + \omega^2)^3}$, $(Rs > 0)$.

2. (a) $2e^{2t}(1 + t) - \dfrac{1}{2}\sin 2t$.

(c) $3 - 2\cos t + 2[(t-4)\sin(t-4)]U(t-4)$, where $U(t)$ is the unit step function.

8. (a) $-1 + 2t + 2t^2 + e^{-2t}$. (b) $t^3 + \dfrac{1}{20}t^5$. (c) $1 + 2te^t$.

(d) $\dfrac{1}{2}t^2$. (e) $4 + \dfrac{5}{2}t^2 + \dfrac{1}{24}t^4$.

INDEX

Abel's formula, 182
Adjugate of a matrix, 89, 90, 91
Algorithm, 50
Arbitrary constants, 63
Associativity of addition, 20
Attractor, 244, 262
Auxilary equation, 184
  (*see also* Characteristic equation)

Basis of a vector space, 126–140, 142–147, 149, 171, 172, 217, 218, 329
  coordinates with respect to, 131, 149
Bernoulli equation, 159
Bessel equation, 166, 303, 304
Binomial theorem, 209
Block diagonal matrix, 329, 330, 331
Bounded function, 191
Bounded solution, 191

Canonical form of a matrix, 327
Capacitance, 286, 287, 289
Cauchy convergence criterion, 306, 307
Center, 244, 262

Characteristic equation, 184, 185, 189, 190
Characteristic polynomial, 184, 192, 214, 220, 248, 253, 256, 285, 329, 330
Closure under addition, 20
Cofactors, 72, 74
  alien, 89
Column space of a matrix, 139
Column vector (*see* Vector)
Commutativity of addition, 20
Comparison theorem for series, 317
Complex Euclidean space, n dimensional, 106
Complex numbers, 26
Complex-valued solutions, 186
Condenser, 13
Convergence of a series of matrices, 209, 306, 307
Convolution, 277
Cramer's rule, 89, 93, 94, 101, 296
Critical point, 236–244
Current, 159, 206

Damping
  critical, 207
  light, 207
  over, 207

343

344 · Index

Determinantal polynomial, 214
Determinants, 71–102
  of an elementary matrix, 80
  expansion by cofactors, 72, 74, 75, 85, 86–89
  properties, 75, 77
  of transposed matrix, 82
  Vandermonde, 99, 100
Dimension, 126, 134–140, 148, 171, 328
Dimensionless product, 6
Dirac delta function, 290
Direct sum, 325, 326, 330, 333

Eigenvalues, 16, 208–264, 327–334
  complex conjugate, 233–235
  double, 224
  of multiplicity k, 214, 329
  simple, 214
Eigenvectors, 16, 208–264, 327–334
  linearly independent, 217, 218, 219, 220, 224, 231, 245, 246, 259, 324
Electrical circuits, 13, 159, 256, 259, 261, 286, 287, 289
Elementary column operations, 76, 79, 80, 86
Elementary row operations, 41, 42, 43, 45, 46, 50, 51, 76, 79, 86
Equivalent systems of algebraic equations, 38, 47, 48
Existence of solutions, 15, 165–169, 312–319
Exponential growth at infinity, 267, 322
Exponential of a matrix, 208–212, 305–307

Flexibility, 4
Force
  external, 195
  of friction, 11
Frequency
  applied, 8
  natural, 8
Fundamental matrix, 175–179, 195–196, 198, 200–201, 208, 210–213, 224–225, 227-228, 234, 236, 253, 256, 261, 263, 296, 300–301
  real, 226
Fundamental set of solutions (*see* Linear homogeneous systems of differential equations)
Fundamental theorem of calculus, 157, 158, 321

Gain function, 288–289, 291
Generalized eigenvector, 327-334
Gram-mole, 7
Growth of solutions, 282

Hooke's law, 3, 5, 9, 12
l'Hospital's rule, 159
Hyperplane, 114

Index, 328
Inductance, 13, 206, 286–287, 289
Induction, proof by, 74, 76–77, 83, 85, 313, 315–316, 330
Infinite sequences, 135
Initial condition, 10, 156, 160, 197
Initial value problem, 10, 11, 13, 156, 160, 161, 162, 164, 166
  for differential equations of order n, 167
  for second-order differential equations, 167
Integral equations, 312, 318
Intersection of subspaces, 324–325
Interval of existence, 166
Invariant subspace, 328, 331

Inverse Laplace transform, 275–281
  linearity of, 280
  uniqueness of, 280

Jordan canonical form, 333–334

Kinetic energy, 300
Kirchhoff's law, 14

Laplace transformation, 16, 265–304, 321
Leading entry, 49, 56
Linear algebraic equations, 2
  systems of (*see* Linear systems of algebraic equations)
Linear combinations, 115, 118, 136, 139, 170
Linear dependence and independence, 115–139, 148, 171, 176, 193–194, 218–219, 224, 231, 234, 327–334
Linear differential equations, 10
  with constant coefficients, 16, 183–195, 226
  of first order, 156–158, 161
  general solution, 185, 188, 191, 194, 202–203, 204, 205, 229, 230, 263
  of order n, 16, 174
  of second order, 10, 151, 162, 163, 173, 200, 202, 226, 254
  solution of, 10, 11
  systems of (*see* Linear systems of differential equations)
Linear differential operator, 183
Linear homogeneous systems of differential equations, 170–183, 195
  algebraic structure of solutions, 170
  fundamental set of solutions, 172, 181, 183, 205
  linearly independent solutions, 172, 174, 176, 180, 181, 183, 184, 187, 189, 190, 191, 192, 195, 204
Linear nonhomogeneous systems of differential equations, 170, 195–206, 227
Linear operator 268
Linear systems of algebraic equations, 5, 15, 16, 29, 32, 34, 37, 49, 69, 176
  consistent, 60, 61, 62, 63, 117, 139
  homogeneous, 55, 63, 213
  inconsistent, 60
  in matrix-vector form, 35
  nonhomogeneous, 64, 65
  nontrivial solution of, 59, 60, 66
  solution of, 36, 37, 39, 63, 71, 137, 138
  solution set of, 39, 59, 60, 138
  trivial solution of, 55, 59
Linear systems of differential equations, 13, 14, 15, 16, 151–207
  with constant coefficients, 16, 208–264, 292
  solution of, 155–207
  triangular, 161, 162
  two-dimensional, 230–244
Linear transformations, 140–147, 149, 150, 327, 330
  matrix of, 142–147, 333
  nilpotent, 329–333
  range of, 328

Main diagonal of a matrix, 32
Mass, 8
Mass spring system, 11, 260
Mathematical model, 3, 14

Matrices, 17, 18, 71, 75, 108
  addition of, 19, 26
  augmented, 39, 40, 41, 48, 49, 117
  coefficient, 61.
  column of, 18
  column rank of, 139
  diagonal, 32, 74, 218, 219, 223, 259
  elementary, 41, 42, 43, 45, 47, 55, 66, 68, 79, 84
  identity, 28, 74, 209
  inverse, 30, 45, 54, 65, 68, 89, 90, 91
  multiplication of, 21, 23, 24, 26, 97
  multiplication by scalars, 20, 26
  nonsingular, 30, 31, 34, 45, 47, 54, 62, 65, 66, 68, 90, 92, 114, 178, 196, 209, 219, 220, 222, 327
  norm of, 305–307
  rank of, 54, 55, 59, 61, 63, 65, 66, 117, 140
  row of, 18
  row-equivalent, 48, 54
  row rank of, 54, 136, 137, 148
  singular, 31
  symmetric, 96, 115
  upper triangular, 32, 54, 62, 65, 66, 68–90
  zero, 20, 74
Matrix functions, 152, 153
  continuity, differentiability, integrability, 154
Mechanical analog, 259
Mechanical system, 260, 261, 300
Molecular weight, 7
Momentum, 9

Newton's second law of motion, 9, 12, 260

Node, 239–242, 262
  improper, 239, 242
  proper, 240
Nonlinear differential equations, 152
Norm, 305–311
  Euclidean, 309

Ohm's law, 14
Orbit, 236–244
Oscillation, 207
  amplitude of, 207

Phase lag, 288, 289
Phase plane, 236–244
Phase portrait, 236–244, 262
Physical assumpions, 3
Physical laws, 3
Pin-jointed framework, 94
Plane, 1, 2, 3,
Potential energy, 300
Product matrix, 24

Range of a linear transformation, 328
Real Euclidean space
  n-dimensional, 103, 105
  3-dimensional, 104
  plane, 105
Real part of a vector, 179
Resistance, 206, 286, 287, 289
Resonance, 8, 207
Resonant frequency, 207
Row-echelon form, 49, 50, 51, 52, 54, 55, 57, 58, 59, 61, 63, 66, 137
Row equivalence, 46, 47, 48, 50, 51

Row space of a matrix, 136
Row vector (*see* Vector)

Saddle point, 241, 262
Similarity of matrices, 147, 150, 218, 219, 223, 224, 259, 334
Simple harmonic motion, 8, 112
Solution matrix, 174–179, 210, 224
Span, 115, 117, 121, 122, 126, 127, 134, 147, 148
Spiral point, 243, 262
Spring, 8, 260
Spring constant, 9, 260
Square-wave function, 271
Steady-state solution, 159, 288
Strain, 3
Stress, 3
Subspaces, 112–118, 217, 222, 246, 247, 250, 325
  invariant, 328, 329
Successive approximations, 313–319

Tension, 4
Transfer function, 288, 289, 291
Transient solution, 159, 288
Transpose of a matrix, 82, 95
Triangle inequality, 309

Uniform convergence, 318
Uniqueness of solutions, 15, 32, 202, 319–321
  of linear algebraic systems, 37, 65, 69, 70
  of linear systems of differential equations, 165–169

Variation of constants, 196–197, 204, 253, 258
  for second-order equations, 200–202
Vector
  addition of, 104, 107
  column, 18
  components of, 56, 58
  functions, 110, 111, 153–156, 159–164, 170, 310
  linear combination of, 58, 59, 60
  multiplication by scalars, 104, 107
  norm of, 308–311
  row, 18
  unit, 105
Vector spaces, 103–150
  finite-dimensional, 134, 135
  of functions, 108–111, 120–123, 135, 136, 141–142, 156, 171
  infinite-dimensional, 134, 171
  of polynomials, 109, 129
Vector sum of subspaces, 324, 325
Velocity, 9
Voltage, 13
  applied, 206, 287

Wronskian, 180, 182, 185, 191